# Energy-Aware System Design

Chong-Min Kyung · Sungjoo Yoo

Editors

# Energy-Aware System Design

## Algorithms and Architectures

 Springer

*Editors*
Prof. Chong-Min Kyung
Electrical Engineering
KAIST
Gwahak-ro 335, Yuseong-gu
305-701, Daejeon
Republic of Korea
kyung@ee.kaist.ac.kr

Prof. Sungjoo Yoo
Embedded System Architecture Lab
Electronic and Electrical Engineering
POSTECH
Hyoja-dong 31, Namgu
790-784, Pohang
Republic of Korea
sungjoo.yoo@postech.ac.kr

ISBN 978-94-007-1678-0       e-ISBN 978-94-007-1679-7
DOI 10.1007/978-94-007-1679-7
Springer Dordrecht Heidelberg London New York

Library of Congress Control Number: 2011931517



*Cover design*: VTeX UAB, Lithuania

Printed on acid-free paper

Springer is part of Springer Science+Business Media (www.springer.com)

# Preface

Up to now the driving force of the development of most information technology (IT) devices and systems has mainly been performance-cost ratio boosting, but this has already begun to change. For some time energy consumption will occupy a growing portion in the design objective function of a large number of IT devices, especially in mobile, health, and ubiquitous applications. Using even the most energy-wise frugal technology, the energy we are spending for logic switching is still at least six orders of magnitude larger than the theoretical limit. The task of reducing that energy gap is not an easy one, but it can be quite effectively carried out if accompanied by a nicely coordinated effort of energy reduction among various design stages in the design process and among various components in the system.

A number of books have already been published that focus on low-energy design in one aspect, i.e., limited to an individual functional block such as on-chip networks, algorithms, processing cores, etc. Instead of merely enumerating various energy-reducing technologies, architectures, and algorithms, this book tries to explain the concepts of the most important functional blocks in typical information processing devices, e.g., memory blocks and systems, on-chip networks, and energy sources, such as batteries and fuel cells.

The most important market for low-energy devices, after the current booming smart phone, is probably energy-aware smart sensors. The variety of applications in the market is truly huge and expanding every year. With more and more traffic (both people and data) on the move, the planet is becoming more dangerous, as well as more exciting. The demand for installing smart sensors on various locations in our society as well as our bodies, i.e., on/in/outside the human body, obviously will grow. The scale and variety of threats against our society and each individual has never been so overwhelming, and this will probably escalate unless we carry out a systematic and coordinated effort toward building a safe society. We believe that the energy-aware smart sensor is one such attempt.

This book tries to show how the design of each functional block and algorithm can be changed by an addition of a new component: energy. Besides explanations of each functional block in early chapters, three application examples are given at the end: data/file storage systems, an artificial cochlea and retina, and a battery-operated surveillance camera. We understand that the coverage is far from complete in terms of the variety of functional blocks, algorithms, and applications. Despite these imperfections, we sincerely hope, through this book, that the readers will gain some perspective and insights into energy-aware IT system design, which will lead us all toward a better, i.e., cleaner and safer society.

Daejeon, Republic of Korea
Pohang, Republic of Korea

Chong-Min Kyung
Sungjoo Yoo

# Contents

# Contributors

**Jung Ho Ahn** Seoul National University, Seoul, Republic of Korea, gajh@snu.ac.kr

**Naehyuck Chang** Seoul National University, Seoul, Republic of Korea, naehyuck@elpl.snu.ac.kr

**Sungwoo Choo** Seoul National University, Seoul, Republic of Korea, choos@snu.ac.kr

**Kyungsu Kang** KAIST, Daejeon, Republic of Korea, kyungsu.kang@gmail.com

**Giwon Kim** KAIST, Daejeon, Republic of Korea, gwkim@vslab.kaist.ac.kr

**Jaemoon Kim** Samsung Electronics, Seoul, Republic of Korea, jaemoon.kim@gmail.com

**John Kim** KAIST, Daejeon, Republic of Korea, jjk12@kaist.edu

**Jungsoo Kim** KAIST, Daejeon, Republic of Korea, jungsoo.kim83@gmail.com

**Sung June Kim** Seoul National University, Seoul, Republic of Korea, kimsj@snu.ac.kr

**Chong-Min Kyung** KAIST, Daejeon, Republic of Korea, kyung@ee.kaist.ac.kr

**Sangkwon Na** Samsung Electronics, Seoul, Republic of Korea, sangkwon.na@gmail.com

**Seongil O** Seoul National University, Seoul, Republic of Korea, swdfish@snu.ac.kr

**Chanik Park** Samsung Electronics, Hwasung-City, Republic of Korea, ci.park@samsung.com

**Youngsoo Shin** KAIST, Daejeon, Republic of Korea, youngsoo@ee.kaist.ac.kr

**Sungjoo Yoo** POSTECH, Pohang, Republic of Korea, sungjoo.yoo@postech.ac.kr

# Chapter 1
# Introduction

**Chong-Min Kyung and Sungjoo Yoo**

**Abstract** Energy efficiency is now an important keyword in everyday life, involving, e.g., $CO_2$ emissions, rising oil prices, longer battery lifetimes for smart phones, and lifelong functioning medical implants. This book addresses energy-efficient IT systems design, especially low power embedded systems design. This chapter discusses how a power-efficient design can be achieved by exploiting various slacks. For instance, temporal slack is utilized for dynamic voltage scaling while thermal slack is exploited for low-leakage operation, both methods thereby enabling low power consumption. This chapter also provides short introductions to the remaining chapters, which address aspects of low power embedded systems design such as low power circuits, memory, on-chip networks, power delivery, and low power design case studies of video surveillance systems, embedded storage, and medical implants.

## 1.1 Energy Awareness

Power consumption has become the most important design goal in a wide range of electronic systems. There are two driving forces toward this trend: continuing device scaling and ever-increasing demand for higher computing power. First, device scaling continues to satisfy Moore's law via a conventional way of scaling (More Moore) and a new way of exploiting vertical integration (More than Moore) [1]. Second, mobile and IT convergence requires more computing power on the silicon chip than ever. Cell phones are now evolving to become mobile PCs. PCs and data centers are becoming commodities in the home and a must in industry. Both the supply enabled by device scaling and the demand triggered by the convergence trend realize more computation on chip (via multi-cores, integration of diverse functionalities

C.-M. Kyung (✉)
KAIST, Daejeon, Republic of Korea
e-mail: kyung@ee.kaist.ac.kr

S. Yoo
POSTECH, Pohang, Republic of Korea
e-mail: sungjoo.yoo@postech.ac.kr

C.-M. Kyung, S. Yoo (eds.), *Energy-Aware System Design*,
DOI 10.1007/978-94-007-1679-7_1, © Springer Science+Business Media B.V. 2011

**Fig. 1.1** Monthly operation
cost of data center [4]

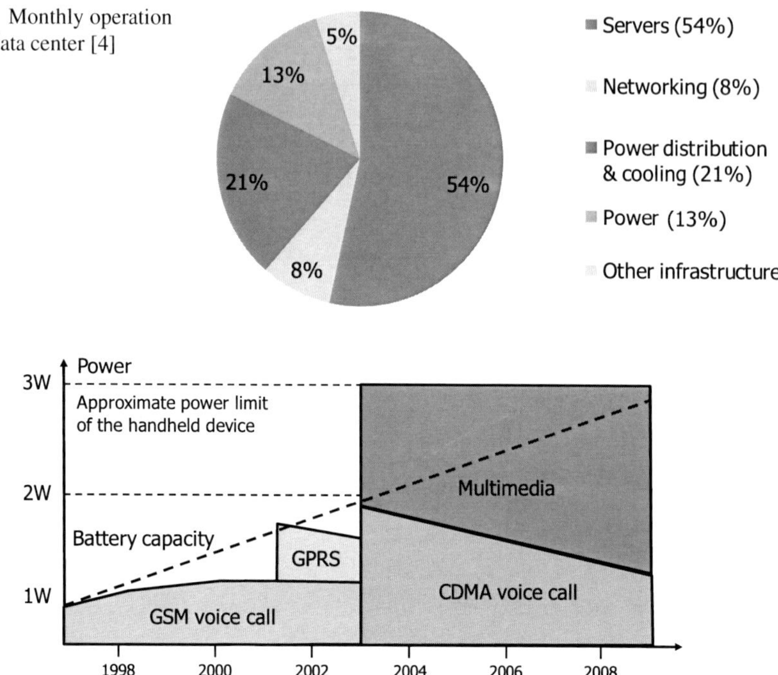


**Fig. 1.2** Battery capacity vs. computational demand [7]. © 2004 IEEE

on mobile SoCs, etc.) and finally more power consumption, incurring power-related
issues and constraints.

We take two examples, the data center and the mobile phone, in order to investigate the impact of the current trend of increasing power consumption. Recently, data centers are becoming a crucial infrastructure in industry and government as well as in everybody's Internet usage. In the United States alone, data centers consumed 61 billion kilowatt-hours (kWh) in 2006, which is 1.5% of the total U.S. electricity consumption and amounts to a total electricity cost of about \$4.5 billion. In 2011, it is expected to reach more than 100 billion kWh [2]. The demand for data centers is projected to increase at 10% compound annual growth rate (CAGR) in the next decade [3].

Figure 1.1 gives the decomposition of the data center operation cost [4]. It shows that about 34% (or more according to other sources) of the data center operation cost is power related. Power dissipation itself occupies 13%, power distribution and cooling 21%. Thus, reducing the power consumption and providing better cooling efficiency become critical in lowering the operation cost. Several approaches are being actively studied, including server rack level power management [5], capping the compute power in an I/O-intensive workload [6], and active liquid cooling [2].

Figure 1.2 shows the trend of computing power requirement and battery capacity in the case of the cell phone [7]. The workload of the cell phone is decomposed into three parts: radio, multimedia, and application. Power consumption in the radio

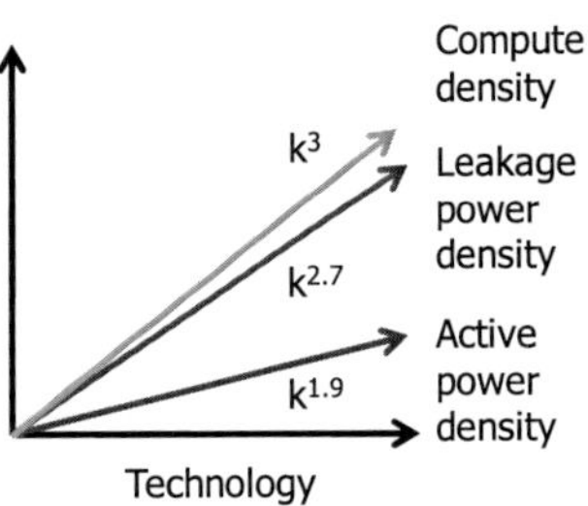

**Fig. 1.3** Power consumption trend [10]



is incurred by the power amplifier during conversation and the receiver in standby for paging. Power consumption of multimedia includes display, camera capture, video decoding/encoding, and three-dimensional (3D) graphics. Applications consume power in, e.g., Internet browsing, email, gaming, and photo handling. The workload of the cell phone increases by an order of magnitude every five years [8].

The battery in a smartphone is allocated only 4–5 cm$^3$ and, thus, offers a maximum of about 2000 mAh as of 2010. However, as shown in Fig. 1.2, the battery capacity improves much more slowly than the increase in computational demand. Thus, battery-operated devices such as smartphones are limited by the energy density of the battery [9].

Figure 1.3 shows the power consumption trend as the technology node advances, where $k$ represents the scaling factor, 1.4. The computing density is defined as the maximally possible number of computations per unit area and time. The figure shows that it increases at a rate of $k^3$. Also according to the figure, as scaling continues, leakage power increases much faster than active power. Note that the trend in Fig. 1.3 assumes that clock frequency continues to rise and voltage scaling slows down. If this trend, i.e., an explosion in power consumption, continues without solution, it will become a roadblock in the IT industry, which has benefited from the rapid increase in computing power during the past decades.

In reality, many low power design methods have been applied to avoid such an explosion in power consumption trend. However, since the power demand will continue to increase in the future, e.g., in cloud computing and smart IT devices (smartphone, smart TV, etc.), the trend itself will continue. The absolute quantity of power consumption continues to rise, even though the currently available low power design methods are applied. Thus, more innovations are required to further reduce power consumption.

Recently, power consumption has become limited by another constraint: $CO_2$ emission. The constraint is general to everyday activities including computing. All personal, industrial, and governmental activities are now evaluated in terms of energy consumption or $CO_2$ emission. Table 1.1 shows examples of energy efficiency measured in terms of number of Google searches [11]. One Google search consumes 1 kJ (0.0003 kWh) of energy on average, which translates into roughly 0.2 g of $CO_2$. As this example shows, energy awareness is expected to spread more widely in our everyday life as well as in the IT industry. The increasing amount of electricity usage of IT technology will result in more pressure to achieve green IT. Low power

**Table 1.1** Energy and $CO_2$ emission in terms of number of Google searches [11]. © Google, Inc.

| Activity | Google searches | Energy consumption |
| --- | --- | --- |
| $CO_2$ emissions of an average daily newspaper (100% recycled paper) | 850 | 850 kJ |
| A glass of orange juice | 1,050 | 1,050 kJ |
| One load of dishes in an EnergyStar dishwasher | 5,100 | 5,100 kJ |
| A five mile trip in the average U.S. automobile | 10,000 | 10,000 kJ |
| A cheeseburger | 15,000 | 15,000 kJ |
| Electricity consumed by the average U.S. household in one month | 3,100,000 | 3,100,000 kJ |

design methods, like those we will present in this book, will contribute to realizing the green IT.

## 1.2 Energy-Aware Design

Energy-aware design, sometimes called energy-efficient design, is the design of a system to meet a given performance constraint with the minimum energy consumption. Figure 1.4 illustrates the relationship between energy and delay. At a given performance constraint, i.e., a delay constraint, an energy-efficient design is one having the minimum energy consumption satisfying the delay constraint. As the figure shows, such energy-efficient designs are called Pareto-optimal designs (e.g., points B and C in the figure) and form a front line (bold line in the figure) in the energy-delay relationship. Pareto-optimal points outperform other design candidates in terms of at least one design metric, e.g., energy or delay.

Energy-efficient design can be achieved in several ways at every level of abstraction, from system level down to transistor device level, as follows:

- System level: energy-aware algorithm (e.g., parallel data structure instead of sequential one), memory-aware software optimization (e.g., utilizing scratch pad memory)
- Architecture: multi-core (including parallel functional units), instruction set selection, dynamic voltage and frequency scaling, power gating
- Logic (or gate level): multi-$V_{\text{th}}/L_{\text{g}}/T_{\text{ox}}/V_{\text{dd}}$ designs (can also be considered in circuit level) instead of worst case design
- Circuit: device sizing, exploiting of transistor stacking to reduce leakage power, hybrid usage of dynamic and static circuit to meet the given delay constraint while minimizing power consumption, differential signaling to reduce voltage swing, etc.
- Device: high $I_{\text{on}}/I_{\text{off}}$ devices, e.g., double-gate or back-gate transistors

Energy awareness means to consider an algorithm (i.e., function) and implementation in terms of a work/energy concept. Thus, energy efficiency is evaluated in terms

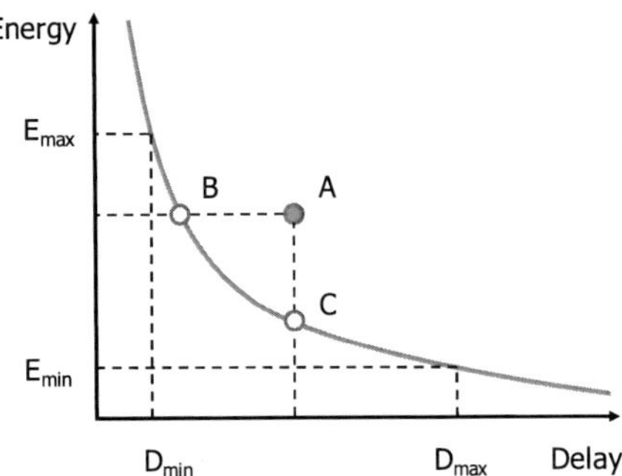

**Fig. 1.4** Energy-delay relationship [12]



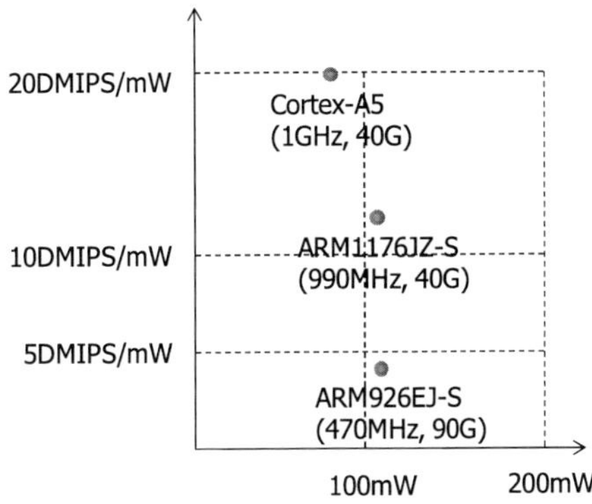

**Fig. 1.5** DMIPS/mW of ARM processors (frequency optimized) [13]



of the amount of work per energy and per unit time, e.g., DMIPS/mW. Figure 1.5 shows an architecture level example where the energy efficiency is improved from one processor generation to another. The figure shows three types of ARM processors, ARM926EJ-S (introduced in the early 2000s), ARM1176JZ-S (mid-2000s), and Cortex-A5 (2010). The figure shows that energy efficiency improves from one generation to the next while the total CPU power consumption remains at a similar level. The energy efficiency improvement results from two factors. One is technology. Note that the power number of ARM926EJ-S is from a 90 nm generic process, while the other two are from a 40 nm generic process. The other factor is architectural improvement. For instance, Cortex-A5 adopts SIMD operations while ARM926EJ-S does not.

Energy-efficient design aims at the best return on investment (ROI), i.e., maximum performance per energy spending. There are several low power design principles for obtaining the best ROI. One representative principle is matching computation and architecture. For instance, it is more energy efficient to run data parallel compute-intensive loops on the DSP instead of running them on the RISC processor. Another example is to utilize asymmetric and/or heterogeneous multi-cores to exploit the fact that single and multi-threaded applications coexist.

Many solutions have been presented at several abstraction levels for energy-aware behavioral and architectural design. Most of the low power design techniques at higher levels than the transistor device level can be considered to exploit "slack" in various forms. In this book, we present several ideas utilizing slack. In the next subsection, we introduce several types of slack and explain how it is utilized for energy-efficient design.

## 1.3 Exploiting Slack Toward Energy-Aware Design

Slack is often called locality, which represents non-uniform, but not random characteristics. There are several types of slack: temporal, spatial, behavioral, architectural, process variation, thermal (2D and 3D), peak power slack, etc. We classify existing low power design methods depending on which type of slack they utilize as follows:

1. Temporal slack: power/clock gating and (conventional) dynamic voltage and frequency scaling, e.g., Intel SpeedStep
2. Spatial slack: multi-core, e.g., ARM Cortex-A9 MP
3. Behavior- and architecture-induced temporal slack: runtime distribution, e.g., Intel Data Center Manager
4. Process variation slack: adaptive voltage scaling, e.g., TI SmartReflex
5. Thermal slack: temperature-aware design, e.g., Intel Turbo Boost
6. Peak power slack: peak power-aware overclocking, e.g., Intel Turbo Boost

### *1.3.1 Temporal Slack*

Figure 1.6 illustrates power gating and dynamic voltage and frequency scaling (DVFS). In Fig. 1.6(a), we assume that the processor has a workload of $N$ clock cycles to be finished by the deadline, $D$. We also assume the quadratic relationship of switching energy/clock cycle $\sim$ voltage$^2$ and a linear relationship between frequency and supply voltage, i.e., frequency $\sim$ voltage. In Fig. 1.6(a), the processor runs at clock frequency $F$ and the execution finishes at time $D/2$. Then, the processor enters an idle state during the slack by shutting off its power until time $D$. Since the power-gated processor consumes negligible power, the total switching energy consumption in this case is $F^2 N$.

Figure 1.6(b) shows the case of applying DVFS to this example. If the workload of $N$ clock cycles is known at time 0, the clock frequency can be set to $F/2$ just to meet the deadline, as shown in the figure. Thus, in this case, the new voltage becomes half the voltage in Fig. 1.6(a), and the energy consumption becomes 25% of that in Fig. 1.6(a), since new energy consumption $\sim$ new_voltage$^2 \sim$ voltage$^2/4$.

As shown in Fig. 1.6, DVFS exploits the slack in order to adjust the frequency and supply voltage such that they are just high enough to serve the current workload. For DVFS to be efficient, accurate workload estimation is critical. Many studies have been presented on workload estimation based on algorithm-specific information, compiler analysis [14], runtime prediction [15], etc. To meet the given deadline constraint, conventional DVFS methods utilize the worst case execution time

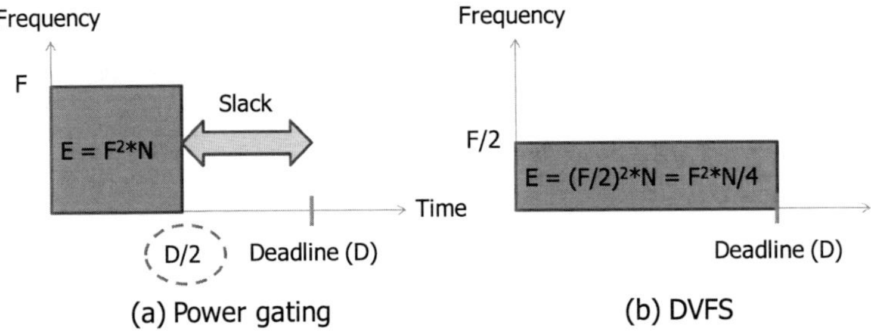



**Fig. 1.6** Power gating vs. dynamic voltage and frequency scaling (DVFS)

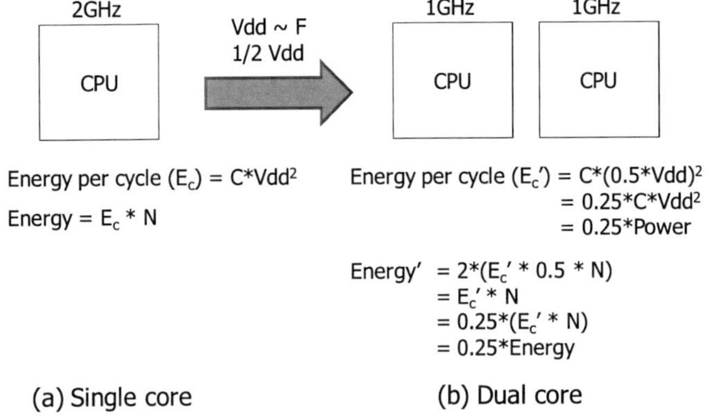



**Fig. 1.7** A benefit of multi-core: reduced energy consumption

(WCET) as the estimated workload and set the frequency to $\text{WCET}/D$. However, this method is pessimistic and loses opportunities for further energy reduction since there is a new slack which is the difference between the worst case and average case execution times. We will explain how to exploit this slack later in this section.

In reality, discrete voltage/frequency levels called operational points, e.g., the $P$-states in the Advanced Configuration and Power Interface (ACPI) [16], are usually applied in DVFS. Frequency change takes a variable latency depending on whether the required frequency is obtained by a simple clock division (a few clocks of latency) or by reconfiguring the PLL (typically, tens of microseconds).

## *1.3.2 Spatial Slack Enabled by Newer Process Technology*

In the past decade, the multi-core technology has proven to be effective in achieving better energy efficiency. One of the driving forces toward multi-core is that new process technology offers more room, i.e., more silicon area to accommodate more cores. Figure 1.7 shows how a multi-core processor improves energy efficiency. In Fig. 1.7(a), assume that a single processor executes a workload of $N$ clock cycles at

2 GHz. In Fig. 1.7(b), assume that two processors are utilized and that each executes half the workload, $N/2$ clock cycles at 1 GHz.

Figure 1.7(b) shows that the switching energy consumption of each core in the dual-core processor is 25% of that in the single-core processor. Since each core executes half the workload, the total energy consumption of the dual-core is 25% of the single-core energy consumption, as shown in the figure.

In addition to spatial slack by new process technology, other factors affect multi-core energy efficiency. On the positive side, the lower operating clock frequency (from 2 GHz to 1 GHz in Fig. 1.7) improves the per-core energy efficiency, e.g., by adopting shallower pipelines [12]. On the negative side, the first hurdle in achieving better energy efficiency is how to expose enough parallelism to fully utilize multiple cores. Another factor is that leakage power becomes more important, since leakage power is proportional to silicon die area (note that spatial slack means more silicon area usage) and the newer process technology incurs more leakage power consumption.

### 1.3.3 Behavior- and Architecture-Induced Temporal Slack: Runtime Distribution

In existing DVFS methods based on the prediction of WCET, the operating frequency, i.e., operating voltage, is set to $WCET/D$ where WCET is the worst case execution time of the remaining workload and $D$ is the time to deadline. In reality, it is rare to encounter the WCET. Instead, the execution time tends to have a distribution. Figure 1.8 illustrates the distribution (probability density function) of the runtime in decoding video clips obtained by running JM8.5 on an ARM946EJ-S processor (SoCDesigner).

Given such a wide runtime variation, by running the processor at the frequency level targeted for the worst case, we will lose opportunities for further reduction in energy consumption. Intuitively, more energy reduction could be obtained by running the processor at a frequency level near the average execution time divided by the time to deadline as long as there is a measure to guarantee the satisfaction of the given deadline constraint [17].

The runtime variation exemplified in Fig. 1.8 comes from two sources. One is the application behavior, which can have loops whose iteration counts are determined by input data. The other is the hardware architecture, the execution time of which varies depending on data values or access patterns. Figure 1.9 illustrates the distribution of the memory stall cycle obtained by running MPEG-4 decoding with 3000 frames of $1920 \times 800$ *Dark Knight* on an LG XNOTE LW25 laptop [18]. Such a wide variation results from access locality in the L2 cache and DRAM. In this case, the worst case assumption on memory stall time leads to losing the opportunities for achieving better energy efficiency.

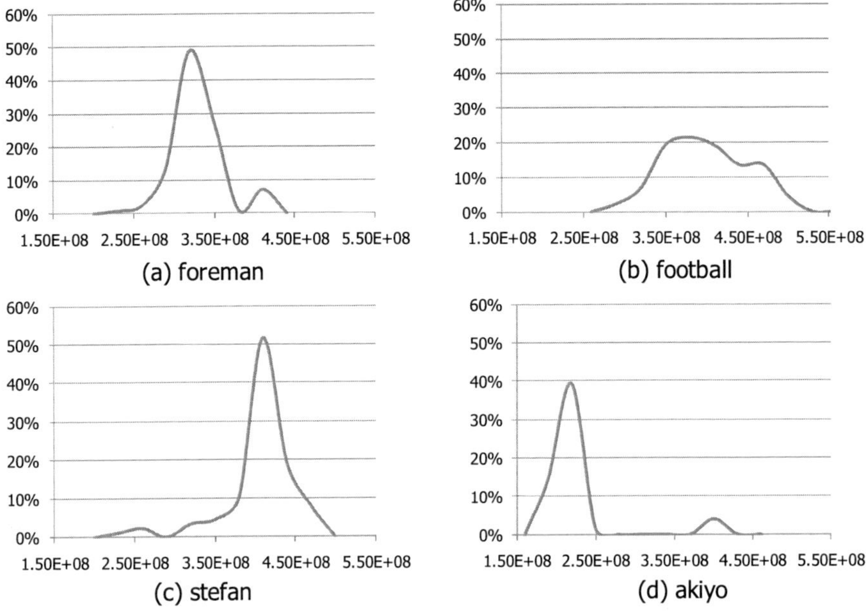



**Fig. 1.8** Runtime variation in decoding different video clips

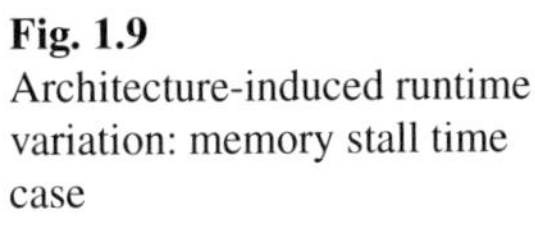

**Fig. 1.9** Architecture-induced runtime variation: memory stall time case

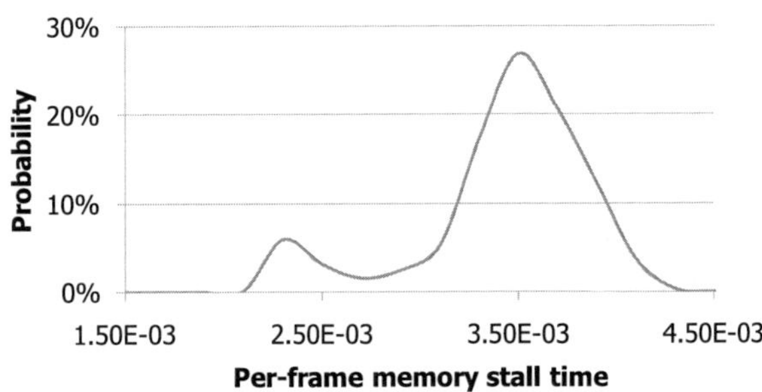



## *1.3.4 Process, Voltage, Temperature, and Reliability Slack*

Determining a design margin is one of the most important issues in designing low power chips. Several kinds of variations are taken into account to determine the timing margin, including process, voltage, and temperature variations, which are called PVT variation. Recently, reliability, e.g., negative bias temperature inversion (NBTI) has also been included in the timing margin. The amount of timing margin to cope with such variations is confidential to each chip manufacturer. Typically, 10–20% of the timing margin is assumed. From the viewpoint of power consumption, a 20% timing margin represents an opportunity cost of 36% ($= 1 - 0.8^2$) reduction in power consumption, as Fig. 1.10(a) shows. The timing margin is used to cope with the worst case of each of the process, voltage, temperature, and relia-

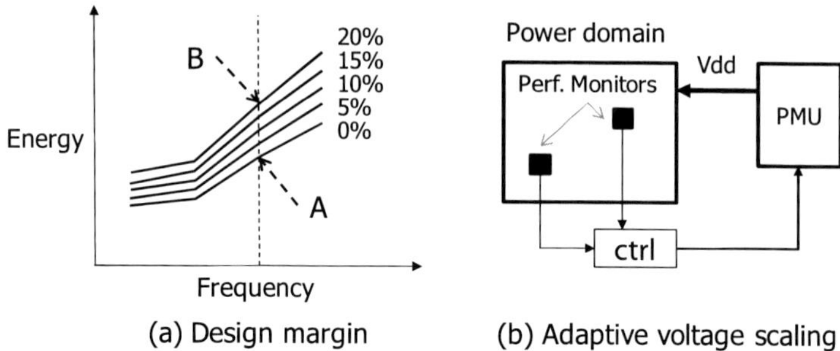


**Fig. 1.10** Coping with PVT variation [19]

bility variations. However, in reality, the worst case occurs only rarely. In addition, the four worst cases may occur at the same time with an extremely low probability. Thus, in normal conditions, the variations will be much smaller than the worst case levels. If we can exploit the slack, i.e., the difference between the worst and nominal conditions, we can recoup the lost opportunity cost.

Figure 1.10(b) illustrates how to exploit the slack dynamically during runtime. The figure shows a feedback loop starting from the performance monitor and going to the voltage regulator. The performance monitor mimics the critical path of the system with replica circuits. Based on the performance evaluation of the replica circuits, the performance monitor identifies the current level of process, voltage, temperature, and reliability variations. Based on the current performance level information, the controller (hardware or software) sets the voltage level to just meet the current operating frequency. Then, the voltage regulator adjusts the supply voltage (and body bias) to the level.

For instance, process variation can yield fast chips which have a lower threshold voltage than the average. The fast chips tend to suffer from high leakage power consumption due to the low threshold voltage. Thus, if the performance monitor reports that at the nominal voltage level (the voltage level obtained from the worst case assumption) the chip can run faster than the nominal frequency, then the supply voltage and/or body bias is reduced. As another example, if the current operating temperature is much lower than the worst case level, then a lower supply voltage than the nominal one is applied, thereby reducing the power consumption while meeting the required operating frequency. This method, called adaptive voltage scaling, is applied by most silicon manufacturers, for example, TI SmartReflex and ARM Intelligent Energy Manager.

## *1.3.5 Temporal and Spatial Thermal Slack*

The increasing demand for computing power and the slow improvement in cooling methods drives the need for thermal management. Thermal management is required for both high performance and mobile computing. In high performance computing,

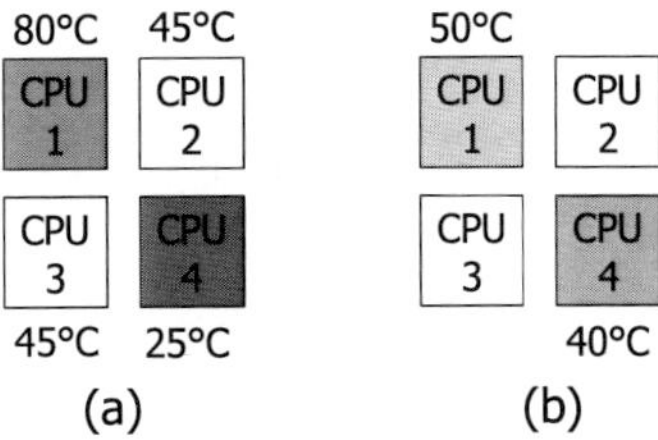



**Fig. 1.11** Exploiting two-dimensional thermal slack

as in the case of a data center, significant efforts are being made to lower the operation cost related to thermal management, i.e., cooling [2, 3, 5]. In mobile computing, thermal issues are considered from the beginning of the industrial design of the product in order to consider air flow inside the mobile devices, e.g., smartphones. In mobile computing, thermal constraints become more important for two reasons. First, there is no active cooling capability in mobile devices due to the small form factor requirement. Second, temperature has a significant impact on leakage power consumption; typically, leakage power consumption is exponentially proportional to temperature.

We can classify thermal slack as temporal and spatial slack. Temporal thermal slack represents the fact that a location on the silicon die can have phases of high and low operating temperature depending on the amount of computation, i.e., power dissipated in that location or nearby. At high temperatures, in order to prevent thermal problems, the execution is throttled or stopped. Thus, the lower the operating temperature, the higher the computing capability of the location. Adaptive voltage scaling, described above, is a way to exploit the temporal thermal slack.

Temperature reading is based on on-die temperature sensors. Multiple sensors monitor the temperature on hot spots (e.g., the instruction decoding stage, ALU, or floating point unit), and their maximum reading is typically interpreted as the core temperature. In real devices, significant temperature gradients exist on the die. For instance, even the intra-core temperature difference between the computing part and the cache exceeds 20°C [20].

Figure 1.11 illustrates how spatial thermal slack can be utilized for better energy efficiency. Figure 1.11(a) shows a quad-core example where CPU1 is the hottest while CPU4 is the coolest. In Fig. 1.11, suppose that a new thread needs to start on one of the four cores. Without considering the temperature gradient and the relationship between leakage power and temperature, any core with available computing power could be selected for the execution of the thread. However, for better energy efficiency, CPU4 needs to be selected because it is the coolest and will consume the least amount of energy by minimizing the leakage power, which is a strong function of temperature.

Temperature can determine the most energy-efficient core on the die, as shown in Fig. 1.11. The same situation occurs in the cases of 3D stacked dies on a small scale and the data center on a larger scale. In 3D stacked dies, the die near the heat sink has better cooling capability and thus is more energy efficient than other dies far from the heat sink. In the data center, computing server racks near the cooling facilities, e.g., at the air flow entrance, have a lower temperature and thus are more energy efficient than those with less cooling capability.

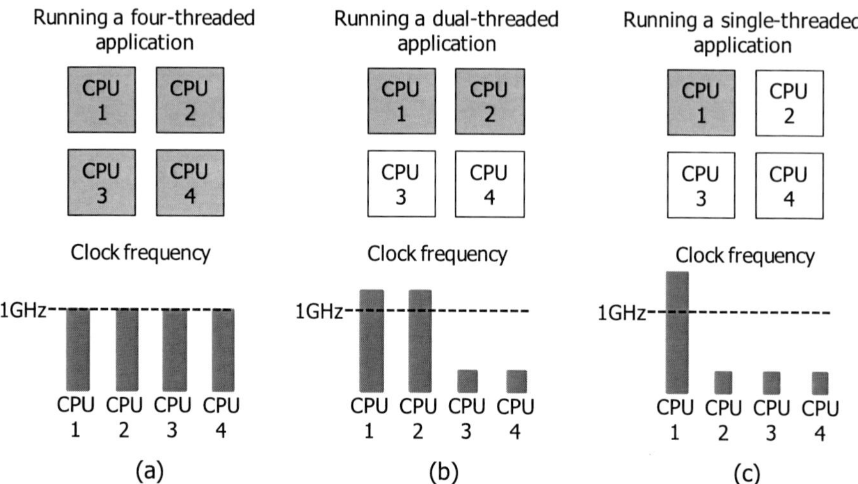


**Fig. 1.12** Peak power slack-aware overclocking [21]

### 1.3.6 Peak Power Slack

The peak power constraint is the maximum power that the power delivery system can provide. The peak power slack is the difference between the peak power constraint and instantaneous power consumption drawn by the silicon die. The peak power slack is often exploited in order to maximize performance while meeting the given peak power constraint. Figure 1.12 shows an example of exploiting the peak power slack.

In Fig. 1.12(a), four CPUs run at 1 GHz a four-threaded application, thus, one thread on a core. In this case, the parallelism of quad-core is fully exploited, thereby giving the best energy efficiency. If there is less parallelism in the computation than that in the underlying architecture, we can exploit the peak power slack to boost the performance of lightly threaded applications. In Fig. 1.12(b), only two threads run on two cores. In this case, the clock frequencies and supply voltages of the running cores can be increased by fully utilizing the peak power constraint. Figure 1.12(c) shows the case of running a single-threaded application at a higher frequency than the nominal level. In terms of drawing maximum performance from the given power budget, such an overclocking is useful, as proven in commercial solutions, e.g., Intel Turbo Boost. However, the energy efficiency of this solution is not yet proven to be better than that of conventional clocking—an interesting issue for the general usage of this method.

### 1.3.7 Holistic Approach for More Slack and Better Ways to Exploit It

Because many methods for low power design utilize slack, discovering new types of slack will open new possibilities for better energy efficiency. We expect that more types of slack and more innovative methods to exploit slack will be studied and applied to real designs. One possible way to discover new types of slack is a holistic approach which allows us to consider a bigger scope than a silicon chip design. One example that will be presented in this book is a surveillance system based on a wireless network where the quality of images captured in the camera subsystem and the transmission rate over the wireless network can be determined in an energy-efficient manner to realize the best quality of service (QoS) for the given budget of energy consumption. Another example of a holistic approach introduced in this book is the low power embedded storage system. In this system, dynamic power management by the storage subsystem only cannot fully exploit the full potential of existing idle time, i.e., slack in the storage subsystem. Instead, collaboration between the host and storage is required to better exploit the slack and thereby run the storage subsystem at lower power states more frequently, thus enabling less energy consumption.

## 1.4  Introduction to Chapters

Chapter 2 introduces various low power circuit techniques. First, methods for power consumption and thermal analysis are presented. Then, low power circuit techniques are explained for reduction in dynamic power, such as clock gating and dual-$V_{dd}$ and static power, e.g., power gating and body biasing.

Chapter 3 explains software-level low power design methods. Low power design of software requires an understanding of the seemingly complex characteristics of software execution cycles, i.e., runtime distribution. In this chapter, a simplified processor power model is first presented. Then, runtime distribution-aware low power design methods are explained which take into account the variations of software execution cycles due to both software program behavior and hardware architecture.

Chapter 4 reviews recent research efforts to improve the performance and energy efficiency of contemporary memory subsystems. First, memory access scheduling policies are explained, including conventional ones for performance and more advanced techniques for effectively managing DRAM power. Research works exploiting emerging technologies, e.g., 3D stacked DRAM and phase-change RAM, are introduced and their impacts on future memory subsystems are analyzed. Then, proposals to modify memory modules and memory device architectures are presented that reflect the memory access characteristics of future manycore systems.

Chapter 5 addresses low power on-chip network design, which is required as manycore design is becoming more popular. It is projected that the on-chip network will be the critical bottleneck of future manycore processors—in terms of both performance and power. In this chapter, we focus on the characteristics of multi-core, manycore on-chip networks and describe how energy-aware on-chip networks can be achieved with different techniques. In particular, we focus on how ideal on-chip networks can be designed such that energy consumption can be minimized and approach the energy consumption of the wires or the channels themselves.

Chapter 6 presents an energy-aware video codec design. It consists of three parts: an implementation of a low power H.264/AVC video codec using embedded compression (EC), an architecture of a power-scalable H.264/AVC video codec, and a power-rate-distortion modeling based on the power scalability of the video codec. The power consumption of the video codec results mainly from the external memory, i.e., DRAM, and the motion estimation (ME). In this chapter, the authors explain low power design techniques to reduce the power consumption of both DRAM and ME to offer about 80% reduction in the power consumption of the video codec.

Chapter 7 explains that efficient power conversion and delivery is equally important for the energy efficiency of an entire system. Specifically, since different types of voltage sources are used in a system for both digital and non-digital parts, the power conversion efficiency of DC-DC converters and linear regulators is crucial to leverage the power efficiency of the entire system. This chapter introduces power conversion subsystems and their efficiency characteristics and discusses system-level solutions to leverage the power conversion efficiency.

3D stacking of silicon dies presents a new low power design challenge coupled with that of temperature. In Chap. 8, the authors present a temperature-aware low power design method for 3D ICs. First, the characteristics of strong vertical thermal coupling in 3D die stacking are exploited to ease function mapping on cores. The authors describe two ideas: instantaneous temperature slack and memory boundedness-aware thread mapping. Instantaneous temperature slack enables one to overcome the conservatism in existing methods based on steady-state temperature, thereby enabling more aggressive utilization of temperature slack during runtime. Memory-bound threads are less sensitive to the core clock frequency change. Thus, the authors propose mapping memory-bound threads on hot and slow cores, which usually lack cooling capability since they are far from the heat sink. This choice enables CPU-bound threads to be mapped on cool and fast cores near the heat sink, thereby improving the total system performance.

Chapter 9 presents a case study of low power solid state disk (SSD) design. The chapter first introduces a multi-channel architecture for a high performance SSD. It then presents a power model of the SSD considering the parallel operations in the multi-channel architecture. It also gives an example in which the SSD power model is used to evaluate time out-based dynamic power management policies.

Chapter 10 gives an energy-aware design example of a wireless surveillance camera (WSC) consisting of image sensor, event detector, video encoder, flash memory, wireless transmitter, and battery. It is based on hierarchical event detection and data management (e.g., local store or remote transmission) to save the energy otherwise wasted on insignificant events. In a WSC, balancing the usage of all resources including battery and flash memory is critical to prolonging the lifetime of the camera, because a shortage of either battery charge or flash memory capacity could lead to a complete loss of events or a significant loss in the quality of the recorded image of events. The authors present a novel method which controls the bit rate of encoded videos and the sampling rate, e.g., the resolution and frame rate, to prolong the lifetime of the WSC.

Chapter 11 discusses two IC design examples for biomedical implantable electronics: cochlear and retinal implants. Energy and power awareness is important in such devices for safety as well as battery lifetime. For example, the pulsed output waveforms for electrical stimulation in such implants should be charge-balanced, because any unbalanced charge, if accumulated beyond a safe limit, can lead to cell damage and electrode corrosion. The chapter also addresses the issue of the long-term reliability of a neural interface whose impedance changes over a long time, therefore requiring monitoring-based adjustment of stimulation parameters to make sure that an optimum amount of electrical charge is delivered to the target neurons.

## References


1. International Technology Roadmap for Semiconductor. http://www.itrs.net
2. US Environmental Protection Agency: Report to congress on server and data center energy efficiency public law 109-431, Aug. 2007
3. McKinsey & Company: Revolutionizing data center energy efficiency, July 2008
4. Hamilton, J.: Data center infrastructure innovation. In: Web Performance and Operations Conference (Velocity), June 2010
5. Meisner, D., Gold, B.T., Wenisch, T.F.: PowerNap: eliminating server idle power. In: International Conference on Architectural Support for Programming Languages and Operating Systems (2009)
6. Intel, Co.: Data center energy efficiency with Intel® power management technologies, Feb. 2010
7. Neuvo, Y.: Cellular phones as embedded systems. In: Proceedings of IEEE International Solid-State Circuits Conference (2004)
8. Van Berkel, C.H.: Multi-core for mobile phones. In: Design Automation and Test in Europe (2009)
9. Wikipedia: Energy density. http://en.wikipedia.org/wiki/Energy_density
10. Flynn, D., Aitken, R., Gibbons, A., Shi, K.: Low Power Methodology Manual: For System-on-Chip Design. Springer, Berlin (2007)
11. Efficient Computing at Google. http://www.google.com/corporate/green/datacenters/index.html
12. Rabaey, J.: Low Power Design Essentials. Springer, Berlin (2009)
13. ARM Ltd.: Processors. http://www.arm.com/products/processors/index.php
14. Azevedo, A., et al.: Profile-based dynamic voltage scheduling using program checkpoints. In: Design Automation and Test in Europe (2002)
15. Bang, S., Bang, K., Yoon, S., Chung, E.Y.: Run-time adaptive workload estimation for dynamic voltage scaling. IEEE Trans. Comput.-Aided Des. Integr. Circuits Syst. **28**(9), 1334–1347 (2009)
16. Naveh, A., et al.: Power and thermal management in the Intel core duo processor. Intel Technol. J. **10**(2) (2006)
17. Hong, S., et al.: Runtime distribution-aware dynamic voltage scaling. In: International Conference on Computer-Aided Design (2006)
18. Kim, J., Yoo, S., Kyung, C.M.: Program phase-aware dynamic voltage scaling under variable computational workload and memory stall environment. IEEE Trans. Comput.-Aided Des. Integr. Circuits Syst. **30**(1), 110–123 (2011)
19. National Semiconductor, Co.: PowerWise Adaptive Voltage Scaling (AVS). http://www.national.com/analog/powerwise/avs_overview
20. Schmidt, R.: Power trends in the electronics industry—thermal impacts. In: IBM Austin Conference on Energy-Efficient Design (2003)
21. Intel, Co.: Inter turbo boost technology 2.0. http://www.intel.com/technology/product/demos/turboboost/demo.htm

# Chapter 2
# Low-Power Circuits: A System-Level Perspective

Youngsoo Shin

**Abstract** Popular circuit techniques for reducing dynamic and static power consumption are reviewed. The emphasis is on the implication when they are applied, e.g., area increase, because this may serve as important information during system-level design. The estimation of power and temperature is also reviewed.

## 2.1 Introduction

During the architectural design (or system-level design, broadly speaking), a lot of what-if questions are likely to be raised and answered. For example, in a network processor, the designers may consider employing two Ethernet controllers, instead of a single one, to improve throughput, but may also want to validate the choice in terms of chip area [1].

Due to the growing importance of power consumption, it is now tempting to assess the design choice in terms of power: what happens if clock gating is applied to block A, which contains many synchronous memory elements; what happens if body biasing is used in block B, which stays in standby mode for most of its operation time? These questions should be answered after the implication of applying each circuit technique is precisely understood from a system-level perspective; e.g., how much does the circuit area increase when clock gating is applied to A, and what is the latency to put B in standby mode and bring it back to active mode?

This chapter is organized to review various low-power circuit techniques from a system-level perspective. A technique to estimate power consumption is discussed in Sect. 2.3; thermal analysis, which has become very important, is also addressed. Power consumption can be categorized into dynamic power during operational time and static power during standby periods. Representative circuit techniques to reduce the dynamic component, i.e., clock gating and dual-$V_{dd}$, as well as other techniques are reviewed in Sect. 2.4. Techniques to reduce the static component, such as power gating and body biasing, are presented in Sect. 2.5.

Y. Shin (✉)
KAIST, Daejeon, Republic of Korea
e-mail: youngsoo@ee.kaist.ac.kr

C.-M. Kyung, S. Yoo (eds.), *Energy-Aware System Design*,
DOI 10.1007/978-94-007-1679-7_2, © Springer Science+Business Media B.V. 2011

## 2.2 CMOS Power Consumption

To understand the nature of power consumption of CMOS circuits, consider the chip floorplan illustrated in Fig. 2.1. The overall operation of a floorplan block can be classified as being in active or in standby mode. Active mode refers to the period of time when the block is actively computing to produce valuable output; the remaining period is called standby mode. In active mode, there are two components of power consumption: dynamic and static power. *Dynamic power* is consumed while a transistor is switching. The length of time that it switches is usually a small proportion of a clock cycle; for the remaining time, the transistor consumes *static power*. Standby mode, which does not involve any transistor switching, consists of static power alone (assuming that there is also no switching activity in a clock). It is important to understand that the static power in active mode is a transient one, while that in standby mode is a static one; therefore, their amounts are very different, as we address later in this section.

### 2.2.1 Dynamic Power

While the output of the CMOS inverter shown in Fig. 2.1 makes a pair of rising and falling transitions, the amount $C_L V_{dd}^2$ of energy is dissipated, half of it by the pMOS transistor and the other half by the nMOS transistor. $C_L$ is the load capacitance, which models the gate capacitance of fanout gates, the wire capacitance, and the intrinsic capacitance of the inverter itself.

The average power consumption due to the switching, which is the total energy dissipation during a particular period of time divided by the length of that period, is given by the well-known expression

$$P_{sw} = \alpha C_L V_{dd}^2 f, \tag{2.1}$$

where $f$ is the clock frequency and $\alpha$[1] is the probability of the output making a pair of rising and falling transitions in a single clock cycle. Note that $\alpha \leq 0.5$ for any combinational gate unless there is a glitch; in practical circuits, $\alpha$ turns out to be very low, typically less than 0.05. Gates that are driven by a clock, for example those in clock buffers, have $\alpha = 1.0$.

Another component of dynamic power consumption, denoted by $P_{sc}$, is caused by *short-circuit current*. This is the current that flows while both the nMOS and pMOS transistors are turned on for a short period of time when the input signal makes a transition (from 0 to 1 or 1 to 0). Interestingly, $P_{sc}$ decreases with increasing $C_L$ [2], because the output, which changes its value more slowly when heavily loaded, keeps the short-circuit current from increasing. $C_L$, however, cannot be arbitrarily increased due to increased circuit delay. $P_{sc}$ is usually pre-characterized

---

[1] Some people use $\alpha$ as a probability that the output makes a transition (either rising or falling) rather than a pair of transitions. With this definition, $1/2\alpha$ would be used instead of $\alpha$ in (2.1).

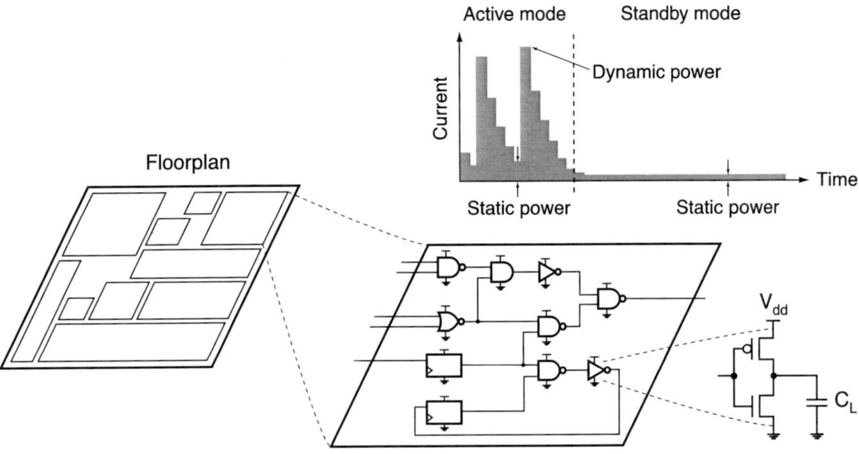


**Fig. 2.1** Power consumption of an architectural block

when each gate is designed and is available during power estimation. In practical circuits, $P_{sc}$ is a small proportion of the total dynamic power consumption $P_{dyn}$; e.g., $P_{sc}/P_{dyn}$ is estimated to be about 10% [3].

### *2.2.2 Static Power*

The static power consumption is a result of the device leakage current, which originates from various physical phenomena [4]. Three components of leakage (subthreshold, gate tunneling, and junction leakage) get more attention than the other ones due to their large proportion in the total static power. The relative importance of these components differs with the technology, the temperature, the style of the circuit, and so on. For instance, gate leakage is important in static random access memory (SRAM) circuits since they typically rely on devices of larger gate length to reduce random dopant variations, while subthreshold leakage is dominant in logic circuits [5].

The subthreshold leakage occurs when the gate-to-source voltage of a transistor is below its threshold voltage ($V_{th}$), i.e., when a device is presumed to be turned off. It is well known that this leakage component increases exponentially with decreasing $V_{th}$, increasing temperature, and increasing gate-to-source voltage. This implies the growing importance of subthreshold leakage as CMOS technology scales down, since $V_{th}$ tends to decrease to maintain circuit speed. It also implies that any quantitative result on static power should be carefully understood; e.g., the value may be very different for different temperatures.

The standby leakage (leakage in standby mode) of the 2-input NAND gate shown in Fig. 2.2(a) for the different inputs is given in the second column of Table 2.1. It is well known that this leakage is lowest when the input is 00, as Table 2.1 confirms.

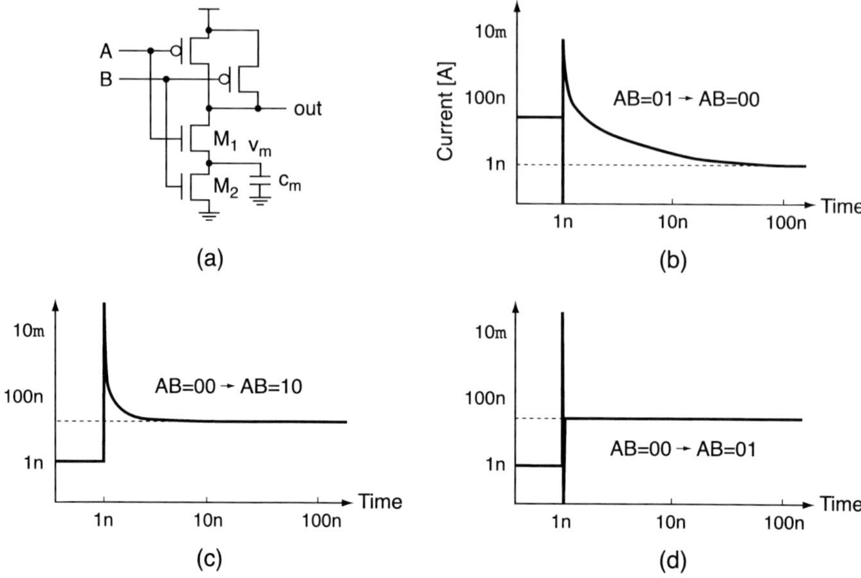

Fig. 2.2 (a) A 2-input NAND gate: active leakage for input transitions (b) from 01 to 00, (c) from 00 to 10, and (d) from 00 to 01

Table 2.1 Standby and active leakage of a 2-input NAND gate in 45-nm technology

| Input (AB) | Standby leakage (nA) | Active leakage (nA) | | |
|---|---|---|---|---|
| | | 1 ns | 5 ns | 10 ns |
| 00 | 1.0 | 11.2 | 3.0 | 1.9 |
| 01 | 16.6 | 16.2 | 16.6 | 16.6 |
| 10 | 10.3 | 17.7 | 10.4 | 10.3 |
| 11 | 26.4 | 26.4 | 26.4 | 26.4 |

The reason is that there is a positive voltage $v_m$ which builds up between $M_1$ and $M_2$ and turns $M_1$ off strongly, due to a negative gate-to-source voltage; this voltage also raises the effective threshold voltage of $M_1$. The whole phenomenon is called the stacking effect [6], because the leakage shrinks as stacked MOS transistors are turned off.

We now turn our attention to active leakage (leakage in active mode). When the input is maintained at 01, the internal node capacitance $c_m$ is fully discharged. If the input is changed to 00 after 1 ns, as depicted in Fig. 2.2(b), the small leakage current through $M_1$ starts to charge $c_m$. As $v_m$ rises, the leakage through $M_1$ falls further due to the stacking effect. But this transition takes a long time, as shown in Fig. 2.2(b). The effect on leakage of a change of input from 00 to 10 is shown in Fig. 2.2(c). The large turn-on current through $M_1$ initially charges $c_m$; however, as $v_m$ rises, $M_1$ turns off, but then its leakage current takes over and continues to charge $c_m$, even though the leakage is gradually falling. If $M_2$ is turned on, for instance by

**Fig. 2.3** Comparison of three components of power consumption in 45-nm technology; leakage is measured assuming 125°C

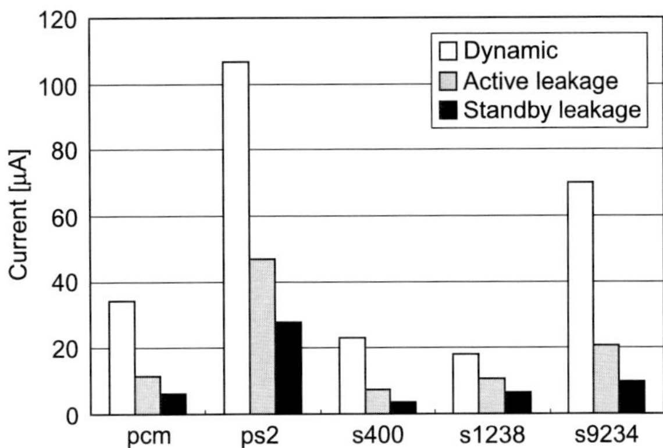


the change of input from 00 to 01 shown in Fig. 2.2(d), the corresponding leakage transition is virtually spontaneous since $c_m$ is quickly discharged.

The average active leakage over different periods after the change of input value is given in the last three columns of Table 2.1. Each value is also averaged over all the transitions that lead to the inputs shown in the first column: thus the first row covers transitions from 01 to 00, from 10 to 00, and from 11 to 00.

The standby and the active leakage are about the same when a 1 is applied to input B (01 and 11 of Table 2.1), which turns on $M_2$. The leakages for 10 and, especially, 00, are significantly different, particularly for the period immediately after the transition, implying a higher operating frequency.

## 2.2.3 Analysis

There are now three components of power consumption: dynamic, active leakage, and standby leakage. The first two components are sources of active-mode power consumption; the last defines standby-mode power consumption. Experiments were performed in 45-nm technology to understand the relative measure of the components; the results are shown in Fig. 2.3. Example circuits were taken from International Symposium on Circuits and Systems (ISCAS) benchmarks as well as from OpenCores [7]. The current was obtained by applying 100 random vectors; the clock period was arbitrarily assumed at 5 ns.

Active leakage represents, on average, 28% of the total active-mode power consumption; it is as high as 37% in s1238 and as low as 23% in s9234. Note that the leakage was measured in conditions where it becomes as large as possible, i.e., a fast process corner in which $V_{th}$ is smallest and at the highest operating temperature. Since dynamic power is scarcely affected by these parameters, the proportion of active leakage will become smaller in different conditions. For example, its proportion decreases to 14% in a nominal process corner with the same temperature.

The average standby leakage is 54% of the average active leakage. The variation in the leakage ratio between circuits can be explained by the extent of the stacking

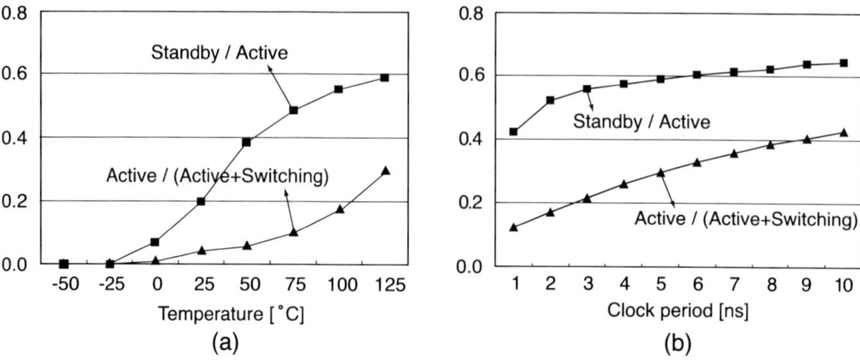



**Fig. 2.4** Ratio of standby to active leakage, and the proportion of active leakage in circuit ps2: with (**a**) varying temperature and (**b**) varying clock period

effect in each circuit. When there are more gates that exhibit the stacking effect in standby mode, we expect the difference between active and standby leakage to increase. This can be confirmed by counting the number of inverters and flip-flops, which are representative of the gates without the stacking effect.

The proportion of active leakage decreases with temperature, as shown in Fig. 2.4(a). The ratio of standby to active leakage also declines, as Fig. 2.4(a) shows, suggesting that the importance of active leakage grows as the temperature drops. When this happens, the transient change in active leakage due to a transition (see Fig. 2.2) takes longer because of its reduced magnitude, which means that $c_m$ is charged more slowly: this increases the difference between active and standby leakage.

As the clock frequency increases and the clock period decreases, the magnitude of the active leakage will increase while the standby leakage remains the same. This is evident from the decreasing ratio between the standby and active leakage shown in Fig. 2.4(b). The total switching current is independent of the clock period, as long as that period is sufficient to accommodate all the switching required. While the average switching current and the active leakage both increase as the clock period decreases, the average switching current increases more rapidly. Thus, the active leakage comes to represent a lower proportion of the total active-mode current, as we see in Fig. 2.4(b).

The contribution of the three components in energy dissipation is determined by the amount of time for which each component is responsible. This in turn is dependent on the fraction of time a circuit stays in active mode, i.e., the duty cycle $D$. Let dynamic power and active leakage be 72% and 28% of the active-mode power consumption, respectively, and active leakage be 1.87 times the standby leakage. Figure 2.5 illustrates the contribution of the three components with different values of $D$, e.g., $4.80D/(5.67D + 1)$ for dynamic power. When $D = 0.1$ such as in a cell phone, 58% of the energy is due to standby leakage, while 31% and 11% are due to dynamic and active leakage. It is apparent that most of the energy is dissipated by dynamic power as $D$ increases, which arises in stationary devices such as servers.

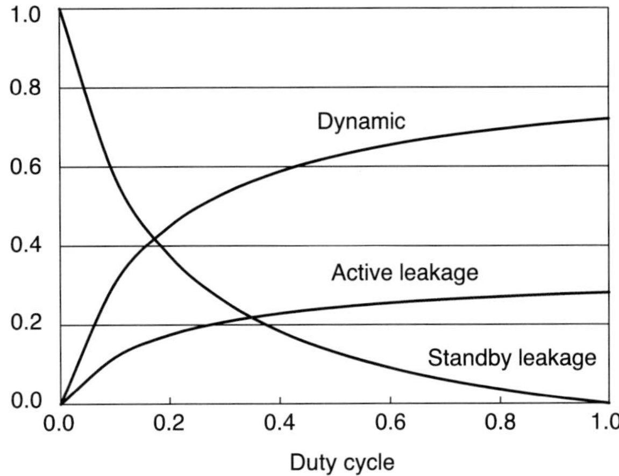


**Fig. 2.5** Contribution of three components in energy dissipation with varying duty cycle $D$

## 2.3 Estimation of Power Consumption

The biggest part of answering what-if questions during architectural design is the ability to estimate power consumption, before and after a particular circuit technique is applied; this is a subject of this section. We also address temperature estimation because the main quantity that determines temperature is power consumption and because temperature has become a roadblock in technology scaling.

### 2.3.1 Dynamic Power

Expression (2.1) suggests that the estimation of $P_{sw}$ comes down to estimating $\alpha$ of each node, once $C_L$ is extracted. This is done either by simulation or by probabilistic analysis.

Different gate delay models can be used in a simulation approach. The simplest model assumes zero gate delay for the sake of simulation time. Each gate can have at most one transition per input vector, since all transitions occur at the same time. If real delay is used, each gate may have different delay resulting in different arrival times at the gate inputs, which causes more than one transition per input vector. But, this takes more time than simulation under zero delay. Gate-level simulation is reported to yield an error of $\pm15\%$ compared to circuit-level simulation, which exhibits $\pm5\%$ error [8]. Another issue is the preparation of input vectors. This is either done by designer-specified use scenarios, or is based on generating a sequence of random vectors. The interesting question here is the number of vectors that should be provided for reasonable accuracy. Experimental study [8] states that using any 100 or 10 consecutive vectors guarantees an error within $\pm5\%$ or $\pm15\%$ (compared to using the whole sequence of vectors from use scenarios), which implies that 10 should be enough for the accuracy of gate-level simulation.

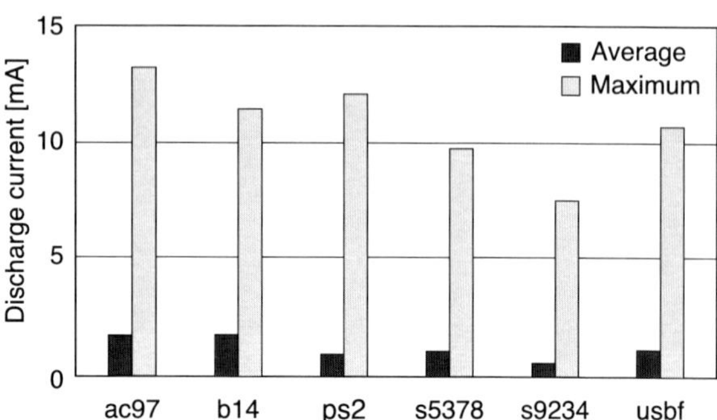

**Fig. 2.6** Comparison of average and maximum discharge current



Probabilistic analysis is more convenient from a designer's perspective. One may simply assume a signal probability (probability of signal being at 1) at each circuit input, quite often 0.5. The probabilities are then propagated toward circuit outputs. The propagation of independent signals is straightforward; e.g., the signal probability at the output of an AND gate is 0.25 when the signal probability of both inputs is 0.5. However, in general circuits, many signals are not independent due to reconvergent fanout; i.e., the same fanout converges at the same gate after going through different paths to the gates. The propagation in this case becomes more difficult, although several methods have been proposed [9].

Note that these power estimation methods target average power consumption. The maximum power consumption, which is necessary for designing a power distribution network, is significantly larger than the average one. This is quantitatively shown for several circuits in Fig. 2.6, in which the difference ranges from 6 to 7 times.

**Accuracy of Estimation**    The important issue in power estimation is its accuracy. This is affected by several factors such as delay model, wire model, and test vectors, but, more importantly, by the design stage in which power estimation is performed. During system-level design, many blocks are in a register transfer level (RTL) description. The description then goes through logic synthesis, in particular technology mapping, to obtain a technology-mapped netlist; some optimizations are then performed, and the layout is finally obtained. Before layout design, the inaccuracy of power estimation ranges $\pm 15\%$; a similar inaccuracy is observed in power estimation before optimization. However, the error of power estimation before technology mapping (an estimation without actual netlist) can reach a factor of 4 or 10, which invalidates any estimation effort at that early stage [8].

### 2.3.2 Static Power

For a given gate-level netlist, estimating leakage power is generally more difficult than estimating switching power. Switching power is weakly dependent on device parameters and operating environments. However, leakage power is strongly affected by the variations of process parameters (e.g., gate length, oxide thickness, and channel dose), variations of operating environment (temperature and $V_{dd}$), and different input patterns.

The dependency of leakage current on process variations is the strongest; e.g., for $3\sigma$ die-to-die $V_{th}$ variation of 30 mV in 180-nm CMOS technology, the leakage current can vary by a factor of 20, while the frequency varies only by 20% [10]. Die-to-die variations are typically taken into account by using process corners; i.e., we can estimate leakage current by assuming one particular set of deterministic device parameters. However, within-die variations, which are occupying an increasing proportion of total process variations with technology scaling, can only be captured by statistical estimation. The dependency of leakage on operating environments is also strong, although less strong than for process variations in practice. Leakage has a superlinear dependency on temperature, e.g., a 30°C change of temperature causes leakage to increase by 30%, and its dependency on supply voltage is exponential, e.g., a 20% fluctuation of $V_{dd}$ causes leakage to change by a factor of 2 or more [11]. Therefore, for accuracy, leakage estimation should be coupled with an analysis of temperature and $V_{dd}$ distribution. The dependency of leakage on input vectors is strong in individual gates, but becomes very weak in whole circuits, especially as circuits have more levels due to lack of controllability.

**Static Estimation**  For leakage analysis or simulation, each gate in the library must be characterized in its leakage. For example, for a 2-input NAND gate, the leakage for each input combination can be characterized: $L_{00}, L_{01}, L_{10}, L_{11}$, where $L_{ij}$ indicates leakage when the inputs take $i$ and $j$. Alternatively, for simplicity, its leakage could be characterized by the average value.

If the leakage of all the gates is characterized, the leakage of an individual gate can be obtained if we know the signal probability of each input. For example, the leakage of the 2-input NAND gate is given by $(1 - p_1)(1 - p_2)L_{00} + (1 - p_1)p_2L_{01} + p_1(1 - p_2)L_{10} + p_1p_2L_{11}$, where $p_1$ and $p_2$ are the signal probabilities of two inputs. The leakage of the whole circuit can then be obtained by summing all the leakages. Thus, the key step is to derive the signal probability of all internal nodes given the signal probability of the primary input, which is the same process as in dynamic power estimation.

**Statistical Estimation**  There are two methods to incorporate within-die process variation in leakage analysis: Monte Carlo simulation (simulation with repeated random sampling of variation source) or statistical estimation. Figure 2.7 illustrates typical leakage histograms after Monte Carlo simulation with 45-nm technology [12], in which $\sigma$ of $V_{th}$ is assumed to be 10% of its normal value. The histogram roughly follows a lognormal distribution.

In statistical estimation, the leakage of each gate is modeled as a lognormal, i.e., $\alpha e^{Y_i}$ [13], where $Y_i$ is a function of process parameters such as gate length and gate oxide thickness, and approximated as a normal distribution. It is shown that both subthreshold and gate tunneling leakage follow this model. The full-chip leakage is then a sum of lognormals, which can be approximated as another lognormal

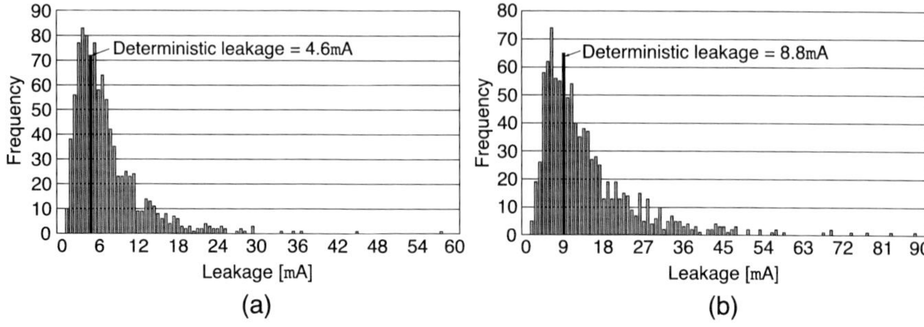

Fig. 2.7  Monte Carlo simulation of leakage: (**a**) c432 and (**b**) c1350

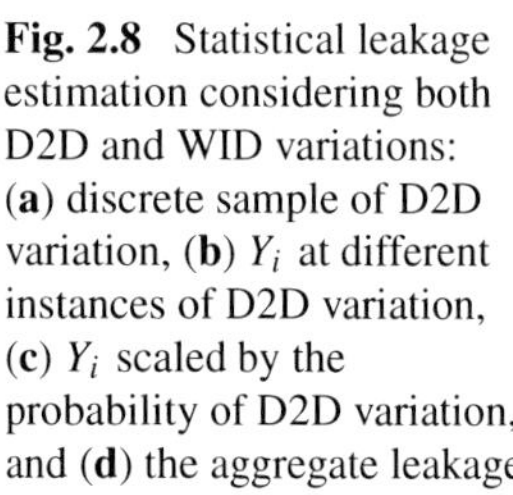
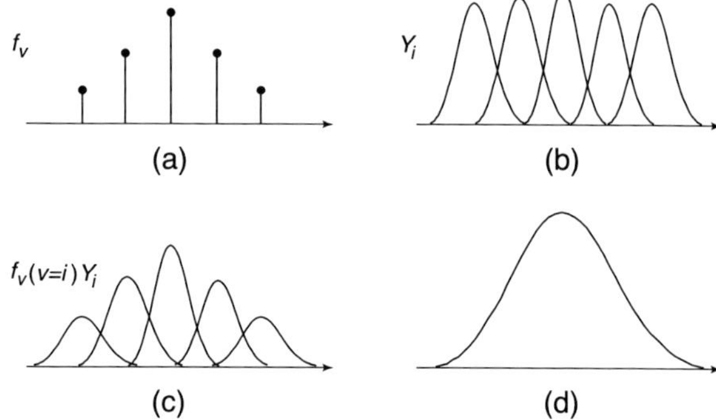

**Fig. 2.8**  Statistical leakage estimation considering both D2D and WID variations: (**a**) discrete sample of D2D variation, (**b**) $Y_i$ at different instances of D2D variation, (**c**) $Y_i$ scaled by the probability of D2D variation, and (**d**) the aggregate leakage

or, more accurately, as an inverse-gamma distribution [14]. If leakage is estimated from a layout, a spatial correlation of device parameters must be taken into account. In other words, $Y_i$ and $Y_j$ are highly correlated if gates $i$ and $j$ are closely located. A chip is divided into an imaginary grid, and a correlation coefficient is defined between a pair of grids, which is then incorporated into the leakage estimation [13].

Statistical leakage estimation considering both die-to-die (D2D) and within-die (WID) variations can be done, as illustrated in Fig. 2.8 [15]. D2D variation is sampled at discrete points (a). Each sampled value becomes a mean of a corresponding normal distribution of $Y_i$ (b). Each $Y_i$ is scaled by the corresponding probability of the sample from D2D space (c). Statistical leakage estimation is done for each $Y_i$ and aggregate leakage is obtained (d).

## 2.3.3 Temperature Estimation

Temperature changes because of the convection of heat. Therefore, it is reasonable to expect to produce temperature change by adjusting the location of hotter and colder blocks, i.e., by trying different floorplans. It is reported that different floor-

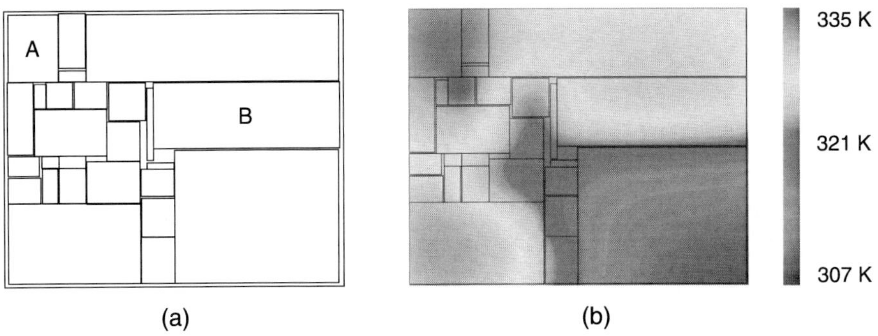


**Fig. 2.9** (**a**) Floorplan of an example chip and (**b**) thermal map

plans of microprocessors can yield a difference of maximum temperature of as high as 37°C [16].

For a given floorplan, as shown in Fig. 2.9(a), the following heat conduction equation is solved to generate a thermal map, shown in Fig. 2.9(b):

$$\rho C_{\mathrm{p}} \frac{\partial T(x, y, z, t)}{\partial t} = \nabla \big[ \kappa(x, y, z, t) \nabla T(x, y, z, t) \big] + g(x, y, z, t), \qquad (2.2)$$

where $T$ is the temperature which we try to obtain, $g$ is the power density of a heat source, and $\kappa$ denotes thermal conductivity; $\rho$ and $C_{\mathrm{p}}$ are material-dependent parameters. Physically, (2.2) implies that the energy stored in a volume $V$ (left-hand side) is equal to the sum of the heat entering $V$ through its boundary surface and the heat generated by itself (right-hand side).

In general, steady-state temperature is of importance because, once a chip reaches that state, the temperature does not respond to an instantaneous change of power consumption. This is due to the relatively large time constant of heat conduction (a few milliseconds) compared to that of a clock cycle (some picoseconds). In a steady state, in which there is no change of temperature over time, the following equation can be solved:

$$\nabla^2 T(x, y, z) = -g(x, y, z)/\kappa, \qquad (2.3)$$

where $\kappa$ is approximated to be constant. Note that $g$ is typically given for each block, say A and B of Fig. 2.9(a); in other words, we approximate the power density of A to be homogeneous—this can be a source of error when the block is very big. Average power consumption (over some period of time) is used for $g$ of (2.3), which can be another source of error, particularly when we try to obtain the maximum temperature. These limitations should be kept in mind when temperature is referred to after estimation.

There are several methods to solve (2.2) or (2.3). Numerical methods include the finite difference method (FDM) or finite element method (FEM), both of which discretize the continuous space domain into a finite number of grid points. But these methods are very slow, usually taking tens or hundreds of minutes; thus, it is not practically possible to use them in any optimization loop.

Fast estimation methods do exist. A notable one is to use a *thermal RC circuit* [17]. This is a circuit built based on the analogy between heat transfer and electrical current: heat flow can be described as a current flowing through a thermal resistance, thus yielding a temperature difference analogous to voltage. Thermal resistance and capacitance are modeled on a per-block basis or, more accurately, on a per-grid basis, in which a chip is divided into a number of imaginary grids. Another fast method to solve (2.3) is to use a Green's function. It can be readily shown that (2.3) is equivalent to

$$T(\mathbf{r}) = \int_{-\infty}^{\infty} G(\mathbf{r}, \mathbf{r}_0)\left(-\frac{g(\mathbf{r}_0)}{\kappa}\right) d\mathbf{r}_0, \qquad (2.4)$$

where $\mathbf{r}$ is $(x, y, z)$ and $\mathbf{r}_0$ is a particular value of $\mathbf{r}$. $G$ satisfies $\nabla^2 G(\mathbf{r}, \mathbf{r}_0) = \delta(\mathbf{r} - \mathbf{r}_0)$ and is called a Green's function; i.e., $G$ is a Green's function if its Laplacian is a delta function. Instead of solving partial differential equation (2.3), we can use (2.4) to directly give $T$ once $G$ is known. The product of cosine functions [18] and the division of hyperbolic functions have been used for $G$.

## 2.4 Circuits to Reduce Dynamic Power

Many circuit techniques have been proposed to reduce dynamic power consumption. Two of them, namely clock gating and dual-$V_{dd}$, deserve attention because of their popularity and effectiveness, and are reviewed in this section in detail. Other techniques are summarized in Sect. 2.4.3.

### 2.4.1 Clock Gating

It is well known that a clock distribution network takes a large portion of total power consumption, e.g., 18% to 36% for processors and 40% for ASICs [19]. This is because the elements of the network including flip-flops (or latches) and clock buffers, as shown in Fig. 2.10, are always triggered. A simple way to reduce this consumption is to gate the clock to a flip-flop, say A, when its input and output are the same. If a clock to A and B can be gated at the same time, we may try to gate the buffer C instead, or higher stage buffers if more flip-flops can be gated together.

Conceptually, clock gating can be implemented as shown in Fig. 2.11(a). The block called clock gating logic determines when the combinational logic does not perform its computation (EN = 0) and when it does (EN = 1). Two things should be noted in regard to clock gating logic. It is an extra logic, which causes an increase of circuit area and power consumption; it therefore should be kept small as much as possible. Clock gating logic itself is a combinational logic, and it thus may generate a hazard; in particular, a static 1-hazard (a change of logic value from

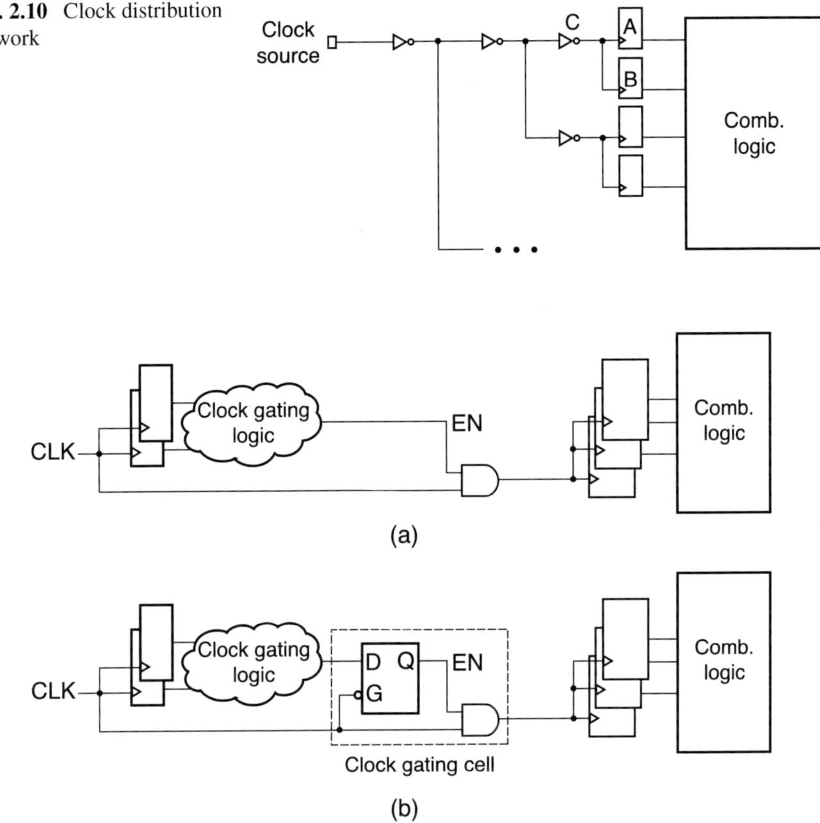



**Fig. 2.10** Clock distribution network

**Fig. 2.11** Clock gating: (**a**) concept and (**b**) implementation

1 to 0 and back to 1, for a short period of time) while $CLK = 1$ makes the flip-flops capture their inputs when they are not supposed to. This is resolved by using a negative sensitive latch, as shown in Fig. 2.11(b). When $CLK = 1$, the latch is opaque and thus blocks any hazard from clock gating logic. The latch together with an AND gate are typically called a clock gating cell. Note that a positive sensitive latch and an OR gate are used if the flip-flops are falling edge triggered ones.

From the designer's perspective, the challenge is to design the clock gating logic such that flip-flops are gated as often as possible while the gating logic is kept small. This is done either manually by human designers or automatically by CAD tools. A generic form of digital circuit consists of a data path and controller, as illustrated in Fig. 2.12. Designers should know when each functional unit is idle from a scheduled data flow description, which could guide them to design clock gating logic. The controller is typically modeled as a finite state machine (FSM) such as the one shown in Fig. 2.12; self-loops associated with states A and B correspond to the mo-

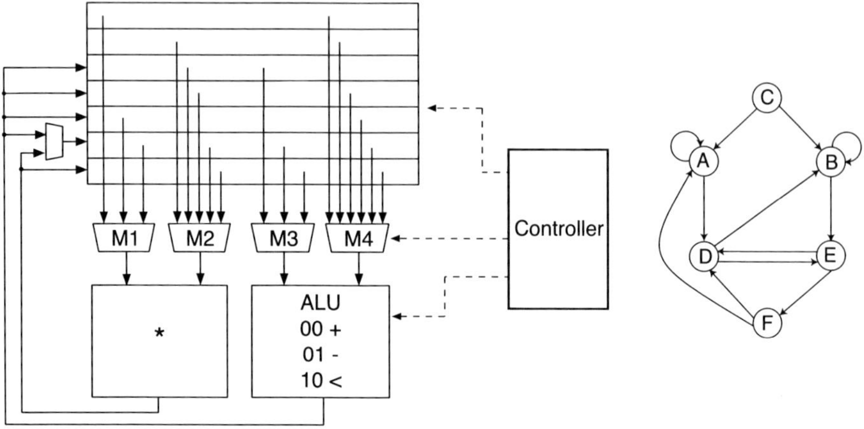

**Fig. 2.12** Digital circuit consists of data path and controller

ment when the clock can be gated; e.g., if state A is assigned binary representation 000, the register content does not change before and after the self-loop.

Automatic synthesis of clock gating logic is a bottom-up approach. Let $\delta_i(\cdot)$ be the Boolean expression for the input of the $i$th flip-flop and let $S_i$ be its output. The clock can be gated when these two take the same value (either 1 or 0); the clock gating logic is thus given by

$$g_i = \overline{\delta_i(\cdot) \oplus S_i}. \tag{2.5}$$

Implementing each $g_i$ as a separate logic incurs too much overhead. Some $g_i$'s thus should be merged as much as possible. Note that if $g_1$ and $g_2$ are merged, flip-flops 1 and 2 are gated only when both can be gated, thereby reducing the gating probability. Therefore, merging $g_i$'s is a trade-off between clock gating logic and gating probability. There are other techniques to reduce the complexity of clock gating logic, e.g., using don't cares, logical approximation, and so on [20].

Figure 2.13 shows the result of clock gating synthesis using a commercial CAD tool [21]. Interestingly, the area starts to decrease as clock gating is applied to some flip-flops, less than 400 flip-flops in Fig. 2.13. This happens because a feedback multiplexer, which is attached to a flip-flop to retain its current value when it needs to, can be removed when clock gating is applied; i.e., if the clock is gated, the value of the flip-flop is retained anyway. This benefit is outweighed by an increasing amount of clock gating logic as more flip-flops are gated. The power consumption monotonically decreases, indicating that extra power consumption from clock gating logic is not large enough (due to the low switching activity of combinational logic) to mask the reduced power consumption of the flip-flops. Figure 2.13 suggests a trade-off between area and power consumption, which can be exploited in architectural design.

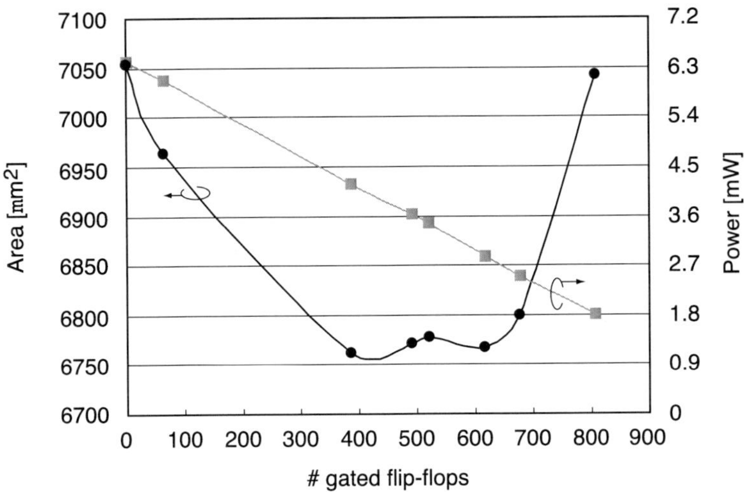



**Fig. 2.13** Clock gating synthesis for wbdma [7], which contains 987 flip-flops in total

## 2.4.2 Dual-$V_{dd}$

From (2.1), it is clear that the best way to reduce dynamic power is to reduce $V_{dd}$. However, reducing $V_{dd}$ comes at the cost of reduced circuit speed. The alternative approach is to use two $V_{dd}$'s, a higher one $V_{ddh}$ and a lower one $V_{ddl}$, so that $V_{ddl}$ is used for gates that do not affect the overall circuit speed. There are several challenges in the implementation of dual-$V_{dd}$:

- Layout architecture: gate placement is restricted in dual-$V_{dd}$, which causes an increase of area and wire length.
- Level conversion: a $V_{ddl}$ gate needs a level converter when it drives a $V_{ddh}$ one.
- Dual-$V_{dd}$ allocation: automatic allocation of $V_{ddh}$ and $V_{ddl}$ must be done for the CAD tool.
- Selection of $V_{ddl}$: $V_{ddh}$ is typically mandated by technology; $V_{ddl}$ is thus a design parameter whose value should be carefully selected.

**Layout Architecture**  Figure 2.14(a) shows the simplest architecture [22], where each row is dedicated to either $V_{ddh}$ or $V_{ddl}$ cells. Standard placement tools can be used for this architecture; once each cell is tagged for a type of row ($V_{ddh}$ or $V_{ddl}$), it can be placed. The wire length typically increases substantially [23], which makes timing closure difficult to achieve; this limits the application of this architecture. Figure 2.14(b) [24] is a layout architecture, which is less constrained as far as cell placement is concerned. However, in order to minimize well isolation (see Fig. 2.15), each of the $V_{ddh}$ and $V_{ddl}$ cells must be grouped as much as possible; this may be performed as a refinement step after the initial placement [24], but carefully, so that initial placement is not perturbed too much. This step must be coordinated

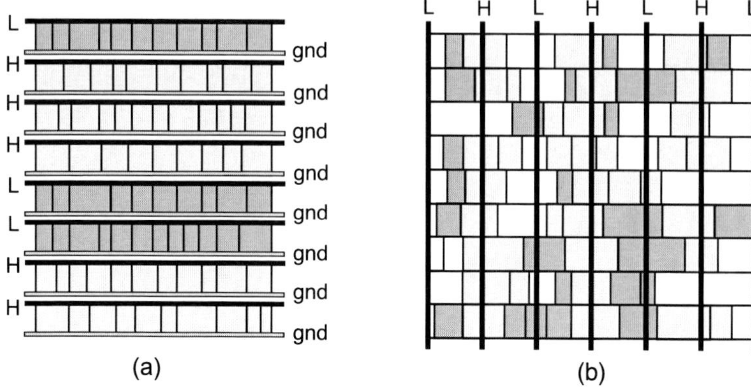

**Fig. 2.14** (**a**) Row-based layout architecture and (**b**) architecture that allows less constrained placement

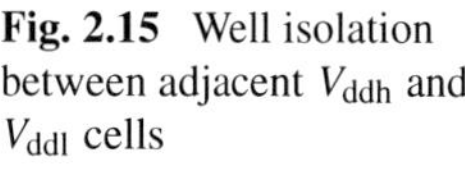

**Fig. 2.15** Well isolation between adjacent $V_{ddh}$ and $V_{ddl}$ cells

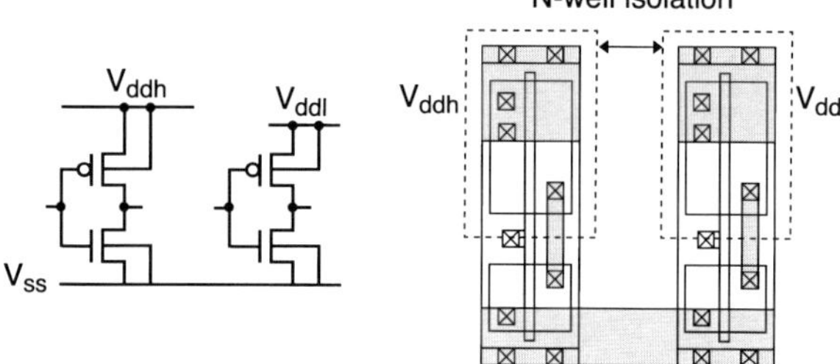

**Fig. 2.16** Dual supply rail standard cells

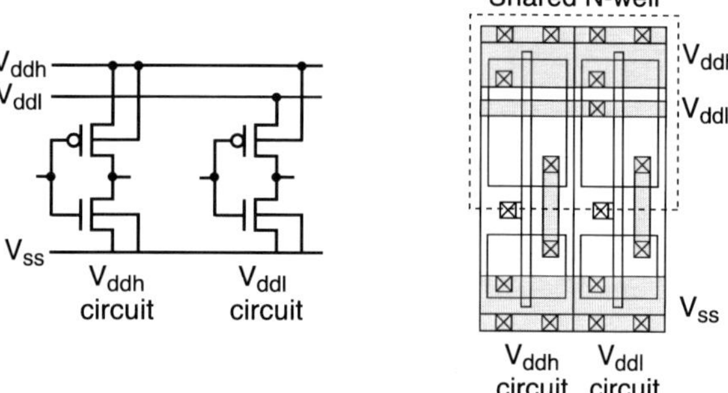

with the topology of the generating power grid, so that each group can be placed directly below (or very close to) corresponding supply voltage rails.

Cell placement is restricted (to a different degree) in both layout architectures of Fig. 2.14. This is unavoidable as long as standard cells developed for single $V_{dd}$ are used for dual-$V_{dd}$ circuits. The restriction can be completely removed in the approach illustrated in Fig. 2.16 [25, 26], where each cell has dual rails for supply voltage: one for $V_{ddh}$ and another for $V_{ddl}$. This of course comes at the cost of developing a custom cell library. In $V_{ddh}$ cells, the n-well is biased to $V_{ddh}$, which

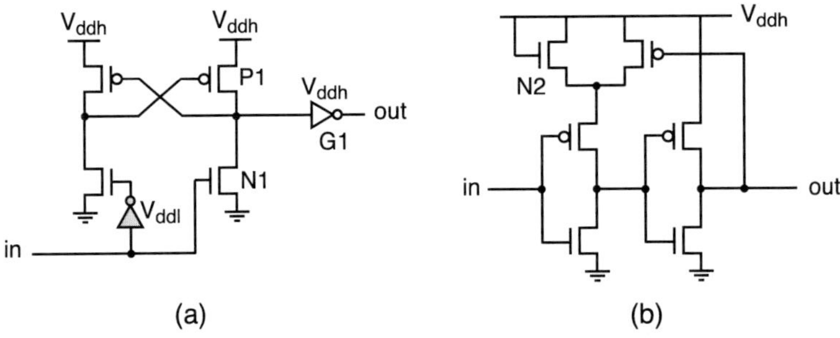



**Fig. 2.17** Level converters: (**a**) pMOS cross-coupled level converter (CCLC) [22] and (**b**) single-supply level converter (SSLC) [24]

is also a voltage supply. The n-well bias of $V_{ddl}$ cells also must be made to $V_{ddh}$, so that adjacent $V_{ddh}$ and $V_{ddl}$ cells can share their n-well for area efficiency. In this setting, the pMOS transistor of $V_{ddl}$ cells is made slower due to its negative body bias (e.g., 18% speed degradation for $V_{ddl} = 1.2$ V with $V_{ddh} = 1.8$ V [26]); however, its subthreshold leakage is reduced substantially. The cell height increases, as it must, but only marginally (e.g., 2.7% [25]). This is because the $V_{ddh}$ rail can be made thinner as the amount of charge it has to deliver becomes less; the $V_{ddl}$ rail can be made even thinner, as $V_{ddl}$ cells are typically a small proportion of the total cells.

**Level Conversion**    The key component of dual-$V_{dd}$ design is a level converter. If $V_{ddl}$ is directly applied to the input of an $V_{ddh}$ inverter, both nMOS and pMOS transistors will be turned on, if $V_{ddl} - V_{ddh}$ is smaller than the threshold voltage of pMOS; this causes huge amount of short-circuit current. Even if pMOS is not turned on, it is only weakly turned off and incurs a large amount of subthreshold current. Therefore, a level converter must be used whenever a $V_{ddl}$ gate drives a $V_{ddh}$ one.

Figure 2.17(a) shows a pMOS cross-coupled level converter (CCLC), which needs both a $V_{ddh}$ and a $V_{ddl}$ voltage supply. This can be designed as a $V_{ddh}$ cell with $V_{ddl}$ supplied through a pin connection; in the layout architecture shown in Fig. 2.14(a), the CCLC must be placed in $V_{ddh}$ rows that are adjacent to a $V_{ddl}$ row, so that a short connection to $V_{ddl}$ can be made. Figure 2.17(b) shows a single-supply level converter (SSLC), which uses a threshold voltage drop across N2 to provide a virtual $V_{ddl}$ to the input inverter. The SSLC is more flexible in placement than the CCLC as it uses only $V_{ddh}$.

In 45-nm technology with $V_{ddh} = 1.1$ V and $V_{ddl} = 0.77$ V, the CCLC and SSLC exhibit 63 ps and 49 ps of delay, respectively, with four $1\times$ inverters as load; these delay numbers represent about 1.7 and 1.3 times the $V_{ddh}$ NAND2 delay. The area of both converters is approximately three times the NAND2 area. This suggests that the number of connections from $V_{ddl}$ to $V_{ddh}$ should be minimized in dual-$V_{dd}$ circuits.

**Selection of $V_{ddl}$**    It is important to decide the value of $V_{ddl}$ for a particular $V_{ddh}$, which is given by the technology. Common sense tells us that there would be an optimum value leading to minimum power consumption: if $V_{ddl}$ is close to $V_{ddh}$, many gates will be allocated to $V_{ddl}$ but the amount of power saving from each allocation will be very small; if we choose a value that is much smaller than $V_{ddh}$, the power saving from using $V_{ddl}$ will be great but only a few gates will take advantage of it due to the abrupt change in delay. The important question is whether this optimum value is different for different circuits or whether there exists some universal number. We obviously prefer the latter.

It was experimentally verified [22] that the optimum $V_{ddl}$ indeed exists at 60–70% of $V_{ddh}$, consistently over many circuits. This was also verified in an analytical way [27], although some approximations are involved. One work claims that optimum $V_{ddl}$ can be brought down to 50% of $V_{ddh}$ when dual-$V_{th}$ is used together with dual-$V_{dd}$ [24], because the reduced circuit speed from using lower $V_{ddl}$ can be recovered to some extent by using dual-$V_{th}$.

**Effectiveness of Dual-$V_{dd}$**     Several works in the literature report the effectiveness of dual-$V_{dd}$. In designing multimedia ASIC [22] with 0.3 μm technology, 76% of cells use $V_{ddl}$ which yields a 47% power saving compared to a design that uses $V_{ddh}$ alone. When this technique is applied to a microprocessor design [24] in 0.13 μm technology, the power savings is rather small, 8%. This is estimated to be due to the higher value of $V_{ddl} = 1.2$ V ($V_{ddh} = 1.5$ V) and the smaller number of $V_{ddl}$ gates from the strict requirement on the processor design.

It is reported [28] that the layout architecture shown in Fig. 2.14(a) causes a 13% to 16% increase of the wire length depending on the placement density; Fig. 2.14(b) exhibits about a 5% increase.

### 2.4.3 Other Methods

To reduce the switching activity of the data path during architecture design, registers based on a dual-edge-triggered flip-flop (DETFF) can be considered. Since the DETFF is triggered at both the rising and falling edges of the clock, the clock of half the frequency can be used for the same throughput, which allows the power consumption of the clock network to be cut in half. Using latch or pulsed-latch as a register is another method to reduce clock power consumption.

A notable approach to reduce the switching activity of the data path is precomputation [29]. Two architectures to implement precomputation are shown in Fig. 2.18. In Fig. 2.18(a), the blocks $g_1$ and $g_2$ take some input bits and predict the output $f$; $g_1 = 1$ and $g_2 = 0$ implies $f = 1$, $g_1 = 0$ and $g_2 = 1$ implies $f = 0$, and $g_1 = 0$ and $g_2 = 0$ implies that $f$ cannot be predicted ($g_1 = 1$ and $g_2 = 1$ are not allowed). Therefore, if $g_1 = 1$ or $g_2 = 1$ (but not both), the input bits are disabled from loading, which yields no switching activity in the combinational block; the corresponding output $f$ is determined by $g_1$ or $g_2$. The input bits are loaded when $g_1 = g_2 = 0$ for normal computation. This architecture has a limitation due to the extra circuitry and increased critical path delay. The architecture shown in Fig. 2.18(b) is similar to

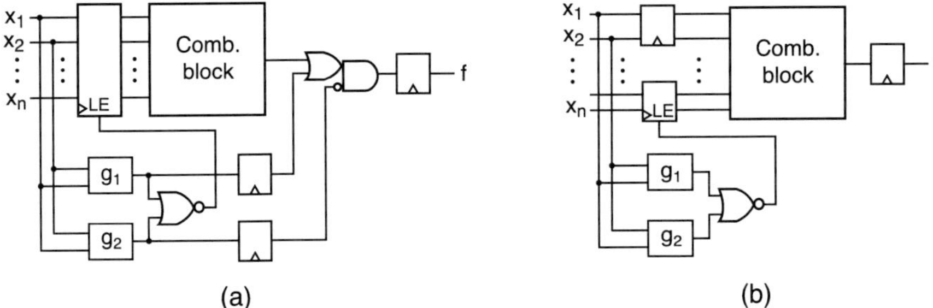

**Fig. 2.18** Precomputation architectures

that of Fig. 2.18(a) except that some input bits are always loaded. Thus, if $g_1 = 1$ or $g_2 = 1$ (but not both), computation depends on only those bits that are loaded while the remaining bits are disabled from loading.

Voltage scaling is the most effective way to reduce dynamic power consumption. Designing or synthesizing a circuit such that its delay is minimized and then lowering $V_{dd}$ until the clock period is just met could be an effective way, even though this typically comes at a cost of increased circuit area. The amount of voltage scaling is limited in ASIC design due to the fixed range of $V_{dd}$ that is used for library characterization.

Every step of the design affects power consumption to some extent [30], e.g., technology mapping, cell and wire sizing, and floorplanning and placement. A notable power savings can be achieved if power consumption is explicitly considered during these design steps.

It is often said that there is more chance to reduce power consumption at higher abstraction levels because less is determined at that point. But the amount of power increase or decrease caused by a design decision at those levels is hard to declare in precise amounts, as discussed in Sect. 2.2.1. Any power savings at higher abstraction levels should be considered very rough and only relative.

## 2.5 Circuits to Reduce Static Power

The two most widely used techniques to reduce static power consumption are reviewed: power gating in Sect. 2.5.1 and body biasing in Sect. 2.5.2. Some other circuit techniques are also reviewed in Sect. 2.5.3.

### 2.5.1 Power Gating

Power gating[2] has become one of the most widely used circuit design techniques for reducing leakage current. Its concept, illustrated in Fig. 2.19, is essentially very

---

[2]This section is also available in [31].

**Fig. 2.19** Power gating circuit [31]. © 2010 ACM

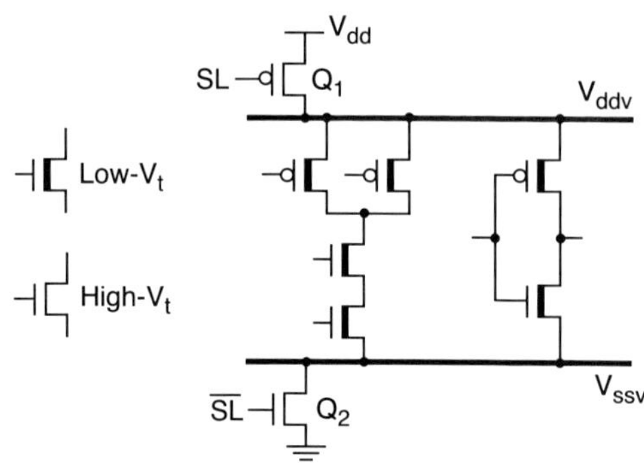



**Fig. 2.20** Master-slave retention flip–flop [31]. © 2010 ACM

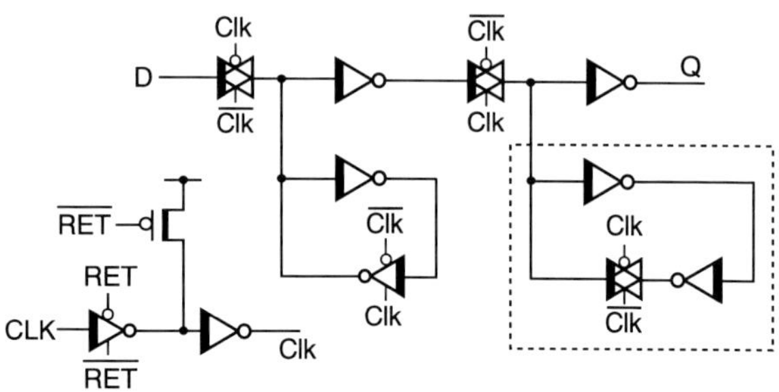



simple. Current switches $Q_1$ and $Q_2$ are shared by all the gates in a circuit, which are thus connected to virtual power lines $V_{ddv}$ and $V_{ssv}$, instead of the real lines, $V_{dd}$ and $V_{ss}$. Low-$V_{th}$ is used within the circuit itself to achieve high performance, and high-$V_{th}$ is used in the switches to achieve low subthreshold leakage. In active mode, SL is kept low, $Q_1$ and $Q_2$ are turned on, and $V_{ddv}$ and $V_{ssv}$ are maintained close to $V_{dd}$ and $V_{ss}$ respectively. In standby mode, SL is kept high, $Q_1$ and $Q_2$ are turned off, and $V_{ddv}$ and $V_{ssv}$ float; leakage from the low-$V_{th}$ circuit is thus limited by high-$V_{th}$ switches. These days, only one switch, a header $Q_1$ or a footer $Q_2$, is usually employed for simplicity.

**Implementation**    Implementing power gating involves several circuit components. Virtual power lines ($V_{ddv}$ or $V_{ssv}$) float once the current switches are turned off, which implies that storage elements lose their values. This can be resolved by using retention registers. There are several different implementations of retention registers; Fig. 2.20 shows one example [32]. The slave latch within the dotted box is directly connected to $V_{dd}$ and $V_{ss}$, while the remainder of the circuit is connected to current switches. In sleep mode, Clk is forced to 0, which isolates the slave latch and thus allows it to retain data with RET set to 1 (0 in active mode). The latch uses a pull-up pMOS. The use of retention flip-flops invariably involves an increase in area as well as an increase in the sequencing overhead, the number of wires, and the power consumption—thus, the use of retention flip-flops should be limited. Decisions on how to provide retention are mainly made intuitively and, in any case,

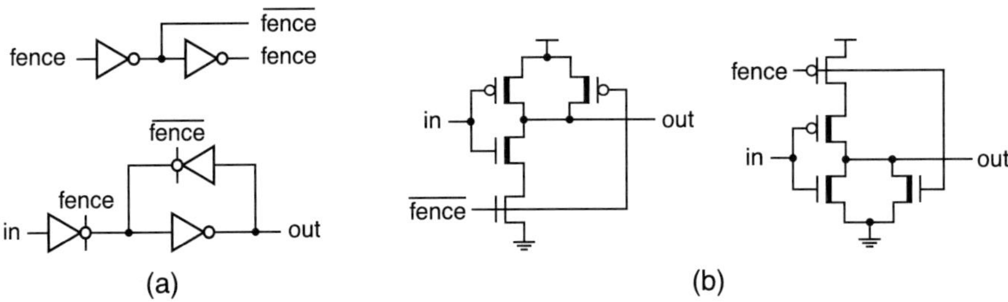

**Fig. 2.21** (**a**) Output isolation circuit and (**b**) output hold circuit [31]. © 2010 ACM

retention is only required for a small fraction of the data in the system [33]. Nevertheless, some designs inherently require many retention flip-flops, such as finite state machine (FSM) controllers or register files.

After the current switches are turned off, the outputs of a circuit start to float, very slowly. This causes a large amount of short-circuit current in the blocks that are connected. A circuit that prevents the output from floating must be inserted to solve these problems. In Fig. 2.21(a) [34], out is decoupled from in once the fence is asserted; the latch then delivers the output value, which it stores in itself. However, it has limited use because it causes delay and also uses extra area and power. A simple pMOS or nMOS can be used instead, as shown in Fig. 2.21(b), if the only objective is to prevent the output from floating, and the output value itself does not need to be preserved. When a footer is used for power gating, a helper pMOS sets out to high; if a header is used, a helper nMOS sets out to low; these helper devices remain off in active mode. In practical designs, isolation circuits (Fig. 2.21(a)) are used for some outputs while the remainder simply use hold circuits (Fig. 2.21(b)).

The layout of power gating circuits using standard cells should be carefully designed. The power network should consist of three power rails $V_{dd}$, $V_{ssv}$, and $V_{ss}$ when footers are used. But these requirements are not uniform: combinational cells need $V_{dd}$ and $V_{ssv}$, retention elements need all three rails, and footer cells use $V_{ssv}$ and $V_{ss}$. Similar requirements are imposed when headers are used. Figure 2.22 shows an example layout design. The $V_{ssv}$ rails are regarded as local $V_{ss}$ rails, while the real $V_{ss}$ rails, on vertical layers, are regularly spaced, as shown in the figure. The required number of footer cells are now dispersed over the placement region, in locations immediately below the $V_{ss}$ rails. The connection of $V_{ss}$ to the retention flip-flops and isolation cells is achieved by signal routing, as we can see in the figure.

**Area and Wire Length**    We have seen that power gating introduces extra circuitry and wires. It is important to understand the extent of the increase in area and wire length that must be tolerated in designing power gating circuits, and what factors exactly contribute to that increase; this is essential information during the early stages of design. We selected the six example circuits summarized in Table 2.2 for quantitative analysis. The first two circuits are purely combinational, while the other four are sequential. Each circuit was synthesized in 1.1 V 45-nm technology. Output

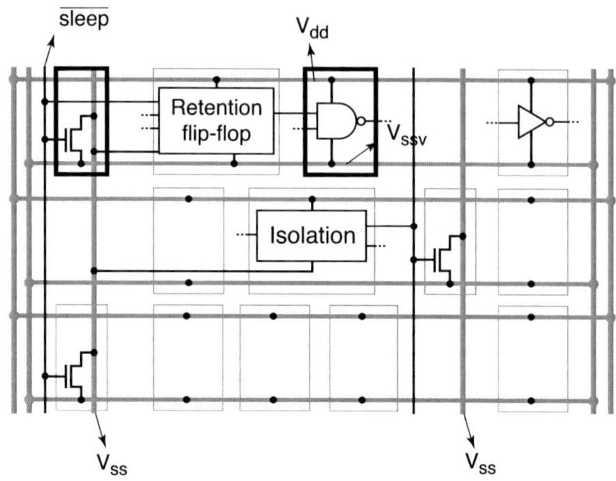



**Fig. 2.22** Layout design of power gating circuit [31]. © 2010 ACM

**Table 2.2** Increase in area and wire length of power gating circuits

| Name | # Comb. | # FFs | # POs | $\Delta$area (%) | | $\Delta$wire length (%) | |
|---|---|---|---|---|---|---|---|
| | | | | Header | Footer | Header | Footer |
| c5315 | 642 | 0 | 123 | 41 | 39 | 28 | 27 |
| c6288 | 1211 | 0 | 32 | 14 | 10 | 12 | 18 |
| s5378 | 605 | 176 | 49 | 41 | 38 | 38 | 36 |
| b21 | 16441 | 490 | 22 | 8 | 6 | 11 | 8 |
| perf_des | 43044 | 8808 | 64 | 28 | 26 | 28 | 22 |
| vga_lcd | 35448 | 17052 | 108 | 36 | 34 | 26 | 24 |

isolation was assumed to be achieved at all primary outputs (POs) by a circuit similar to that shown in Fig. 2.21(a), and we also assumed that data would be retained at all flip-flops by circuits similar to that shown in Fig. 2.20. The $\Delta V$ across the header or footer was set to 1% of $V_{dd}$, and this gives us the total gate width $W$. If we then divide $W$ by the gate width of a single header or footer cell, we obtain the number of header or footer cells required. The layout architecture of Fig. 2.22 was used as the basis for automatic placement and routing [35], using metal layer M3 for the $V_{dd}$ or $V_{ss}$ rails. However, the header or footer cells were manually placed in a regular pattern, and fixed in their locations, followed by automatic placement of the remaining cells.

The last four columns of Table 2.2 show the increase of area and wire length over the same designs without power gating. The area corresponds to the sum of the areas of all the cells. During routing, we forced the cells into about 70% of the placement region to achieve a tight placement, and metal layers up to M4 were allowed for automatic routing. In Fig. 2.23 the extra requirements of power gating are analyzed in more detail by showing the contribution of various components.

There is a fairly large increase in area (41%) when headers are introduced into c5315. This happens because it has large number of POs (considering the number of

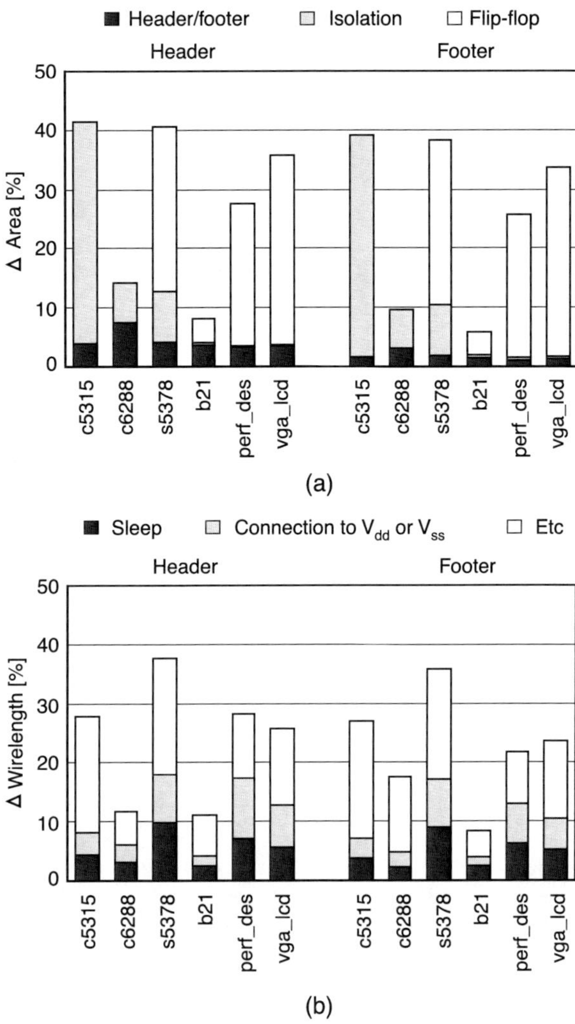

**Fig. 2.23** Increase in (**a**) area and (**b**) wire length caused by power gating [31]. © 2010 ACM



combinational gates), and the same number of isolation circuits must be introduced; this is apparent from the first bar in Fig. 2.23(a). The circuit c6288 has fewer outputs, and so the increase in area is less, although 14% is still sizable; this is largely attributable to headers, as we can see in Fig. 2.23(a). The increase in area due to current switches (headers) alone is about 4% and 8% in these two circuits; this is reduced to 2% and 3% when footers are used (similar results have been reported in industrial examples [33, 36]), and the design using footers also results in lower increases in total area (39% and 10%), as shown in column 6 of Table 2.2. In the four sequential circuits, the growth in area is dominated by the retention flip-flops (see the white portions of the bars in Fig. 2.23(a)); each retention flip-flop that we used is 53% bigger than its non-power-gated counterpart. Table 2.2 shows that the increases in the area of these four sequential circuits are largely determined by the

**Table 2.3** Transition energy and minimum idle time of several example circuits implemented in 65-nm technology

| Name | # Comb. | # FFs | # POs | $E_{tr}$ (pJ) | $T_{min\_idle}$ (μs) |
|------|---------|-------|-------|---------------|----------------------|
| b12  | 855     | 119   | 6     | 15.5          | 9.1                  |
| b13  | 240     | 53    | 10    | 9.9           | 17.0                 |
| irda | 160     | 32    | 33    | 10.0          | 8.4                  |
| ic2  | 312     | 49    | 8     | 6.3           | 0.9                  |
| mc   | 122     | 17    | 32    | 9.0           | 11.8                 |
| wb   | 604     | 274   | 362   | 130.0         | 0.6                  |

number of flip-flops, as a proportion of the total number of circuit elements. There are three components to the increase in wire length, as shown in the last two columns of Table 2.2 and in Fig. 2.23(b): extra wires for routing the `sleep` signal, the wires connecting the data retention elements to the $V_{dd}$ or $V_{ss}$ rails, and the increase in the lengths of all the other signal wires due to increased congestion.

The values in Table 2.2 should be considered as extreme, since data retention was assumed to be required at all POs and flip-flops. In practice, the increase in area and wire length would depend on the proportion of POs and at which data needed to be retained.

**Mode Transition**    Turning on and off the current switches dissipates extra energy, denoted by $E_{tr}$. Therefore, unless a switch is turned off long enough, power gating may increase power consumption rather than decrease it. We may thus define the minimum idle period, at which power gating starts to benefit energy dissipation, $T_{min\_idle}$. Table 2.3 shows its value for example circuits taken from International Test Conference (ITC) benchmarks and OpenCores [7], and experimentally implemented in 1.2 V 65-nm technology. It should be noted that $T_{min\_idle}$ is not directly proportional to circuit size. The value of $T_{min\_idle}$ for this particular technology is of the order of microseconds, which is equivalent to 100 clock cycles in 100 MHz and 1000 cycles at 1 GHz.

Basic power gating results in a slow wake-up process. $V_{ssv}$ is close to $V_{dd}$ when the footers are turned off (close to $V_{ss}$ when the headers are turned off) during sleep mode, which implies that all the nets internal to a circuit are charged up close to $V_{dd}$, irrespective of their logic states. Those nets whose original logic value was 0 simultaneously start to discharge once the footers are turned on. This can cause two problems: a high current peak at $V_{ss}$ may cause malfunctions in adjacent blocks that are still in active mode; and a large rush current combined with the parasitic of the power network and package yields a ground bounce at $V_{ss}$, which delays the time at which a circuit can start operation again, by increasing the wake-up delay.

Several approaches have been proposed to reduce the rush current and the wake-up delay. The VRC [37], shown in Fig. 2.24(a), produces a faster wake-up than basic power gating, since $V_{ddv}$ and $V_{ssv}$ are clamped to the built-in potential of a diode, and so both the charging and discharging currents during wake-up are reduced. Since the rush current is caused by turning on all the switches at the same time, a natural solution is to turn them on one by one, which can be physically implemented by

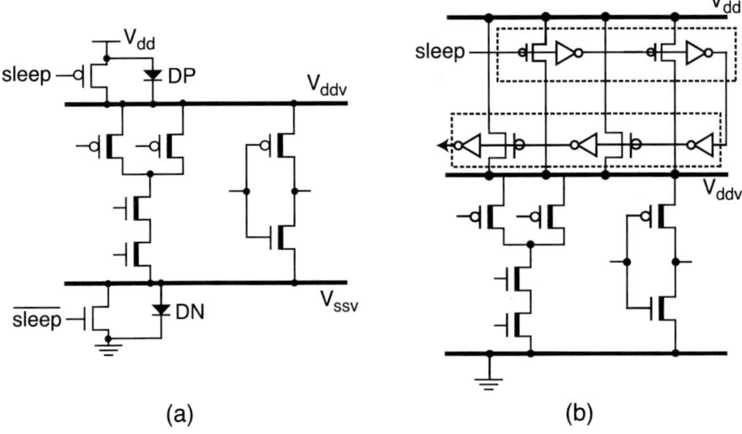

**Fig. 2.24** (**a**) Virtual power/ground rail clamp (VRC) and (**b**) two-pass turn-on [31]. © 2010 ACM

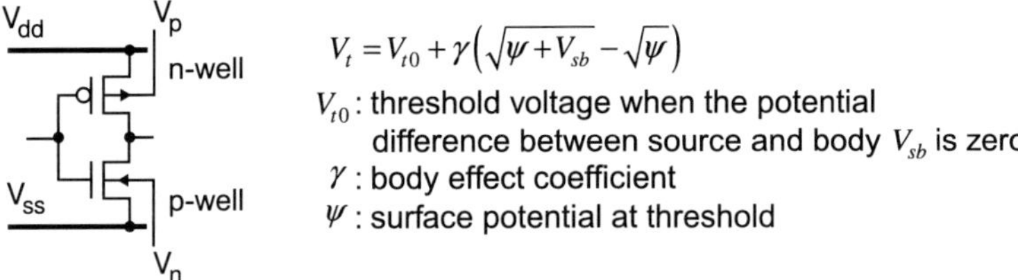

$$V_t = V_{t0} + \gamma\left(\sqrt{\psi + V_{sb}} - \sqrt{\psi}\right)$$

$V_{t0}$ : threshold voltage when the potential difference between source and body $V_{sb}$ is zero

$\gamma$ : body effect coefficient

$\psi$ : surface potential at threshold

**Fig. 2.25** Threshold voltage of MOS transistor

connecting the switches in a daisy-chain. The rush current is still likely to be too large unless the daisy-chain operates very slowly [38], so that the virtual rails and internal nodes have enough time to charge or discharge. Figure 2.24(b) illustrates a circuit scheme called two-pass turn-on [39], in which a header consists of two components: a series of weak switches followed by a series of strong ones. The first pass turns on the weak switches one by one, in such a way that $V_{ddv}$ is fully restored (to a voltage close to $V_{dd}$) when the last switch is turned on. The limited current flow through the weak pMOS switch constrains the rush current. The weaker the first set of switches, the lower the current peak, but turn-on delay increases. In the second pass, the strong switches are turned on, and then all the switches supply the current needed in active mode.

### 2.5.2 Body Biasing

The threshold voltage ($V_{th}$) of a MOS transistor is illustrated in Fig. 2.25. Typically, the substrate (or p-well) of an nMOS is tied to $V_{ss}$, i.e., $V_n = V_{ss}$; similarly, the n-well of a pMOS is tied to $V_{dd}$, i.e., $V_p = V_{dd}$. In body biasing, $V_p$ and $V_n$ are controlled (dynamically or statically), rather than being tied. In reverse body biasing (RBB), a voltage smaller than $V_{ss}$ is applied to $V_n$ and a voltage larger than $V_{dd}$ is applied to $V_p$. As expected from Fig. 2.25, this causes $V_{th}$ to increase since $V_{sb}$ becomes positive; as a result, the device becomes slower and subthreshold leakage is exponentially reduced. The opposite technique is called forward body biasing (FBB).

Body biasing is a useful technique to counteract process variations. The threshold voltage always fluctuates due to imperfections in the manufacturing process and to variations in temperature and $V_{dd}$. When $V_{dd}$ is high enough, this does not affect circuit delay that much since the difference between $V_{dd}$ and $V_{th}$, which determines circuit delay, is large enough compared to $\Delta V_{th}$. But, as $V_{dd}$ becomes lower as technology is scaled down, the delay may vary a lot, which severely affects yield. In this scenario, FBB may be applied to chips that are manufactured with $V_{th}$ larger than target, and RBB may be applied to those with unnecessarily high $V_{th}$ thereby causing too much leakage.

**Body Biasing for Low Leakage** The second application of body biasing is to reduce leakage current. There are several scenarios for this purpose:

- RBB can be applied in standby mode to reduce subthreshold leakage [40].
- A circuit may be designed with $V_{th}$ higher than its original target value, and then FBB is applied in active mode to compensate for the large initial delay, while zero body bias (ZBB) is applied in standby mode [41] to reduce the leakage.
- A circuit may be designed with $V_{dd}$ lower than its initial target value, and then FBB is used in active mode to compensate (this helps reduce dynamic power because of the lower $V_{dd}$ even though active leakage will be larger), while ZBB is applied in standby mode.
- A circuit may be designed so that some gates use FBB while the remainder use ZBB [41]. This technique emulates dual-$V_{th}$ (FBB for low-$V_{th}$ and ZBB for high-$V_{th}$) but without using any extra masks.

It is important to understand that RBB becomes less effective with technology scaling due to a steepening increase in junction leakage [42] with the extent of the biasing. This also implies that there is an optimal level of RBB for minimizing leakage, which is about 500 mV in sub-100 nm technology. Another technology-related problem is that RBB is impaired by increasing the within-die variation of $V_{th}$ [43]. When body biasing is applied in dual-$V_{th}$ circuits, care must be taken. The body effect coefficient ($\gamma$ in Fig. 2.25) is different for different $V_{th}$, which implies that the amount of $V_{th}$ shift is different for the same body bias.

**Implementation** The layout methodology of body biasing circuits is shown in Fig. 2.26, where a double-back layout pattern is employed. Standard cells are designed without taps, so that $V_n$ and $V_p$ are kept without any connection. A special cell, called a tap cell, must be inserted to provide the potentials for $V_n$ and $V_p$, which come from body bias generator circuit. The tap cells are inserted in a regular fashion; the columns of the tap cells are separated by some predetermined distance, e.g., 50 μm [44], which is determined by the well impedance.

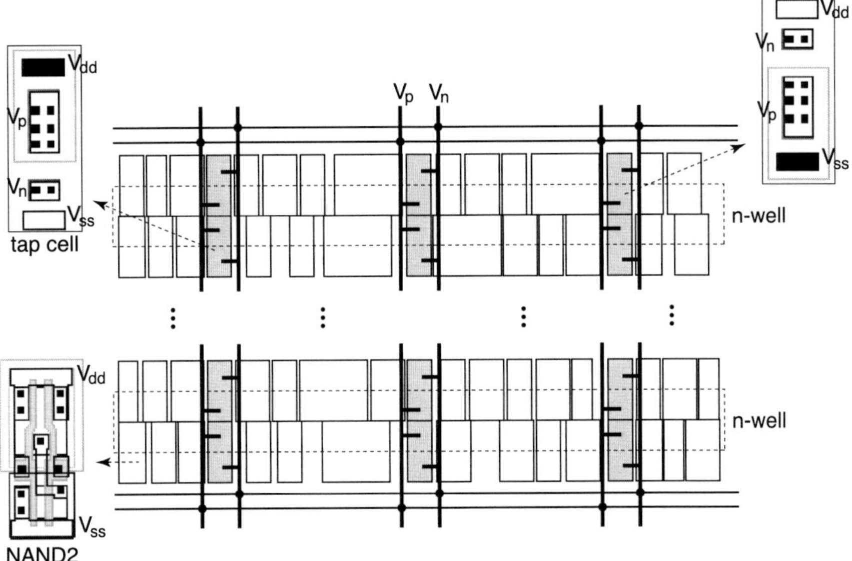

**Fig. 2.26** Layout methodology of body biasing circuits [45]. © 2007 IEEE

## 2.5.3 Other Techniques

Many circuit design techniques have been proposed and used to control leakage. The multiple-$V_{dd}$, multiple-$V_{th}$, and multiple-gate-length approaches are static techniques; the use of two voltages, in dual-$V_{dd}$ schemes, is most common. Dynamic techniques include input vector control (IVC), as well as power gating and body biasing. Static techniques are used during the design process with the support of a tool that determines the $V_{dd}$ or $V_{th}$ levels and the gate lengths of gates or blocks; the conventional synthesis-based design flow is almost unaffected. These approaches reduce leakage in both active and sleep modes, but the amount of leakage that can be saved is relatively small (e.g., 42% with dual-$V_{dd}$ [26] and 30% using dual-gate-length [39] techniques). Dynamic techniques rely on explicit control of the leakage-reducing circuits at run time, and are typically more complicated, so that the conventional design flow has to be customized. Dynamic techniques only reduce the leakage in sleep mode, but the savings is substantial (e.g., $40\times$ using power gating [39]). The IVC technique is an exception, in that its design is not very complicated and the leakage savings is not great, even though it must be categorized as a dynamic technique.

Multiple-$V_{dd}$ techniques have been widely used to reduce switching power, which has a quadratic dependence on $V_{dd}$, but they are even more effective in reducing leakage because subthreshold and gate leakage are roughly proportional to $V_{dd}^3$ and $V_{dd}^4$, respectively [46]. The use of multiple-$V_{th}$ configurations is prevalent in contemporary CMOS technologies; they are sometimes combined with multiple-$V_{dd}$ or multiple-gate-length techniques.

IVC applies a predetermined input vector, called a minimum leakage vector (MLV), during sleep mode or a short idle period. Its effectiveness is largely determined by the controllability of the circuit, which is negatively correlated with logic depth. As this increases, fewer gates can usually be put in the lowest leakage state. Controllability can be improved by adding multiple control points within a circuit or by modifying some gates. Several methods have been proposed to find a good MLV, including random search and heuristics based on gate controllability or iterative dynamic programming [47], and there is an exact method using Boolean satisfiability formulation [48].

## 2.6 Conclusion

Architectural or system-level design involves a trade-off between one design parameter and another. Power consumption is no exception. Applying power gating, for example, helps save a substantial amount of standby leakage, but comes at a cost of extra area and wire length. Unless the trade-off is taken into account early on, an architecture may be hard to implement or may not be implementable at all. Understanding the implementation details of various low-power circuit techniques is thus very important; this has been the subject of this chapter.

## References


1. Bergamaschi, R., Shin, Y., Dhanwada, N., Bhattacharya, S., Dougherty, W., Nair, I., Darringer, J., Paliwal, S.: SEAS: a system for early analysis of SoCs. In: Proc. Int. Conf. on Hardware/Software Codesign and System Synthesis, Oct. 2003, pp. 150–155 (2003)
2. Veendrick, H.: Short-circuit dissipation of static CMOS circuitry and its impact on the design of buffer circuits. IEEE J. Solid-State Circuits 468–473 (1984)
3. Nose, K., Sakurai, T.: Analysis and future trend of short-circuit power. IEEE Trans. Comput.-Aided Des. **19**, 1023–1030 (2000)
4. Roy, K., Mukhopadhyay, S., Mahmoodi-Meimand, H.: Leakage current mechanisms and leakage reduction techniques in deep-submicrometer CMOS circuits. Proc. IEEE **91**(2), 305–327 (2003)
5. Narendra, S., Chandrakasan, A. (eds.): Leakage in Nanometer CMOS Technologies. Springer, Berlin (2005)
6. Ye, Y., Borkar, S., De, V.: A new technique for standby leakage reduction in high-performance circuits. In: Proc. Symp. on VLSI Circuits, June 1998, pp. 40–41 (1998)
7. OpenCores [Online]. Available: http://www.opencores.org/
8. Brand, D., Visweswariah, C.: Inaccuracies in power estimation during logic synthesis. In: Proc. Int. Conf. on Computer Aided Design, Nov. 1996, pp. 388–394 (1996)
9. Ercolani, S., Favalli, M., Damiani, M., Olivo, P., Riccó, B.: Estimate of signal probability in combinational logic networks. In: Proc. European Test Conf., Apr. 1989, pp. 132–138 (1989)
10. Borkar, S., Karnik, T., Narenda, S., Keshavarzi, A., De, V.: Parameter variations and impact on circuits and microarchitecture. In: Proc. Design Automation Conf., June 2003, pp. 338–342 (2003)
11. Su, H., Liu, F., Devgan, A., Acar, E., Nassif, S.: Full chip leakage estimation considering power supply and temperature variations. In: Proc. Int. Symp. on Low Power Electronics and Design (2003)

12. Zhao, W., Cao, Y.: New generation of predictive technology model for sub-45nm design exploration. In: Proc. Int. Symp. on Quality Electronic Design, Mar. 2006, pp. 585–590 (2006)
13. Chang, H., Sapatnekar, S.: Full-chip analysis of leakage power under process variations, including spatial correlations. In: Proc. Design Automation Conf., June 2005
14. Acar, E., Agarwal, K., Nassif, S.: Characterization of total chip leakage using inverse (reciprocal) gamma distribution. In: Proc. Int. Symp. on Circuits and Systems, May 2006
15. Rao, R., Srivastava, A., Blaauw, D., Sylvester, D.: Statistical estimation of leakage current considering inter- and intra-die process variation. In: Proc. Int. Symp. on Low Power Electronics and Design, Aug. 2003
16. Han, Y., Koren, I.: Simulated annealing based temperature aware floorplanning. J. Low Power Electron. 3(2), 141–155 (2007)
17. Huang, W., Stan, M., Skadron, K., Sankaranarayanan, K., Ghosh, S., Velusamy, S.: Compact thermal modeling for temperature-aware design. In: Proc. Design Automation Conf., June 2004, pp. 878–883 (2004)
18. Zhan, Y., Sapatnekar, S.: A high efficiency full-chip thermal simulation algorithm. In: Proc. Int. Conf. on Computer Aided Design, Nov. 2005, pp. 635–638 (2005)
19. Chinnery, D., Keutzer, K.: Closing the power gap between ASIC and custom: an ASIC perspective. In: Proc. Design Automation Conf., June 2005, pp. 275–280 (2005)
20. Arbel, E., Eisner, C., Rokhlenko, O.: Resurrecting infeasible clock-gating functions. In: Proc. Design Automation Conf., July 2009, pp. 160–165 (2009)
21. Synopsys. IC Compiler User Guide, Dec. 2008
22. Usami, K., Igarashi, M., Minami, F., Ishikawa, T., Kanzawa, M., Ichida, M., Nogami, K.: Automated low-power technique exploiting multiple supply voltages applied to a media processor. IEEE J. Solid-State Circuits 33(3), 463–472 (1998)
23. Shin, I., Paik, S., Shin, Y.: Register allocation for high-level synthesis using dual supply voltages. In: Proc. Design Automation Conf., July 2009, pp. 937–942 (2009)
24. Puri, R., Stok, L., Cohn, J., Kung, D., Pan, D., Sylvester, D., Srivastava, A., Kulkarni, S.: Pushing ASIC performance in a power envelope. In: Proc. Design Automation Conf., June 2003, pp. 788–793 (2003)
25. Wang, J., Shieh, S., Wang, J., Yeh, C.: Design of standard cells used in low-power ASIC's exploiting the multiple-supply-voltage scheme. In: Proc. ASIC Conf., Sept. 1998, pp. 119–123 (1998)
26. Shimazaki, Y., Zlatanovici, R., Nikoli, B.: A shared-well dual-supply-voltage 64-bit ALU. In: Proc. IEEE Int. Solid-State Circuits Conf., Feb. 2003, pp. 104–105 (2003)
27. Kuroda, T., Hamada, M.: Low-power CMOS digital design with dual embedded adaptive power supplies. IEEE J. Solid-State Circuits 35(4), 652–655 (2000)
28. Shin, I., Paik, S., Shin, D., Shin, Y.: HLS-dv: a high-level synthesis framework for dual-$V_{dd}$ architectures. IEEE Trans. Very Large Scale Integr. (2011)
29. Alidina, M., Monteiro, J., Devadas, S., Ghosh, A., Papaefthymiou, M.: Precomputation-based sequential logic optimization for low power. IEEE Trans. Very Large Scale Integr. 2(4), 426–436 (1994)
30. Chinnery, D., Keutzer, K. (eds.): Closing the Power Gap Between ASIC & Custom. Springer, Berlin (2007)
31. Shin, Y., Seomun, J., Choi, K.-M., Sakurai, T.: Power gating: circuits, design methodologies, and best practice for standard-cell VLSI designs. ACM Trans. Des. Autom. Electron. Syst. 15(4), 28:1–28:37 (2010)
32. Mair, H., Wang, A., Gammie, G., Scott, D., Royannez, P., Gururajarao, S., Chau, M., Lagerquist, R., Ho, L., Basude, M., Culp, N., Sadate, A., Wilson, D., Dahan, F., Song, J., Carlson, B., Ko, U.: A 65-nm mobile multimedia applications processor with an adaptive power management scheme to compensate for variations. In: Proc. Symp. on VLSI Circuits, June 2007, pp. 224–225 (2007)
33. Lueftner, T., et al.: A 90-nm CMOS low-power GSM/EDGE, multimedia-enhanced baseband processor with 380-MHz ARM926 core and mixed-signal extensions. IEEE J. Solid-State Circuits 42(1), 134–144 (2007)

34. Won, H.-S., Kim, K.-S., Jeong, K.-O., Park, K.-T., Choi, K.-M., Kong, J.-T.: An MTCMOS design methodology and its application to mobile computing. In: Proc. Int. Symp. on Low Power Electronics and Design, Aug. 2003, pp. 110–115 (2003)
35. Synopsys. Astro User Guide, Mar. 2007
36. Kanno, Y., et al.: Hierarchical power distribution with power tree in dozens of power domains for 90-nm low-power multi-CPU SoCs. IEEE J. Solid-State Circuits **42**(1), 74–83 (2007)
37. Kumagai, K., Iwaki, H., Yoshida, H., Suzuki, H., Yamada, T., Kurosawa, S.: A novel powering-down scheme for low Vt CMOS circuits. In: Proc. Symp. on VLSI Circuits, June 1998, pp. 44–45 (1998)
38. Shi, K., Howard, D.: Challenges in sleep transistor design and implementation in low-power designs. In: Proc. Design Automation Conf., July 2006, pp. 113–116 (2006)
39. Royannez, P., Mair, H., Dahan, F., Wagner, M., Streeter, M., Bouetel, L., Blasquez, J., Clasen, H., Semino, G., Dong, J., Scott, D., Pitts, B., Raibaut, C., Ko, U.: 90 nm low leakage SoC design techniques for wireless applications. In: Proc. IEEE Int. Solid-State Circuits Conf., Feb. 2005, pp. 138–139 (2005)
40. Seta, K., Hara, H., Kuroda, T., Kakumu, M., Sakurai, T.: 50% active-power saving without speed degradation using standby power reduction (SPR) circuit. In: Proc. IEEE Int. Solid-State Circuits Conf., Feb. 1995, pp. 318–319 (1995)
41. Narendra, S., Keshavarzi, A., Bloechel, B., Borkar, S., De, V.: Forward body bias for microprocessors in 130-nm technology generation and beyond. IEEE J. Solid-State Circuits **38**(5), 696–701 (2003)
42. Keshavarzi, A., Narendra, S., Borkar, S., Hawkins, C., Roy, K., De, V.: Technology scaling behavior of optimum reverse body bias for standby leakage power reduction in CMOS IC's. In: Proc. Int. Symp. on Low Power Electronics and Design, Aug. 1999, pp. 252–254 (1999)
43. Narenda, S., Antoniadis, D., De, V.: Impact of using adaptive body bias to compensate die-to-die Vt variation on within-die Vt variation. In: Proc. Int. Symp. on Low Power Electronics and Design, Aug. 1999, pp. 229–232 (1999)
44. Clark, L., Morrow, M., Brown, W.: Reverse-body bias and supply collapse for low effective standby power. IEEE Trans. Very Large Scale Integr. **12**(9), 947–956 (2004)
45. Choi, B., Shin, Y.: Lookup table-based adaptive body biasing of multiple macros. In: Proc. IEEE Int. Symp. on Quality Electronic Design, ISQED, Mar. 2007, pp. 533–538 (2007)
46. Krishnamurthy, R., Alvandpour, A., De, V., Borkar, S.: High-performance and low-power challenges for sub-70nm microprocessor circuits. In: Proc. Custom Integrated Circuits Conf., May 2002, pp. 125–128 (2002)
47. Cheng, L., Deng, L., Chen, D., Wong, M.: A fast simultaneous input vector generation and gate replacement algorithm for leakage power reduction. In: Proc. Design Automation Conf., July 2006, pp. 117–120 (2006)
48. Aloul, F., Hassoun, S., Sakallah, K., Blaauw, D.: Robust SAT-based search algorithm for leakage power reduction. Lect. Notes Comput. Sci. **2451**, 253–274 (2002)

# Chapter 3
# Energy Awareness in Processor/Multi-Processor Design

**Jungsoo Kim, Sungjoo Yoo, and Chong-Min Kyung**

**Abstract**   A processor is one of the major energy consumers in a state-of-the-art system on chip (SoC); thus, reducing the energy consumption of processors is an important issue in realizing an energy-efficient SoC design. Various techniques have been proposed, such as dynamic voltage and frequency scaling (DVFS), adaptive body biasing (ABB), power gating, and instruction throttling. Recently, dynamic power management harnessing multiple low-power techniques has proven to be an effective method, since it enables us to have all operating conditions complement each other in reducing the energy consumption. The dynamism of a software program running on a processor makes it difficult to determine the right action to take; without care, an increase in energy consumption can be the result. Both leakage as well as dynamic energy consumption should be considered, since leakage energy consumption becomes comparable to or can even surpass dynamic energy consumption in the deep submicrometer regime. In this chapter, we describe state-of-the-art dynamic power management methods, adjusting processor states with proper consideration of the dynamic behavior of the software program in order to minimize both dynamic and leakage energy consumption.

## 3.1 Introduction

A system on chip (SoC) consists of multiple functional blocks such as processor, bus, memory, and peripherals. Among them, the processor is one of the largest energy consumers in the system. Various dynamic power management techniques, such as dynamic voltage and frequency scaling (DVFS), clock/power gating, and adaptive body biasing (ABB), have been applied to reduce the energy consumption

J. Kim (✉) · C.-M. Kyung
KAIST, Daejeon, Republic of Korea
e-mail: jungsoo.kim83@gmail.com

C.-M. Kyung
e-mail: kyung@ee.kaist.ac.kr

S. Yoo
POSTECH, Pohang, Republic of Korea
e-mail: sungjoo.yoo@postech.ac.kr

C.-M. Kyung, S. Yoo (eds.), *Energy-Aware System Design*,
DOI 10.1007/978-94-007-1679-7_3, © Springer Science+Business Media B.V. 2011

of the processor. However, the complex behavior of the software program running on a processor makes it a big challenge to reduce the processor energy consumption. In this chapter, we first present a simplified processor power model to study the major energy sources in processors in order to reduce the energy consumption. We then show the complex behavior of software programs, and explain some techniques to reduce the energy consumption by exploiting this complex behavior.

## 3.2 Processor Power Model

The power consumption of a processor consists largely of dynamic and leakage power consumption. Dynamic power is the amount of power delivered from the supply voltage to charge and discharge the load capacitance when transistors switch, and is modeled as follows:

$$P_{\text{dyn}} = C_{\text{eff}} V_{\text{dd}}^2 f \tag{3.1}$$

where $C_{\text{eff}}$ is the average switched capacitance per cycle, which can be extracted from extensive simulation or measurement, and $f$ and $V_{\text{dd}}$ are the clock frequency and supply voltage, respectively.

As processors are implemented in advanced process technologies, the leakage power consumption drastically increases. There are various leakage mechanisms: (1) subthreshold conduction, (2) gate direct tunneling current, (3) junction tunneling leakage, (4) gate induced drain leakage (GIDL), (5) hot carrier injection current, (6) punchthrough current, etc., as described in [1]. Among them, the three major components are subthreshold, gate, and junction leakage currents. Thus, the leakage power consumption of the unit circuit $P_{\text{leak}}$ can be expressed as follows [2]:

$$P_{\text{leak}} \simeq P_{\text{sub}} + P_{\text{gate}} + P_{\text{junc}} = V_{\text{dd}} I_{\text{sub}} + V_{\text{dd}} I_{\text{gate}} + |V_{\text{bs}}| I_{\text{j}} \tag{3.2}$$

where $P_{\text{sub}}$, $P_{\text{gate}}$, and $P_{\text{junc}}$ are the subthreshold, gate, and junction leakage power consumption, respectively, $I_{\text{sub}}$, $I_{\text{gate}}$, and $I_{\text{j}}$ are the subthreshold, gate, and junction leakage currents, and $V_{\text{dd}}$ and $V_{\text{bs}}$ are the supply and body bias voltages, respectively.

The subthreshold current, $I_{\text{sub}}$, is the current which flows from the drain to the source of a transistor by the mechanism of diffusion of the minority carriers through the channel when a transistor is off ($V_{\text{gs}} < V_{\text{th}}$). The subthreshold leakage current depends exponentially on both the gate-to-source voltage and threshold voltage as follows:

$$I_{\text{sub}} = \left(\frac{W}{L}\right) I_{\text{s}} \left[1 - e^{-V_{\text{dd}}/V_{\text{T}}}\right] e^{\frac{-(V_{\text{th}}+V_{\text{off}})}{n V_{\text{T}}}} \tag{3.3}$$

where $W$ and $L$ are the width and height of the device, $I_{\text{s}}$, $n$, and $V_{\text{off}}$ are empirically determined constants for a given process, and $V_{\text{T}}$ is the thermal voltage.

A reverse bias applied at the substrate (or body), $V_{\text{bs}}$, increases $V_{\text{th}}$. The relationship between $V_{\text{bs}}$ and $V_{\text{th}}$ is modeled as follows:

$$V_{\text{th}} = V_{\text{th0}} + \gamma \left(\sqrt{\phi_{\text{s}} - V_{\text{bs}}} - \sqrt{\phi_{\text{s}}}\right) - \theta_{\text{DIBL}} V_{\text{dd}} \tag{3.4}$$

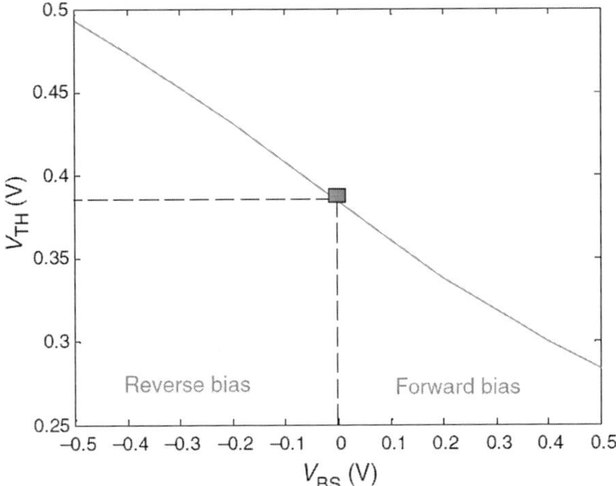


**Fig. 3.1** Relationship between $V_{\text{th}}$ and $V_{\text{bs}}$ [3]

where $V_{\text{th0}}$ is the zero-bias threshold voltage, and $\phi_s$, $\gamma$, and $\theta_{\text{DIBL}}$ are constants for a given technology. $V_{\text{bs}}$ is the voltage applied between the body and source of the transistor. Equation (3.4) shows that $V_{\text{th}}$ has a linear dependence on $V_{\text{dd}}$. Furthermore, if $|V_{\text{bs}}| \simeq \phi_s$, then $\sqrt{\phi_s - V_{\text{bs}}} - \sqrt{\phi_s}$ can be linearized as $k V_{\text{bs}}$, which yields

$$V_{\text{th}} \simeq V_{\text{th1}} - K_1 V_{\text{dd}} - K_2 V_{\text{bs}} \tag{3.5}$$

where $K_1$, $K_2$, and $V_{\text{th1}}$ are constants. Figure 3.1 illustrates experimental results showing the relationship between $V_{\text{th}}$ and $V_{\text{bs}}$ [3].

$V_{\text{off}}$ in (3.3) is typically small; thereby, $1 - \exp(-V_{\text{dd}}/V_T)$ in (3.3) is nearly 1. Then, by substituting (3.5) into (3.3), we can approximate $I_{\text{sub}}$ in (3.3) in a simpler form as follows:

$$I_{\text{sub}} \simeq K_3 e^{K_4 V_{\text{dd}}} e^{K_5 V_{\text{bs}}} \tag{3.6}$$

Gate direct tunneling current is due to the tunneling of electrons (or holes) from the bulk silicon and source/drain overlap region through the gate oxide potential barrier into the gate (or vice versa). As the process technology scales down, the tunneling current increases exponentially with a decrease in the oxide thickness and an increase in the potential drop across the oxide. $I_{\text{gate}}$ has an exponential relation with $V_{\text{dd}}$ and is approximately expressed as follows:

$$I_{\text{gate}} \simeq K_6 e^{K_7 V_{\text{dd}}} \tag{3.7}$$

where $K_6$ and $K_7$ are fitting parameters.

Finally, junction leakage is current which flows through the junction due to tunneling of electrons from the valence band of the $p$-region to the conduction band of the $n$-region (band-to-band tunneling (BTBT)) by a high electric field across a reverse biased $p$–$n$ junction. In a nanoscale device the junction leakage current becomes significant because of higher doping at the junctions. The junction tunneling current depends exponentially on the junction doping and the reverse bias across the junction.

Putting all these components together, we can express the total power consumption of a processor as follows [2, 4]:

$$P_{\text{tot}} = P_{\text{dyn}} + P_{\text{leak}} \tag{3.8}$$

$$= C_{\text{eff}} V_{\text{dd}}^2 f + L_{\text{g}} \left( V_{\text{dd}} K_3 e^{K_4 V_{\text{dd}}} e^{K_5 V_{\text{bs}}} + K_6 e^{K_7 V_{\text{dd}}} + |V_{\text{bs}}| I_{\text{j}} \right) \tag{3.9}$$

where $C_{\text{eff}}$ and $N_{\text{g}}$ are the effective capacitance and effective number of gates of the target processor, respectively. They can be extracted from extensive simulation at transistor, gate, or register-transfer (RT) levels. By using the power model, the energy consumption per clock cycle can be expressed as

$$e_{\text{cycle}} = \frac{P_{\text{tot}}}{f} = C_{\text{eff}} V_{\text{dd}}^2 + \frac{L_{\text{g}}}{f} \left( V_{\text{dd}} K_3 e^{K_4 V_{\text{dd}}} e^{K_5 V_{\text{bs}}} + K_6 e^{K_7 V_{\text{dd}}} + |V_{\text{bs}}| I_{\text{j}} \right) \tag{3.10}$$

## 3.3 Workload Characteristics Running on a Processor

### 3.3.1 Inter-Tasks Workload Characteristics and Classification

The main operation of tasks is to load instructions and data from memory, execute operations to manipulate data, and store results into memory. Thus, the workload of tasks (in terms of the number of processor clock cycles measured at a processor) running on a processor, i.e., $W$, can basically be decomposed into the two components: (i) on-chip workload, $W_{\text{cpu}}$, caused by executing instructions to manipulate data and (ii) off-chip workload, $W_{\text{mem}}$, caused by accessing external components, e.g., DRAM, PCI peripherals etc., as follows:

$$W = W_{\text{cpu}} + W_{\text{mem}} \tag{3.11}$$

Usually, processor and memory (e.g., DRAM, HDD, SSD) are mapped into different voltage and frequency islands. This means that processor and memory are operated with different voltage and frequency levels, and the number of clock cycles for accessing external memory can be varied according to the operating frequency level of the processor. For instance, when the memory access time is 100 ns, each off-chip memory access takes 100 and 200 processor clock cycles at 1 GHz and 2 GHz, respectively. Since the memory access time, called memory stall time, is invariant to the processor clock frequency, the number of processor clock cycles spent for memory access grows as the clock frequency increases. That is, the number of stall cycles spent fetching data from memory increases as the processor frequency increases when a processor and a memory are operated with different voltage and frequency levels. The dependency of the memory stall cycle and processor operating frequency can be approximately modeled as the product of processor frequency,

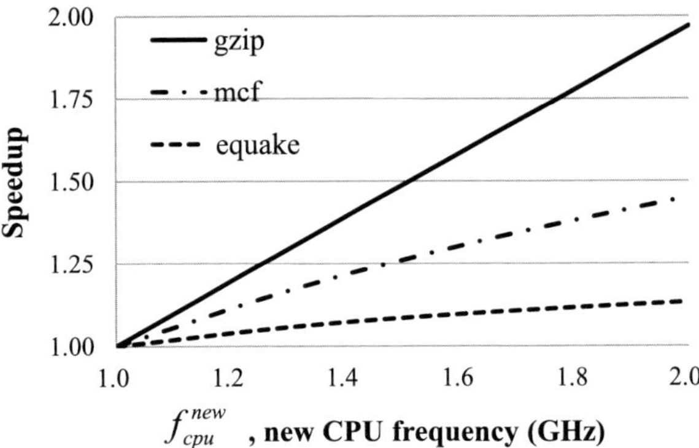



**Fig. 3.2** Speed-up vs. new frequency as clock frequency of processor changes

$f_{cpu}^{new}$ **, new CPU frequency (GHz)**

$f_{cpu}$, and the amount of time spent by accessing external memory in running a task, $T_{mem}$, as follows:

$$W_{mem} = f_{cpu} T_{mem} \qquad (3.12)$$

The workload characteristics vary for the tasks running on a processor. According to the relative ratio of processor computation and external memory access, tasks are classified into CPU-bound or memory-bound tasks. For CPU-bound tasks, $W_{cpu}$ is much larger than $W_{mem}$, whereas $W_{mem}$ is much larger than $W_{cpu}$ in memory-bound tasks. When a processor is operated at a fixed frequency level, workload character-istics of tasks are usually classified with the number of executed instructions per clock cycle (IPC). However, the relative ratio of $W_{cpu}$ and $W_{mem}$ varies according to the processor frequency when a processor supports various voltage and frequency levels, since the memory stall cycle increases as the processor frequency increases. The variations are different over task characteristics such that the rate of increase of the memory stall cycle is higher in memory-bound tasks than in CPU-bound tasks. Therefore, we also classify the task characteristics with respect to the amount of external memory access in a task. The amount of external memory access can be measured as the relative amount of execution time reduction as we increase the processor frequency, which is called speed-up (SU) and is expressed as follows:

$$SU(f) = \frac{T(f^{ref})}{T(f)} \qquad (3.13)$$

where $f^{ref}$ is a reference processor frequency (in this case, 1 GHz). $T(f)$ represents the amount of execution time until the completion of a task when a processor is oper-ated at the frequency level, $f$. Thus, a bigger speed-up indicates that we can achieve a larger reduction in task execution time as we increase the processor frequency.

Figure 3.2 shows different speed-ups of three tasks in a SPEC2000 benchmark, where the $x$- and $y$-axes represent the processor frequency and corresponding speed-up, respectively. As shown in Fig. 3.2, speed-up increases almost linearly as the processor frequency increases. However, the slopes of the tasks are different. For instance, "gzip" shows the steepest increase in speed-up as the processor frequency

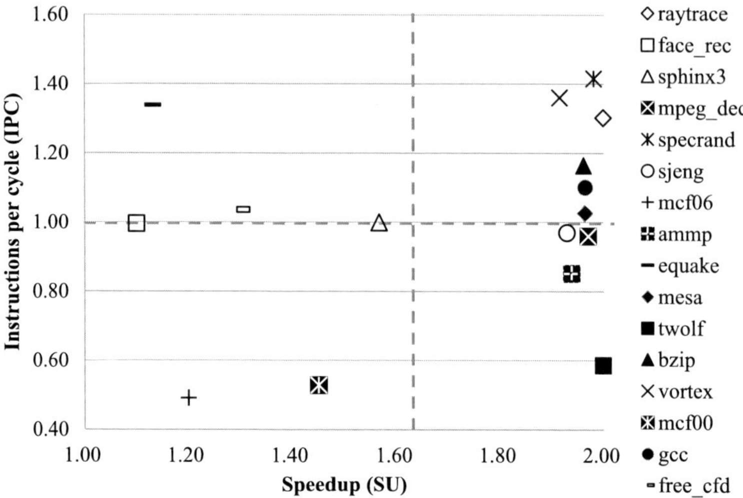

**Fig. 3.3** IPC and SU of programs in SPEC2000/2006, ALPBench, and CFD benchmark

increases, while the increase of speed-up in "equake" is the slowest. The differences come from the different memory access characteristics in each task.

Figure 3.3 shows extensive results of various tasks (obtained from SPEC2000/ 2006, ALPBench, and CFD benchmarks) with respect to IPC and speed-up.

## *3.3.2 Intra-Task Workload Characteristics*

Even within a task, the workload characteristics vary greatly due to the dynamism of the workload. The variation is largely caused by the following three factors [5]:

1. Data dependency: data-dependent loop counts
2. Control dependency: if/else, switch/case statement
3. Architectural dependency: cache hit/miss, translation lookaside buffer (TLB) hit/miss, etc.

Figures 3.4(a) and (b) show the trace and distribution of the per-frame workload of an MPEG-4 decoder, respectively. The results were obtained from decoding 1000 frames of a 1920 × 800 movie clip (an excerpt from *Dark Knight*) on an *LG XNOTE LW25* laptop.[1]

---

[1]The *LG XNOTE LW25* laptop consists of a 2 GHz Intel Core2Duo T7200 processor with 128 KB L1 instruction and data cache, 4 MB shared L2 cache, and 667 MHz 2 GB DDR2 SDRAM.

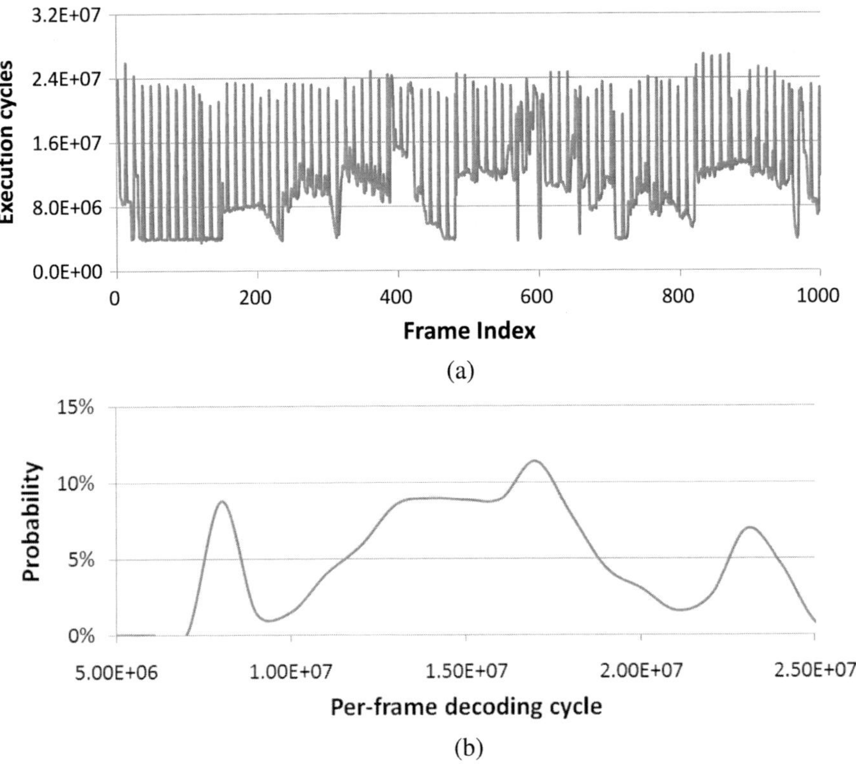

**Fig. 3.4** An example of (**a**) trace and (**b**) distribution of workload (in terms of the number of clock cycles)

### 3.3.3 Workload Profiling and Modeling

The workload characteristics can be profiled and modeled using performance counters which are usually implemented in most modern processors to monitor events occurring in the processor. The following events can be monitored using performance counters [6]:

- Cycle count
- Integer/floating point instruction count
- Instruction count and load/store count
- Branch taken/not taken count and branch mispredictions
- Pipeline stalls due to memory subsystem and resource conflicts
- I/D cache misses for different levels
- TLB misses and invalidations.

The amount of time spent accessing external components, i.e., $T_{mem}$, which is invariant to the processor clock frequency, can be obtained by counting the number of events related to external component access, e.g., last-level cache miss, TLB

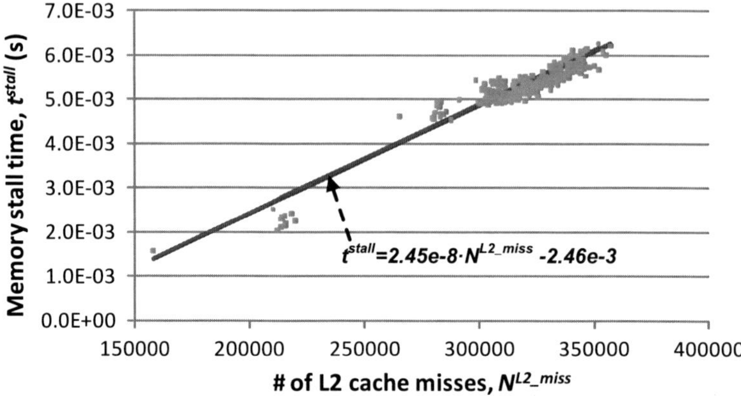



**Fig. 3.5** Memory stall time vs. number of L2 cache misses as approximated by a *straight line*

miss, etc. The following equation shows a linear regression model which fits the relationship between $T_{\mathrm{mem}}$ and the number of last-level cache misses,

$$W_{\mathrm{mem}} = a_{\mathrm{p}} N_{\mathrm{L2\_miss}} + b_{\mathrm{p}} \qquad (3.14)$$

where $a_{\mathrm{p}}$ and $b_{\mathrm{p}}$ are fitting parameters. Figure 3.5 illustrates that (3.14) (solid line) tracks the measured memory stall time quite well (dots) when running the H.264 decoder program in FFmpeg [7].

The accuracy can be improved by using more performance counters related to external memory access, e.g., TLB miss, interrupt, etc. However, the number of events simultaneously monitored in a processor is usually limited (in the experimental platform, two events), and the overhead to access the counter increases as we use larger sets of performance counters. Thus, we need to carefully trade off the accuracy and the overhead in the modeling.

By using the regression model, we can decompose the total workload into $W_{\mathrm{cpu}}$ and $T_{\mathrm{mem}}$ as follows. After the execution of a code section (or a task), we can obtain the total workload, $W$, by the performance counter monitoring of the cycle count and the amount of time accessing external components, $T_{\mathrm{mem}}$, by counting the last-level cache misses and using (3.14). Then, from (3.11), $W_{\mathrm{cpu}}$ is calculated with $W$ and $T_{\mathrm{mem}}$ as follows:

$$W_{\mathrm{cpu}} = W - W_{\mathrm{mem}} = W - f_{\mathrm{cpu}} T_{\mathrm{mem}} \qquad (3.15)$$

## 3.4 Basics of Dynamic Voltage and Frequency Scaling

Dynamic voltage and frequency scaling (DVFS) is a widely used power management technique in processors. In DVFS, voltage and frequency are increased or decreased depending upon the circumstances of the processor, e.g., workload running

**Table 3.1** Supported performance states for the *Intel Pentium M* processor [9]

| Frequency | Voltage |
| --- | --- |
| 1.6 GHz | 1.484 V |
| 1.4 GHZ | 1.420 V |
| 1.2 GHz | 1.276 V |
| 1.0 GHz | 1.164 V |
| 800 MHz | 1.036 V |
| 600 MHz | 0.956 V |

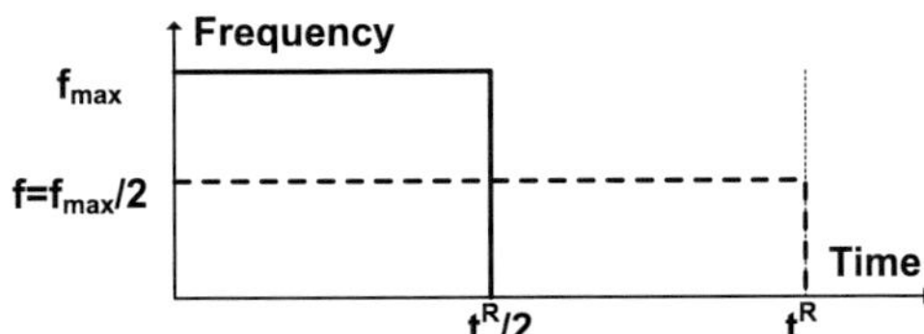


**Fig. 3.6** Concept of dynamic voltage and frequency scaling (DVFS)

on a processor, temperature, etc., based on the following relationship of dynamic power consumption, $P_{dyn}$, with respect to supply voltage, $V_{dd}$, and frequency, $f$:

$$P_{dyn} \propto V_{dd}^2 f \tag{3.16}$$

As shown in (3.16), dynamic power consumption decreases quadratically with supply voltage, $V_{dd}$, and linearly with frequency, $f$. Thus, we can reduce the power consumption by adapting the voltage and frequency level according to the required performance of the processor. When performance demand is low, we can obtain a substantial power reduction by lowering the clock frequency along with the supply voltage to deliver minimum power consumption at the required frequency level. The relationship between supply voltage and operating frequency is modeled as follows [8]:

$$f \propto \frac{(V_{dd} - V_{th})^\alpha}{V_{dd}} \tag{3.17}$$

where $V_{th}$ is the threshold voltage and $\alpha$ is a measure of velocity saturation whose value range is $1 \leq \alpha \leq 2$. Assuming that $V_{dd} >> V_{th}$ and $\alpha \simeq 2$, we can approximate the minimum supply voltage to be proportional to the required frequency level, i.e., $V_{dd} \sim f$.

The supply voltage scaling in accordance with the frequency allows us to reduce dynamic power consumption in a cubic manner by lowering the required frequency level, i.e., $P_{dyn} \propto f^3$. Table 3.1 shows the supported pairs of voltage and frequency levels in the Intel Pentium M processor [9].

Figure 3.6 illustrates how we set the frequency level on a processor, where the $x$- and $y$-axes represent time and frequency, respectively. Let us assume that there is a task running on a processor which must be finished by its time deadline, $t^R$, and is completed at half of its deadline, $t^R/2$, when it runs with maximum frequency

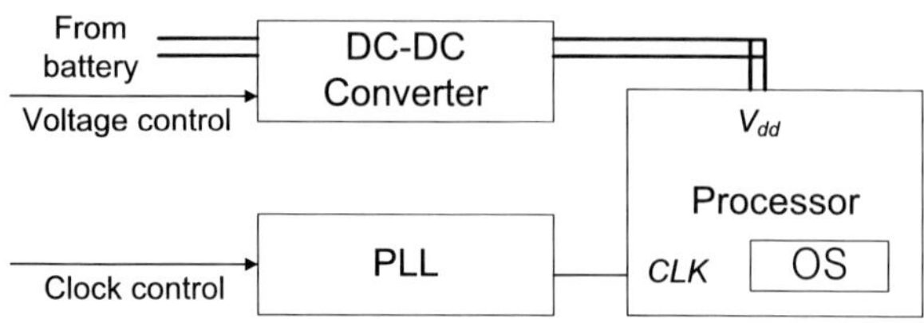

**Fig. 3.7** Implementation of DVFS for a processor [10]



level, $f_{max}$. In this case, we can save power consumption by lowering the frequency along with the supply voltage of the processor to half of $f_{max}$ so that the time slack from the task completion time to the deadline is removed; i.e., the completion time of the task becomes $t^R$. The energy consumed by running the task is calculated as the product of power consumption and execution time. Therefore, we can obtain about a four times reduction in the energy consumption by applying DVFS, which is simply calculated as follows:

$$\text{Before DVFS:} \quad E_{dyn} \propto f_{max}^3 \frac{t^R}{2} = \frac{f_{max}^3 t^R}{2} \tag{3.18}$$

$$\text{After DVFS:} \quad E_{dyn} \propto \left( \frac{f_{max}}{2} \right)^3 t^R = \frac{f_{max}^3 t^R}{8} \tag{3.19}$$

Thus, the frequency is set to the ratio of the remaining workload of a task until the completion of the task (in terms of number of clock cycles) to the remaining time to deadline. At the moment of voltage and frequency setting, the remaining workload of a task until its completion is uncertain, whereas the remaining time to deadline can be accurately measured. Therefore, the effectiveness of DVFS is strongly dependent on the accuracy of the remaining workload prediction. Indeed, most of the recently proposed DVFS methods focus primarily on accurate prediction of the remaining workload.

Implementing DVFS in a processor includes three key components as shown in Fig. 3.7:

1. A microprocessor that can operate over a wide voltage and frequency range.
2. A DC-DC converter and phase-locked loop (PLL) which generate the minimum voltage required for the desired speed.
3. A power management unit which can vary the processor speed.

Leakage power as well as dynamic power are consumed in a processor. The portion of leakage power consumption in the total power consumption increases rapidly as the technology process scales down. Consequently, the energy consumption per clock cycle can increase as we lower the voltage and frequency since the increase in the leakage energy consumption is larger than the decrease in the dynamic power consumption. Figure 3.8 shows the components of the energy consumption per cycle for the 70 nm technology [11]. In this figure, $E_{AC}$, $E_{DC}$, and $E_{on}$ are, respectively, the dynamic, leakage, and inherent power cost of keeping the processor on (PLL circuitry, I/O subsystem, etc.). The $x$- and $y$-axes represent supply voltage, $V_{dd}$, and energy consumption per cycle, $E_{total}$, respectively. As shown in Fig. 3.8, the energy

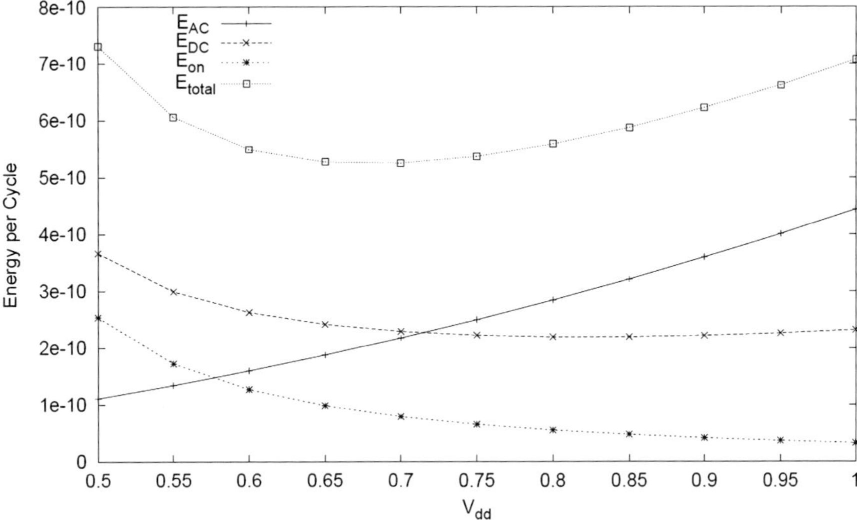

**Fig. 3.8** Total energy consumption with respect to supply voltage in deep submicrometer regime [11]

consumption per cycle increases when the supply voltage is lower than a certain voltage level, called the *critical speed*. Therefore, in the deep submicrometer region, we usually use power gating along with DVFS to lower the voltage and frequency down to the *critical speed*, and then apply power gating when the required voltage and frequency are lower than the *critical speed*, since the total energy consumption then increases as we lower the voltage and frequency levels.

Along with supply voltage scaling, threshold voltage scaling is also widely used to reduce processor leakage energy consumption. This technique, which is called adaptive body biasing (ABB), is performed by controlling the voltage applied to the substrate. When we apply negative voltage to the body, i.e., $V_{bs} < 0$, known as reverse body biasing (RBB), $V_{th}$ increases, and thus leakage power consumption, especially subthreshold leakage, decreases, while performance, i.e., maximum clock speed, decreases. However, in forward body biasing (FBB), we apply positive voltage to the body, i.e., $V_{bs} > 0$, which leads to an exponential increase in leakage power consumption and performance improvement. Thus, by adaptively controlling the voltage applied to the body, we can control the leakage and performance of the system. Some hardware schemes have been proposed to apply ABB to reduce the leakage energy consumption of systems [12, 13].

## 3.5  Workload Characteristic-Aware DVFS

As we mentioned in the previous section, the latencies caused by running a task can be classified as on-chip latencies (e.g., data dependency, cache hit, and branch prediction) and off-chip latencies (e.g., external memory latency and PCI latency). On-chip latencies are caused by events that occur inside the processor and are proportionally reduced with increased processor frequency. By contrast, off-chip latencies are independent of the processor frequency and thus are not affected by scaling of the processor frequency. Thus, the total execution time to run a task, $T$, can be expressed as

$$T = T_{\text{cpu}} + T_{\text{mem}} = \frac{W_{\text{cpu}}}{f_{\text{cpu}}} + T_{\text{mem}} \qquad (3.20)$$

When we scale the voltage and frequency levels in DVFS, we must satisfy the time deadline constraint, $T_{\text{d}}$; i.e., we set the voltage and frequency levels such that a task finishes at its time deadline $T = T_{\text{d}}$ in order to achieve the maximum energy savings. To do this, we set the processor frequency as follows:

$$f_{\text{cpu}} = \frac{W_{\text{cpu}}}{T_{\text{d}} - T_{\text{mem}}} \qquad (3.21)$$

Since the voltage and frequency levels are dependent on $W_{\text{cpu}}$ and $T_{\text{mem}}$, an accurate prediction of these values is important to improve the effectiveness of DVFS in terms of energy savings. However, as we mentioned in Sect. 3.3.2, workload characteristics are not deterministic. Many works have proposed methods to accurately predict the workload and thus improve the effectiveness of DVFS [26–38]. According to the complexity and accuracy of predictors, they can largely be classified into the following three categories:

1. Conventional predictor: last value predictor, (weighted) moving average predictor, etc.
2. Control theory-based predictor: PI and PID predictor
3. Adaptive filter: Kalman filter, particle filter

Even with the most complex and accurate predictor, it is impossible to perfectly predict upcoming workloads since the runtime characteristics of a task have two characteristics: nonstationary program phase behavior and runtime distribution within a program phase. The program phase is defined as a time duration whose workload characteristics (e.g., mean, standard deviation, and max value of runtime) are clearly different from other time durations. In the following sections, we explain an intra-task DVFS method that exploits the workload characteristics [5, 14–16].

### 3.5.1 Runtime Distribution-Aware Workload Prediction

#### 3.5.1.1 Solution Overview

Figures 3.9(a), (b), and (c) illustrate three types of input required in the proposed procedure. Figure 3.9(a) shows a software program partitioned into program regions, each shown as a box. A program region is defined as a code section with

**Fig. 3.9** Solution inputs:
(a) software program (or
source code) partitioned into
program regions, (**b**) energy
model (in terms of energy per
cycle) as a function of
frequency, and (**c**) $f$–$v$ *table*
storing the energy-optimal
pairs, $(V_{dd}, V_{bb})$, for $N$
frequency levels

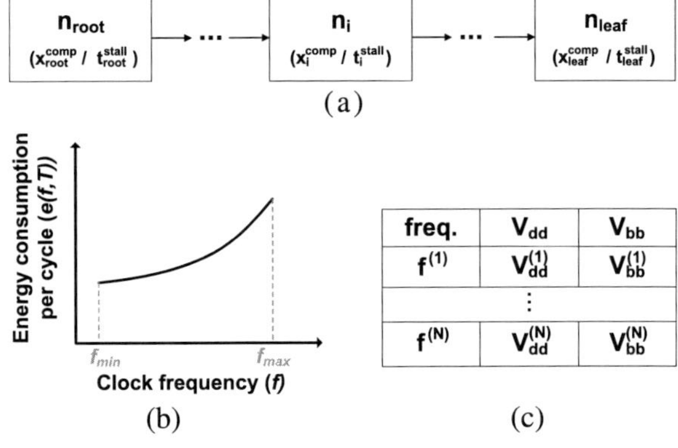

associated voltage/frequency setting. The partition can be performed manually by a designer or via an automatic tool [17] based on execution cycles of code sections obtained by a priori simulation of the software program. The $i$th program region is denoted as $n_i$, and the first and the last program regions are called the root ($n_{\text{root}}$) and leaf ($n_{\text{leaf}}$) program regions, respectively. In this work, we simply focused on a software program which periodically runs from $n_{\text{root}}$ to $n_{\text{leaf}}$ at every time interval. At the start of a program region, the voltage/frequency is set and maintained until the end of the program region. At the end of a program region, the computational cycle and memory stall time are profiled. Then, the joint PDF of the computational cycle and memory stall time is updated. We set frequency as well as $V_{dd}$ and $V_{bb}$, which will be discussed later in this section. Figure 3.9(b) shows an energy model (more specifically, energy-per-cycle vs. frequency). Figure 3.9(c) shows a precharacterized table called $f$–$v$ *table* in which the energy-optimal pair, $(V_{dd}, V_{bb})$, is stored for each frequency level ($f$). When the frequency is scaled, $V_{dd}$ and $V_{bb}$ are adjusted to the corresponding level stored in the table. Note that, due to the dependency of leakage energy on temperature, energy-optimal values of $(V_{dd}, V_{bb})$ corresponding to $f$ vary depending on the operating temperature. Therefore, we prepare $f$–$v$ *table* for a set of quantized temperature levels.

Given the three inputs in Fig. 3.9, we find the energy-optimal workload prediction, i.e., $w_i^{\text{opt}}$, of each program region during program execution. Algorithm 3.1 shows the overall flow of the proposed method. The proposed method is largely divided into workload prediction (lines 1–11) and voltage/frequency ($v/f$) setting (lines 12–14) steps which are invoked at the end and the start of every program region, respectively.

In the workload prediction step, we profile runtime information, i.e., $x_i^{\text{stall}}$ and $t_i^{\text{stall}}$, and update the statistical parameters of the runtime distributions, e.g., mean, standard deviation, and skewness of $x_i^{\text{stall}}$ and $t_i^{\text{stall}}$ (lines 1–2). After the completion of the leaf program region, the number of program runs, i.e., *iter*, is increased (line 4). At every PHASE_UNIT program run (line 5), where PHASE_UNIT is the predefined number of program runs (e.g., 20-frame decoding in MPEG-4), we perform the workload prediction and program phase detection by utilizing the profiled

---

**Algorithm 3.1** Overall flow

---

1: **if** (end of $n_i$) **then**
2:    Online profiling and calculation of statistics
3:    **if** ($n_i == n_{\text{leaf}}$) **then**
4:       iter++
5:       **if** ((iter % PHASE_UNIT)==0) **then**
6:          **for** from $n_{\text{leaf}}$ to $n_{\text{root}}$ **do**
7:             Workload prediction
8:          **end for**
9:          Program phase detection
10:       **end if**
11:    **end if**
12: **else if** (start of $n_i$) **then**
13:    Voltage/frequency scaling with *feasibility check*
14: **end if**

---

runtime information and its statistical parameters (lines 5–10). The periodic work-load prediction is performed in the reverse order of program flow as presented in [5, 14], and [18], i.e., from the end ($n_{\text{leaf}}$) to the beginning ($n_{\text{root}}$) of a program (lines 6–8). In this step, we find the workload prediction of $n_i$ which minimizes total energy consumption. Then, the program phase detection is performed to identify which program phase the current instant belongs to (line 9). In the $v/f$ setting step (lines 12–14), we set voltage and frequency with $w_i^{\text{opt}}$ while satisfying the hard real-time constraint (line 13).

### 3.5.1.2 Energy-Optimal Workload Prediction

For simplicity, we explain the workload prediction method when a task is partitioned into two program regions, $n_1$ and $n_2$. The total energy consumption to run the two program regions is expressed as

$$E_{\text{tot}} = \left(E_1^{\text{cpu}} + E_2^{\text{cpu}}\right) + \left(E_1^{\text{mem}} + E_2^{\text{mem}}\right) \tag{3.22}$$

where $e_i^{\text{cpu}}$ and $e_i^{\text{mem}}$ represent the energy consumption for running the computational workload in a processor and the memory stall workload in the $i$th program region, respectively. $E_1^{\text{cpu}}$ and $E_2^{\text{mem}}$ are expressed as follows:

$$E_i^{\text{cpu}} = e_{\text{cycle}}^{\text{cpu}}(f)x_i^{\text{cpu}} \tag{3.23}$$

$$E_i^{\text{mem}} = e_{\text{cycle}}^{\text{mem}}(f)ft_i^{\text{mem}} \tag{3.24}$$

where $f$ is the processor operating frequency, $x_i^{\text{cpu}}$ and $t_i^{\text{mem}}$ are the computational cycles and memory stall time at the $i$th program region, respectively, and $e_{\text{cycle}}^{\text{cpu}}$ and $e_{\text{cycle}}^{\text{mem}}$ are the energy consumption per cycle when a processor runs computational workload and accesses external memory, respectively. Note that processor energy

**Table 3.2** Energy fitting parameters for approximating the processor energy consumption to the accurate estimation obtained from PT scalar with BPTM high-k/metal gate 32 nm HP model for different temperatures, along with the corresponding (maximal and average) errors

| Temp. (°C) | Fitting parameters | | | | | Max[avg] error (%) |
|---|---|---|---|---|---|---|
| | $a_s$ | $b_s$ | $a_l$ | $b_l$ | $c$ | |
| 25 | $1.2 \times 10^{-1}$ | 1.3 | $4.6 \times 10^{-9}$ | 20.5 | 0.11 | 2.8 [0.9] |
| 50 | $1.2 \times 10^{-1}$ | 1.3 | $2.0 \times 10^{-7}$ | 16.6 | 0.12 | 1.4 [0.5] |
| 75 | $1.2 \times 10^{-1}$ | 1.3 | $2.2 \times 10^{-6}$ | 14.2 | 0.14 | 1.4 [0.4] |
| 100 | $1.2 \times 10^{-1}$ | 1.3 | $1.4 \times 10^{-5}$ | 12.4 | 0.15 | 1.7 [0.7] |

consumption depends on the type of instructions executed in the pipeline path because of the different switching activity and enabled functional blocks [19]. To simply consider the energy dependence on processor state, we classify the processor operation into two states: the computational state for executing instructions and the memory stall state primarily spent for waiting data from external memory. When a processor is in the memory stall state, switching energy consumption is suppressed by applying clock gating, whereas leakage energy consumption remains almost the same as in the computational state. The amount of reduction in switching energy is modeled as the parameter $\beta$, and the per-cycle energy consumption of the two processor states is expressed as follows:

$$e_{\mathrm{cycle}}^{\mathrm{cpu}} = a_s f^{b_s} + a_l f^{b_l} + c \tag{3.25}$$

$$e_{\mathrm{cycle}}^{\mathrm{mem}} = \beta a_s f^{b_s} + a_l f^{b_l} + c \tag{3.26}$$

where $a_s$, $b_s$, $a_l$, $b_l$, and $c$ are fitting parameters that approximate the energy model in (3.10) with a simpler form in order to reduce the complexity of finding workload prediction. The accuracy of the approximation for Intel Core 2-class microarchitecture with the Berkeley Predictive Technology Model (BPTM) high-k/metal gate 32 nm HP model [20] at various temperature levels is shown in Table 3.2.

Since $x_i^{\mathrm{cpu}}$ and $t_i^{\mathrm{stall}}$ have distributions as shown in Fig. 3.9(b), we can express the average energy consumption, $\overline{E_{\mathrm{tot}}}$, as the sum of $E_{\mathrm{tot}}$ with respect to the joint PDFs of two program regions, i.e., $J_1$ and $J_2$. The frequency of each program region, $f_i$ and $f_{i+1}$ in (3.23) and (3.24), is expressed as the ratio of the remaining computational workload prediction ($w_1$ and $w_2$) to the remaining time-to-deadline prediction for running the computational workload, i.e., the total remaining time to deadline ($t_1^R$ and $t_2^R$) minus the remaining memory stall prediction ($s_i$ and $s_{i+1}$), which are set to the average value of the memory stall time of the corresponding program region, as shown in (3.27) and (3.28),

$$f_1 = \frac{w_1}{t_1^R - s_1} \tag{3.27}$$

$$f_2 = \frac{w_2}{t_2^R - s_2} \tag{3.28}$$

where $t^{\mathrm{R}}_{i+1}$ in (3.28) is expressed as

$$t^{\mathrm{R}}_2 = t^{\mathrm{R}}_1 - \frac{x^{\mathrm{cpu}}_1}{f_1} - t^{\mathrm{mem}}_1 \tag{3.29}$$

By replacing $f_1$ and $f_2$ in (3.25) and (3.26), we can express the total energy consumption running a task, $E_{\mathrm{tot}}$, as a function of $w_1$ and $w_2$. The workload prediction is calculated in the reverse order of a task execution, i.e., $w_2 \to w_1$. Note that when a program is the last program region, we should set the workload prediction as the worst-case computational cycles of the program region. In this case, $w_2$ is set to the worst-case execution. Then, $w_1$ is the only unknown variable. $\overline{E_{\mathrm{tot}}}$ is a convex function with respect to $w_1$; thus, $w_1$ minimizing the $E_{\mathrm{tot}}$ can be found by solving the following equation:

$$w_1 \Rightarrow \frac{\partial \overline{E_{\mathrm{tot}}}}{\partial w_1} = 0 \tag{3.30}$$

The complexity for finding the solution can be avoided during runtime by accessing a look-up table (LUT) which precharacterizes the solutions according to runtime distribution [16].

### 3.5.1.3 Voltage and Frequency Setting

The voltage and frequency are set at the start of each program region. The procedure consists of two steps: (i) phase detection and prediction and (ii) voltage and frequency setting under deadline constraint. The program phase (especially in terms of computational cycles and memory stall time), whose time duration is managed on the granularity of PHASE_UNIT (as defined in Algorithm 3.1), is characterized by a salient difference in computational cycle and memory stall time. Conventionally, the program phase is characterized by utilizing only the average execution cycle of basic blocks without exploiting the runtime distributions of the computational cycle and memory stall time [21, 22]. To exploit the runtime distributions in characterizing a program phase, we define a new program phase as a workload prediction for each program region. Note that workload predictions reflect the correlation as well as the runtime distributions of both the computational cycle and memory stall time. Thus, the workload prediction is a good indicator which represents the joint PDF of each program region.

Periodically, i.e., on each PHASE_UNIT, we check to see whether a program phase is changed by calculating the Hamming distance between the workload predictions of the current period and those of the current program phase. When the Hamming distance is greater than the threshold $\theta_{\mathrm{p}}$ (set to 10% of the magnitude of the current program phase vector in our experiments), we evaluate that the program phase is changed, and then check to see if there is any previous program phase whose Hamming distance with the workload prediction of the current period is within the threshold $\theta_{\mathrm{p}}$. If so, we reuse the workload prediction of the matched previous phase

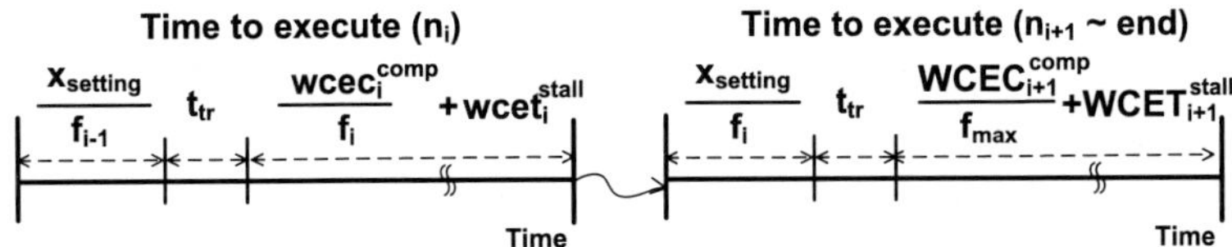


**Fig. 3.10** Calculation of possible worst-case runtime used in feasibility check

as that of the new phase to set voltage/frequency. If there is no previous phase satisfying the condition, we store the newly detected program phase and use the workload prediction of a newly detected program phase to set voltage/frequency until the next program phase detection.

When setting the performance level of each program region, the performance setting function checks to see if the performance level (to be set to $w_i^{opt}/(t_i^R - s_i)$) satisfies the given deadline constraint, even if the worst-case execution time occurs after the frequency is set. Figure 3.10 illustrates how to calculate the worst-case execution time. We decompose the worst-case execution time into two parts. One is the execution time related to $n_i$, and the other is that of the remaining program regions, i.e., $n_{i+1} \sim n_{\text{leaf}}$. As shown in the figure, the worst-case execution time of $n_i$ consists of three components: (1) runtime of frequency setting function ($x_{\text{setting}}/f_{i-1}$), (2) transition time ($t_{\text{tr}}$) for voltage (e.g., DC-DC conversion time) and frequency (e.g., PLL locking time), and (3) the worst-case processor computation cycle ($\text{wcec}_i^{\text{comp}}/f_i$) and memory stall time ($\text{wcet}_i^{\text{stall}}$) of the program region. The worst-case execution time from $n_{i+1}$ to $n_{\text{leaf}}$ also consists of the following three components: (1) runtime of frequency setting function running at $f_i$ ($x_{\text{setting}}/f_i$), (2) transition time ($t_{\text{tr}}$), and (3) worst-case time of processor computation running at $f_{\text{max}}$ ($\text{WCEC}_{i+1}^{\text{comp}}/f_{\text{max}}$) and memory stall ($\text{WCET}_{i+1}^{\text{stall}}$) from $n_{i+1}$ to $n_{\text{leaf}}$. Based on the calculation of worst-case execution time, we check whether the possible worst-case execution time exceeds the remaining time to deadline, i.e., $t_i^R$, as (3.31) shows. If it exceeds $t_i^R$, we increase the frequency of $n_i$, i.e., $f_i$, to the extent that the given deadline constraint is met

$$\left( \frac{x_{\text{setting}}}{f_{i-1}} + t_{\text{tr}} + \frac{\text{wcec}_i^{\text{comp}}}{f_i} + \text{wcet}_i^{\text{stall}} \right)$$

$$+ \left( \frac{x_{\text{setting}}}{f_i} + t_{\text{tr}} + \frac{\text{WCEC}_{i+1}^{\text{comp}}}{f_{\text{max}}} + \text{WCET}_{i+1}^{\text{stall}} \right) \leq t_i^R \tag{3.31}$$

### 3.5.1.4 Experimental Results

We used two real-life multimedia programs, MPEG-4 and H.264 decoder in FFmpeg [7]. We applied two picture sets for the decoding. First, we used, in total, 4200 frames of a 1920 × 1080 video clip consisting of eight test pictures, including *Rush Hour* (500 frames), *Station2* (300 frames), *Sunflower* (500 frames),

*Tractor* (690 frames), *SnowMnt* (570 frames), *InToTree* (500 frames), *Controlled-Burn* (570 frames), and *TouchdownPass* (500 frames) in [23]. Second, we used 3000 frames of a $1920 \times 800$ movie clip (as excerpted from *Dark Knight*). We inserted nine voltage/frequency setting points in each program: seven for macroblock decoding and two for file write operation for decoded image. We performed profiling with PAPI [6] running on LG XNOTE with Linux 2.6.3.

We performed experiments at 25°C, 50°C, 75°C, and 100°C. The energy consumption was calculated by PTscalar [4] and Cacti5.3 with the BPTM high-k/metal gate 32 nm HP model. We used seven discrete frequency levels from 333 MHz to 2.333 GHz with 333 MHz step size. We set 20 µs as the time overhead for switching voltage/frequency levels and calculated the energy overhead using the model presented in [24].

We compared the following four methods:

1. *RT-CM-AVG* [25]: runtime DVFS method based on the average ratio of memory stall time and computational cycle (*baseline*).
2. *RT-C-DIST* [18]: runtime DVFS method which only exploits the PDF of the computational cycle.
3. *DT-CM-DIST* [15]: design-time DVFS method which exploits the joint PDF of the computational cycle and memory stall time.
4. *RT-CM-DIST* [16]: runtime version of DT-CM-DIST.

We modified the original RT-CM-AVG [25], which runs inter-task DVFS without a real-time constraint so that it supports intra-task DVFS with a real-time constraint. In running DT-CM-DIST [15], we performed a workload prediction with respect to 20 quantized levels of remaining time, i.e., bins, using the joint PDF of the first 100 frames in design time.

Tables 3.3(a) and (b) compare the energy consumption for MPEG-4 and H.264 decoder, respectively, at 75°C. The first column shows the name of the test pictures. Columns 2, 3, and 4 represent the energy consumption of each DVFS method normalized with respect to that of RT-CM-AVG.

Compared with RT-CM-AVG [25], RT-CM-DIST offers 5.1%–34.6% and 4.5%–17.3% energy savings for MPEG-4 and H.264 decoder, respectively. Figure 3.11 shows the statistics of used frequency levels when running *SnowMnt* in MPEG-4 decoder. As Fig. 3.11 shows, RT-CM-AVG uses the lowest frequency level, i.e., 333 MHz, more frequently than the other two methods. It also leads to the frequent use of high frequency levels, i.e., levels above 2.00 GHz, where energy consumption drastically increases as frequency rises, in order to meet the real-time constraint. However, by considering the runtime distribution in RT-CM-DIST, high frequency levels incurring high energy overhead are less frequently used because the workload prediction with distribution awareness is more conservative than the average-based method.

Table 3.4 shows energy savings results for one of the test pictures, *SnowMnt*, at four temperatures, 25°C, 50°C, 75°C, and 100°C. As the table shows, more energy savings can be achieved as the temperature increases. The reason is that the energy penalty caused by frequent use of a high frequency level can be more obviously ob-

**Table 3.3** Comparison of energy consumption for test pictures at 75°C: (a) MPEG-4 (20 fps) and (b) H.264 decoder (12 fps)

| Image | RT-C-DIST [18] | DT-CM-DIST [15] | RT-CM-DIST (proposed) |
|---|---|---|---|
| (a) | | | |
| Rush Hour | 1.08 | 0.83 | 0.79 |
| Station2 | 1.34 | 0.97 | 0.95 |
| Sunflower | 0.99 | 0.76 | 0.74 |
| Tractor | 1.01 | 0.78 | 0.75 |
| SnowMnt | 1.02 | 0.90 | 0.81 |
| InToTree | 0.97 | 0.79 | 0.71 |
| ControlledBurn | 0.88 | 0.67 | 0.65 |
| TouchdownPass | 1.15 | 0.91 | 0.86 |
| Average | 1.05 | 0.83 | 0.78 |
| (b) | | | |
| Rush Hour | 1.11 | 0.94 | 0.93 |
| Station2 | 1.05 | 0.90 | 0.83 |
| Sunflower | 1.09 | 0.97 | 0.88 |
| Tractor | 1.18 | 1.00 | 0.96 |
| SnowMnt | 1.03 | 1.03 | 0.84 |
| InToTree | 1.14 | 1.00 | 0.93 |
| ControlledBurn | 1.07 | 0.94 | 0.88 |
| TouchdownPass | 1.10 | 0.99 | 0.94 |
| Average | 1.10 | 0.97 | 0.90 |

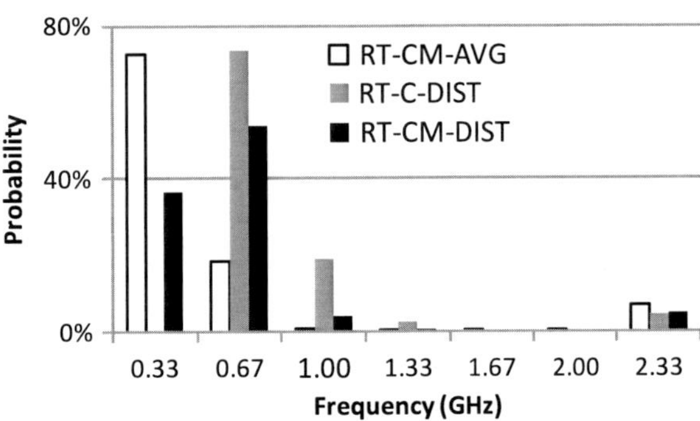


**Fig. 3.11** Statistics of used frequency levels in MPEG-4 for decoding *SnowMnt*

served as the temperature increases, since leakage energy consumption is exponentially increasing according to the temperature. By considering the temperature dependency of leakage energy consumption, RT-CM-DIST sets voltage/frequency so as to use high frequency levels less frequently as the temperature increases, whereas RT-CM-AVG does not consider the temperature increases. Note that, in most cases, the MPEG-4 decoder gives more energy savings than H.264 because, as Fig. 3.12

**Table 3.4** Comparison of energy consumption for *SnowMnt* at four temperature levels

| | Temp (°C) | RT-C-DIST [18] | DT-CM-DIST [15] | RT-CM-DIST (proposed) |
|---|---|---|---|---|
| MPEG-4 decoder | 25 | 1.15 | 0.94 | 0.86 |
| | 50 | 1.10 | 0.92 | 0.84 |
| | 75 | 01.02 | 0.90 | 0.81 |
| | 100 | 0.96 | 0.89 | 0.79 |
| H.264 decoder | 25 | 1.06 | 1.02 | 0.91 |
| | 50 | 1.05 | 1.02 | 0.88 |
| | 75 | 1.03 | 1.03 | 0.84 |
| | 100 | 1.01 | 1.03 | 0.80 |

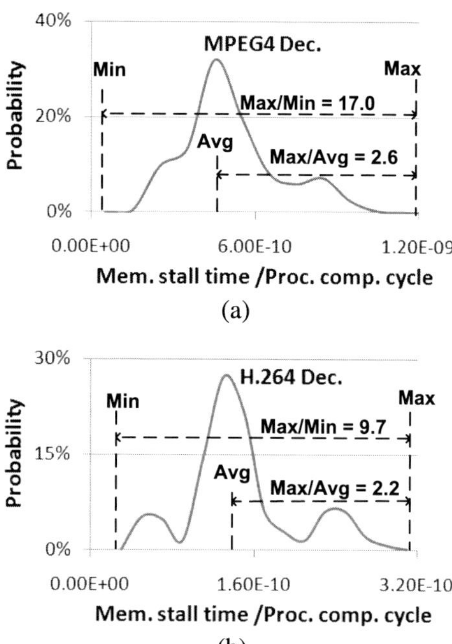


**Fig. 3.12** Distribution of memory boundedness in (**a**) MPEG-4 and (**b**) H.264 decoder

shows, the distribution of memory boundedness (defined as the ratio of memory stall time to computational cycle) of MPEG-4 has a wider distribution than that of H.264 in terms of Max/Avg and Max/Min ratios.

Compared with RT-C-DIST [18], which exploits only the distribution of computational cycle in runtime, RT-CM-DIST provides up to 20.8%–28.9% and 15.1%–21.0% further energy savings for MPEG-4 and H.264 decoder, respectively. The amount of further energy savings represents the effectiveness of considering the distribution of memory stall time as well as the correlation between computational cycle and memory stall time, i.e., the joint PDF of computational cycle and memory

**Table 3.5** Comparison of energy savings for *Dark Knight* at 75°C

|                | RT-C-DIST [18] | DT-CM-DIST [15] | RT-CM-DIST [16] |
| -------------- | -------------- | --------------- | --------------- |
| MPEG-4 decoder | 1.26           | 1.20            | 0.89            |
| H.264 decoder  | 1.16           | 1.16            | 0.89            |

stall time. RT-C-DIST regards the whole number of clock cycles, which is profiled at the end of every program region, as the computational cycle. Thus, RT-C-DIST cannot consider the joint PDF distribution of computational cycle and memory stall time. Consequently, it sets frequency levels higher than required levels, as shown in Fig. 3.11.

In Table 3.3, compared with DT-CM-DIST, which exploits runtime distributions of both computational and memory stall workload in design time, RT-CM-DIST provides 2.1%–10.2% and 1.2%–18.1% further energy savings for MPEG-4 and H.264 decoder, respectively. The largest energy savings can be obtained at *SnowMnt* for both MPEG-4 and H.264 decoder, which has distinctive program phase behavior. Since DT-CM-DIST finds the optimal workload using the first 100 frames (design-time fixed training input), which is totally different from that of the remaining frames (runtime-varying input), it cannot provide proper voltage and frequency setting.

To further investigate the effectiveness of considering complex program phase behavior, we performed another experiment using 3000 frames of a movie clip. The program phase behavior is more obviously observed for a movie clip with a fast-moving scene. Table 3.5 shows the normalized energy consumption at 75°C when decoding the movie clip from *Dark Knight* in MPEG-4 and H.264 decoder, respectively. RT-CM-DIST outperforms DT-CM-DIST by up to 26.3% and 23.3% for MPEG-4 and H.264 decoder, respectively, because complex program phase behavior exists due to the frequent scene changes in the movie clip.

# References


1. Agarwal, A., Kim, C.H., Mukhopadhyay, S., Roy, K.: Leakage in nanoscale technologies: mechanisms, impact and design considerations. In: DAC (2004)
2. Mudge, T., Flautner, K., Vlaauw, D., Martin, S.M.: Combined dynamic voltage scaling and adaptive body biasing for lower power microprocessors under dynamic workloads. In: Proc. ICCAD, pp. 721–725 (2002)
3. Rabaey, J.: Low Power Design Essentials. Series on Integrated Circuits and Systems. Springer, New York (2009)
4. Liao, W., He, L., Lepak, K.M.: Temperature and supply voltage aware performance and power modeling at microarchitecture level. IEEE Trans. Comput.-Aided Des. Integr. Circuits Syst. **24**(7), 1042–1053 (2005)
5. Hong, S., Yoo, S., Jin, H., Choi, K.-M., Kong, J.-T., Eo, S.-K.: Runtime distribution-aware dynamic voltage scaling. In: Proc. ICCAD, pp. 587–594 (2006)
6. PAPI [Online]. http://icl.cs.utk.edu/papi
7. FFmpeg [Online]. http://www.ffmpeg.org

8. Yan, L., Luo, J., Jha, N.K.: Joint dynamic voltage scaling and adaptive body biasing for heterogeneous distributed real-time embedded systems. IEEE Trans. Comput.-Aided Des. Integr. Circuits Syst. **24**(7), 1030–1041 (2007)
9. Enhanced Intel SpeedStep technology for the Intel Pentium M processor. Intel Whitepaper, Mar. 2004
10. Pedram, M.: Power optimization and management in embedded systems. In: Proc. ASP-DAC (2001)
11. Jejurikar, R., Pereira, C., Gupta, R.: Leakage aware dynamic voltage scaling for real-time embedded systems. In: Proc. DAC (2004)
12. Kim, C., Roy, K.: Dynamic VTH scaling scheme for active leakage power reduction. In: Proc. DATE (2002)
13. Nose, K., Hirabayashi, M., Kawaguchi, H., Lee, S., Sakurai, T.: VTH-hopping scheme for 82% power saving in low-voltage processor. In: Proc. IEEE Custom Integrated Circuits Conf. (2001)
14. Hong, S., Yoo, S., Bin, B., Choi, K.-M., Eo, S.-K., Kim, T.: Dynamic voltage scaling of supply and body bias exploiting software runtime distribution. In: Proc. DATE, pp. 242–247 (2008)
15. Kim, J., Lee, Y., Yoo, S., Kyung, C.-M.: An analytical dynamic scaling of supply voltage and body bias exploiting memory stall time variation. In: Proc. ASPDAC, pp. 575–580 (2010)
16. Kim, J., Yoo, S., Kyung, C.-M.: Program phase-aware dynamic voltage scaling under variable computational workload and memory stall environment. IEEE Trans. Comput.-Aided Des. Integr. Circuits Syst. **30**(1), 110–123 (2011)
17. Oh, S., Kim, J., Kim, S., Kyung, C.-M.: Task partitioning algorithm for intra-task dynamic voltage scaling. In: Proc. ISCAS, pp. 1228–1231 (2008)
18. Kim, J., Yoo, S., Kyung, C.-M.: Program phase and runtime distribution-aware online DVFS for combined $V_{dd}/V_{bb}$ scaling. In: Proc. DATE, pp. 417–422 (2009)
19. Kavvadias, N., Neofotistos, P., Nikolaidis, S., Kosmatopoulos, C.A., Laopoulos, T.: Measurement analysis of the software-related power consumption in microprocessors. IEEE Trans. Instrum. Meas. **53**(4), 1106–1112 (2004)
20. BPTM high-k/metal gate 32 nm high performance model [Online]. http://www.eas.asu.edu/ ptm
21. Sherwood, T., Perelman, E., Hamerly, G., Sair, S., Calder, B.: Discovering and exploiting program phases. In: IEEE Micro, Nov. 2003, pp. 84–93 (2003)
22. Sherwood, T., Sair, S., Calder, B.: Phase tracking and prediction. In: Proc. ISCA, pp. 336–347 (2003)
23. VQEG [Online]. ftp://vqeg.its.bldrdoc.gov/
24. Azevedo, A., Issenin, I., Cornea, R., Gupta, R., Dutt, N., Veidenbaum, A., Nicolau, A.: Profile-based dynamic voltage scheduling using program checkpoints. In: Proc. DATE, pp. 168–175 (2002)
25. Choi, K., Soma, R., Pedram, M.: Fine-grained dynamic voltage and frequency scaling for precise energy and performance tradeoff based on the ratio of off-chip access to on-chip computation times. IEEE Trans. Comput.-Aided Des. Integr. Circuits Syst. **24**(1), 18–28 (2005)
26. Shin, D., Kim, J.: Optimizing intra-task voltage scheduling using data flow analysis. In: Proc. ASPDAC, pp. 703–708 (2005)
27. Bang, S.-Y., Bang, K., Yoon, S., Chung, E.-Y.: Run-time adaptive workload estimation for dynamic voltage scaling. IEEE Trans. Comput.-Aided Des. Integr. Circuits Syst. **28**(9), 1334–1347 (2009)
28. Liang, W.-Y., Chen, S.-C., Chang, Y.-L., Fang, J.-P.: Memory-aware dynamic voltage and frequency prediction for portable devices. In: Proc. RTCSA, pp. 229–236 (2008)
29. Dhiman, G., Rosing, T.S.: System-level power management using online learning. IEEE Trans. Comput.-Aided Des. Integr. Circuits Syst. **28**(5), 676–689 (2009)
30. Lorch, J.R., Smith, A.J.: Improving dynamic voltage scaling algorithm with PACE. ACM SIGMETRICS Perform. Eval. Rev. **29**(1), 50–61 (2001)
31. Gu, Y., Chakraborty, S.: Control theory-based DVS for interactive 3D games. In: Proc. DAC, pp. 740–745 (2008)

32. Rixner, S., Dally, W.J., Kapasi, U.J., Mattson, P., Owens, J.D.: Memory access scheduling. In: Proc. ISCA, pp. 128–138 (2000)
33. Seo, J., Kim, T., Lee, J.: Optimal intratask dynamic voltage-scaling technique and its practical extensions. IEEE Trans. Comput.-Aided Des. Integr. Circuits Syst. **25**(1), 47–57 (2006)
34. Govil, K., Chan, E., Wasserman, H.: Comparing algorithms for dynamic speed-setting of a low-power CPU. In: Proc. MOBICOM, pp. 13–25 (1995)
35. Kim, J., Oh, S., Yoo, S., Kyung, C.-M.: An analytical dynamic scaling of supply voltage and body bias based on parallelism-aware workload and runtime distribution. IEEE Trans. Comput.-Aided Des. Integr. Circuits Syst. **28**(4), 568–581 (2009)
36. Xian, C., Lu, Y.-H.: Dynamic voltage scaling for multitasking real-time systems with uncertain execution time. In: Proc. GLSVLSI, pp. 392–397 (2006)
37. Wu, Q., Martonosi, M., Clark, D.W., Reddi, V.J., Connors, D., Wu, Y., Lee, J., Brooks, D.: A dynamic compilation framework for controlling microprocessor energy and performance. In: Proc. IEEE MICRO, pp. 271–282 (2005)
38. Isci, C., Contreras, G., Martonosi, M.: Live, runtime phase monitoring and prediction on real systems with application to dynamic power management. In: Proc. MICRO, pp. 359–370 (2006)

# Chapter 4
# Energy Awareness in Contemporary Memory Systems

**Jung Ho Ahn, Sungwoo Choo, and Seongil O**

**Abstract** As a critical part of a stored-program digital computer, the main memory system supports instructions and data to be consumed and produced by processing cores. High storage density and low manufacturing cost are the main design goals of the dynamic random access memory (DRAM) technology, making DRAM a primary option for main memory storage on contemporary computer systems that demand ever-increasing storage capacity. However, this focus on storage density and low cost leads to slow access time, low bandwidth, and high performance sensitivity on access patterns compared to the static random access memory (SRAM) technology, which is mainly used for on-chip storage. Thus, both the performance and energy efficiency of the main memory system are critical in improving the energy efficiency of the entire computer system. Also, emerging technologies such as 3D through-silicon vias and phase change memory have started to impact the structures and roles of main memory systems. This chapter categorizes and reviews recent research efforts to improve the performance or energy efficiency of contemporary memory systems. First, we describe conventional memory access scheduling policies which generate DRAM commands based on pending memory requests; we then review more advanced techniques that are focused on either improving performance or effectively managing DRAM power. Research works exploiting emerging technologies are introduced, and their impacts on future memory systems are analyzed. Then, we cover the proposals to modify memory modules or memory device architectures that reflect the access characteristics of future manycore systems.

J. Ahn (✉) · S. Choo · S. O
Seoul National University, Seoul, Republic of Korea
e-mail: gajh@snu.ac.kr

S. Choo
e-mail: choos@snu.ac.kr

S. O
e-mail: swdfish@snu.ac.kr

C.-M. Kyung, S. Yoo (eds.), *Energy-Aware System Design*,
DOI 10.1007/978-94-007-1679-7_4, © Springer Science+Business Media B.V. 2011

## 4.1 Introduction

Memory is a critical part of a stored-program digital computer. The computer's central processing unit (CPU) takes instructions and data from the memory, processes the data, and returns them back to the memory. An ideal memory can load and store data instantaneously and hold an infinite amount of data, which is not implementable in reality. Instead, modern computers have multiple levels of data storage: the ones closer to the CPU are smaller in capacity and faster in access time, and those farther from the CPU are bigger in capacity and slower in access time to mimic the ideal memory. The farthest (or lowest-level) one from the CPU is typically called main memory; the smaller and faster ones are called scratchpads or local memories if they have their own address spaces, and are called caches if they don't. Caches hold copies of main memory data that are frequently accessed, and can be managed by either hardware or software. In a broad sense, all levels of data storage can be called a memory system, but in this chapter we limit this. We say that a computer's memory system consists of main memory storage devices and controllers that are connected between the main memory devices and the caches (or scratchpads) to communicate data by sending control signals to the main memory devices in response to requests from the upper-level storage devices.

Dynamic random access memory (DRAM) is used for main memory storage devices in contemporary memory systems since its storage density is much higher than that of static random access memory (SRAM) even though it is slower than SRAM in access time (see Table 4.1). Continuous improvements in process technology enable commodity DRAM chips to hold several gigabits (Gbs) in each these days [1]. However, the primary design goal of the commodity DRAM chips is to lower manufacturing costs by improving area efficiency. To achieve high area efficiency, circuits to control DRAM cells are designed in a smaller area compared to other memory types, and the wires have higher resistance and capacitance. This makes most internal parts of a DRAM device operate slowly; many data bits are transferred in parallel to I/O buffers that are connected to external devices through I/O pins and operate at a higher speed. So modern DRAM chips achieve much better performance on a sequence of accesses to consecutive addresses (sequential accesses) than on a sequence of accesses to random addresses (random accesses). Moreover, the performance of global wires that send commands and transfer data to individual (among the billions of) DRAM cells has improved slowly compared to that of transistors and local wires [2]. As a result, the data transfer rate of a DRAM chip that governs its peak throughput improves more slowly than the rate at which the capacity of the DRAM chip increases, and its random access time improves at an even slower rate, as shown in Fig. 4.1.

The energy efficiency of a computer system can be enhanced by increasing either the performance or the energy efficiency of its memory system. A computer system with a higher-performance memory system will take less time to finish a task. Since a computer system dissipates static power regardless of its activity due to transistor leakage and clocking, this execution time savings will lead to system-wide energy savings. Various types of timing constraints apply in operating a DRAM device:

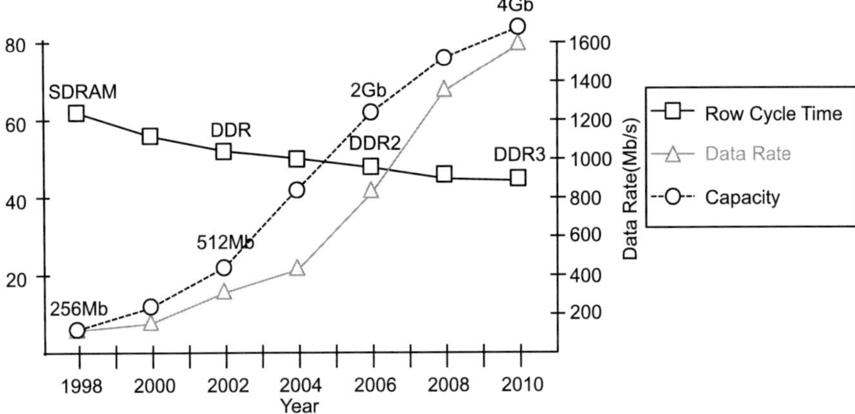

Fig. 4.1   Capacity, bandwidth, and latency trends of popular modern DRAM devices [3–6]

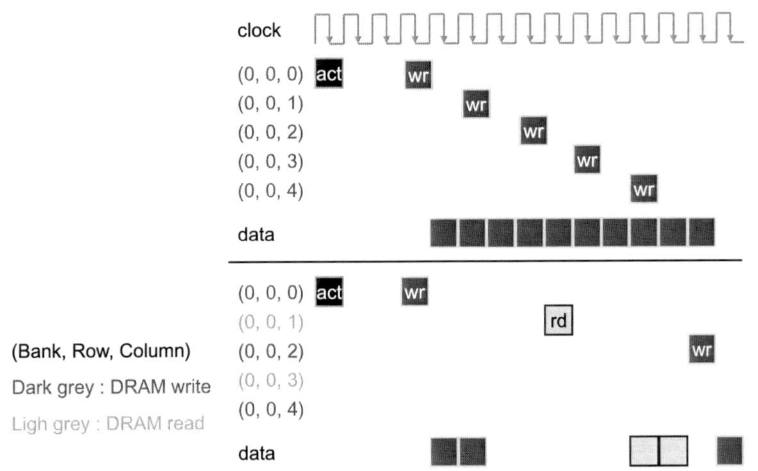

Fig. 4.2   The time to complete a series of memory references varies heavily depending on the order in which each reference is serviced [7]

interconnects are shared between multiple subcomponents of the device, data in DRAM cells must be activated before they can be accessed, and the device must operate under peak power constraints. Hence, the memory controllers in the memory system can affect its performance heavily. Proper reordering and precharge speculation can lead to huge savings in time to serve memory requests, depending on the memory access patterns, as shown in Fig. 4.2 [7].

If the energy efficiency of a memory system is improved without impacting its performance, the entire computer system will have better energy efficiency. However, architectural techniques to make the memory system more energy efficient usually affect the computer system performance, often negatively. As a result, these

**Fig. 4.3** DRAM power breakdown of a Micron 2 Gb DDR3 DRAM chip [8]. row/col means a ratio between row-level commands (ACTIVATE/PRECHARGE) and column-level commands (READ/WRITE) [9]

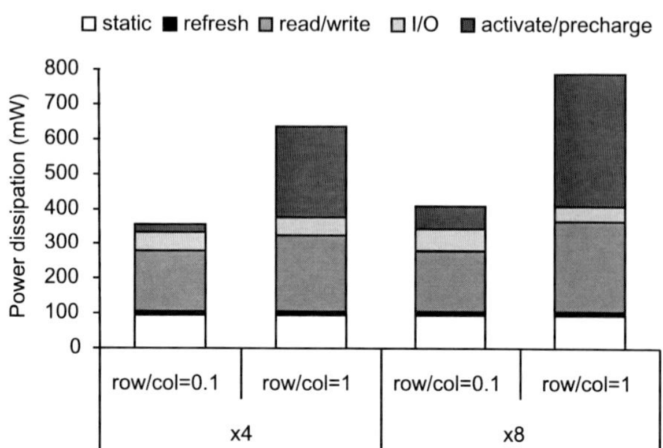



techniques must be exercised carefully. Figure 4.3 shows the power distribution of DDR3 DRAM chips [8], which have been widely used as of 2010. DRAM power consumption can be divided into two groups: static and dynamic power. Regardless of memory access, a DRAM chip consumes power since it needs to keep locking the phase of an external clock, transistors leak, and refresh operations are needed. The amount of dynamic power depends strongly on the access patterns applied to the DRAM chip. If random accesses are applied to the chip, the data in the DRAM cells will need to be activated before it can be read or written in most cases, so activation operations consume more than half of the total DRAM power consumption. By contrast, if sequential accesses are applied, the cost of activation operations can be largely amortized by subsequent reads and writes, and most of the dynamic power is due to the delivery of data through on-chip and off-chip wires. For a computer system that requires high throughput from its memory system, architectural techniques to lower the dynamic power of the DRAM chips with minimal impact on system performance will be effective. If a computer system needs high memory capacity, such as those running in-memory databases or housing multiple virtual machines, static power including refresh power must be minimized.

This chapter reviews recent research efforts to improve the energy efficiency of contemporary computer systems by enhancing the performance or energy efficiency of their memory systems. Section 4.2 introduces a canonical memory system that is a part of a modern computer system, and performance and energy-efficiency trends of DRAM chips and their components. Section 4.3 covers memory access scheduling policies and their implementations to achieve better memory system performance. Section 4.4 focuses on architectural techniques to manage DRAM power consumptions explicitly. Papers to exploit emerging technologies such as 3D die stacking, flash memory, and phase change memory on memory systems are reviewed in Sect. 4.5, and techniques to better place data in multiple memory systems of future manycore computer systems are covered in Sect. 4.6. Section 4.7 explains the proposals to modify memory module or device architectures for higher energy efficiency on future VLSI technologies.

## 4.2 Background

This section describes a memory channel of a canonical memory system, which is composed of multiple DRAM chips orchestrated by a memory controller (Fig. 4.4). A DRAM chip has billions of DRAM cells, each of which holds a bit of information. A DRAM cell, as shown in Fig. 4.5, consists of one access transistor and one capacitor. The capacitor holds data by the amount of stored charges. To read or write data in the cell, the access transistor becomes open to connect the capacitor in the DRAM cell and a bitline external to the cell. Charges flow into or out of the cell capacitor to write "one" or "zero" to the cell. For a read, charges in the cell capacitor and the bitline are redistributed, and the corresponding voltage change in the bitline is sensed by the sense amplifier. A capacitor is a passive device, and the capacitance of the cell capacitor is limited (around 30 fF for years [1]), so if the bitline capacitance is much higher than that of the DRAM cell, the voltage change on the bitline by the charge sharing with cell transistor is too small to be reliably detectable. Since the bitline capacitance is proportional to the number of DRAM cells connected to the bitline, there is an upper bound on the number of cells that can be attached for correct detection. Also, the charges in the DRAM cells leak, so they must be refreshed periodically to retain the values inside.

These DRAM cells are laid out in a two-dimensional (2D) structure called a DRAM mat, as shown in Fig. 4.6. Within a mat, all the access transistors of the cells in the same row are connected by a single wordline and controlled by the single driver chosen by the row decoder in the mat. All the cell transistors of the cells in the same column are attached to a single bitline through access transistors, but the row decoder guarantees that only up to one cell is "connected" to the bitline at any given time. For reliable and fast detection of voltage differences caused by charge redistribution, bitlines are precharged to a certain voltage level (typically to half of the $V_{dd}$ for a DRAM [1]). Once the row decoder chooses a row to access and bitlines are precharged, the cells in the row are connected to the corresponding bitlines. To load data, voltage changes due to charge sharing are detected by the sense amplifiers. The column decoder chooses the column or a group of columns that are affected by the load or store operation. To store data, the specific row is first loaded and the write drivers drive currents to the bitlines selected by the column decoder to charge or discharge cell transistors.

These mats are again laid out in a 2D structure to compose a DRAM chip. All mats are connected by levels of on-chip interconnects (called local I/Os, mid I/Os, and global I/Os on a three-level interconnect structure, for example) to communicate with the memory controller through off-chip I/Os. Due to large fanin and fanout, these on-chip interconnects use sense amplifiers to achieve high operating speed, but they are still slow, with a long access time (on the order of dozens of nanoseconds). Both data and control signals are transferred through the on-chip interconnects. Control signals consist of command signals and address signals. The address signals specify the DRAM mats, the row of the mat, and possibly the column ranges of the row on which the data signals are transferred. The command signals determine the type of operations conducted in the specified mat, row, or

                                                                                                                    

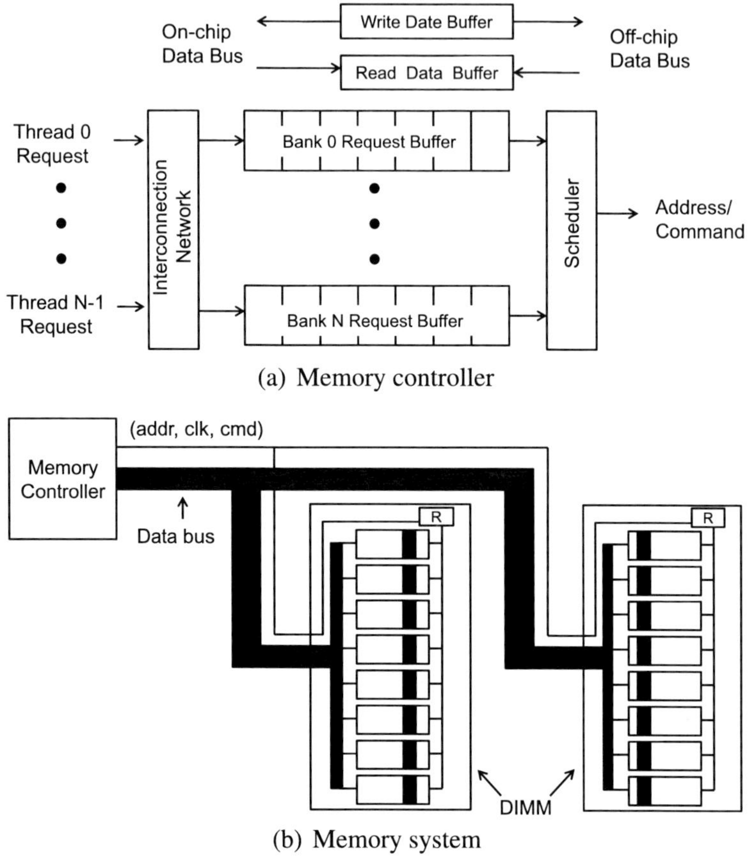


**Fig. 4.4** (**a**) A memory controller and one or more memory modules compose (**b**) a conventional memory system

columns, such as precharging the bitlines of a mat and reading data from a row. To overcome the performance degradation due to the high access latency, access operations are pipelined on multiple DRAM banks in the DRAM chip so that bitlines in a DRAM bank are precharged, rows in DRAM mats of another bank are sensed, and part of the sensed rows in the third bank are transferred through on-chip interconnects, all at the same time. Besides the mats and the interconnects, the DRAM chips have peripheral circuits such as a delay-locked loop [10] and voltage converters that contribute to static power and the targets of DRAM power-down modes, which were first introduced in Rambus DRAM (RDRAM [11]) and followed by other DRAM designs.

To increase the maximum data throughput from a DRAM chip without increasing the operating speed of the mats and banks, the on-chip interconnects become wider and the transfer rate of the off-chip I/Os is increased. The data transfer rate per I/O pin of modern DDR3 DRAM chips, which are mainly for CPUs, is up to 1.6 Gbps

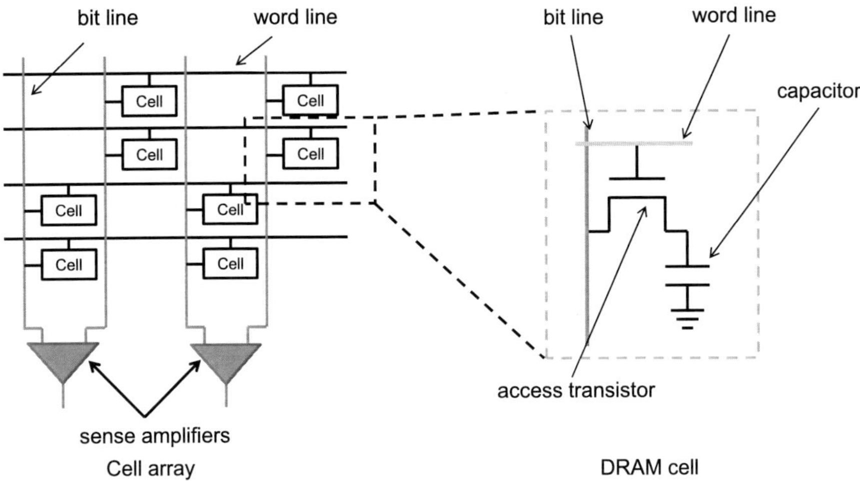



**Fig. 4.5** (**a**) A cell array and (**b**) a DRAM cell

**Fig. 4.6** A DRAM mat consists of DRAM cells laid out in a 2D array, a row decoder, a column decoder, and sense amplifiers

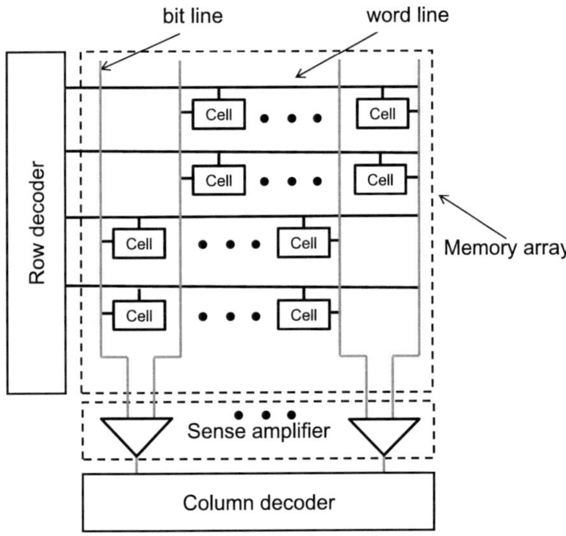



while that of modern GDDR5 DRAM chips, which are mainly for graphics processing units (GPUs), is up to 7 Gbps as of 2010 [12, 13]. One of the reasons why the data transfer rate per I/O pin of GDDR5 DRAM chips is much higher than that of DDR3 DRAM chips is that the connection between the GPU and the GDDR5 chip is point-to-point, whereas multiple DDR3 chips are connected to a memory controller through a shared bus. The electrical properties of the shared bus are much worse than that of a point-to-point link; the bus is lengthy, so the waveforms are more distorted and signals reflect due to impedance mismatch on the nodes connected to the bus. To overcome these problems, differential signaling is applied to high speed DRAM devices such as Rambus XDR DRAM chips [14]. The maximum data throughput from a DRAM chip is determined by multiplying the data transfer rate per I/O pin by the number of data I/O pins. The latter is often called data path width per DRAM chip, and the notation $\times N$ is used to represent an $N$-bit data path.

DRAM chips are grouped together to form a DRAM rank. The DRAM rank is a unit of access from a memory controller in modern computer systems; one or more DRAM ranks are housed in a memory module. Single-inline memory modules (SIMMs) with a 32-bit data path were once popular, but they are being replaced by dual-inline memory modules (DIMMs) to supply higher data throughput. All ranks in a bank share the control and data paths except the rank select signals that specify the target of a certain command. The data path width of a DIMM is twice that of a SIMM, 64, so that 16 $\times 4$ chips, 8 $\times 8$ chips, or 4 $\times 16$ chips typically compose a rank.

Continuous increases in the number of transistors that can be integrated per die, the diminishing returns of increasing the complexity of a single computation core, and the growing interest in energy efficiency make multicore architectures popular, and the number of computation cores per package keeps increasing [15]. Since the amount of cache dedicated per core does not increase much as more cores are integrated per package, the multicore processors need more main memory throughput so that multiple memory controllers are populated per processor. The demands on memory capacity also keep increasing as emerging applications such as in-memory databases and virtual machines [16] become more popular. To support large main memory capacity, multiple ranks can be connected to a single memory controller. As explained above, the shared bus is used to connect multiple DDR3 DRAM ranks (as well as their previous generations). To increase the data transfer rate per I/O pin so that the wire placing a burden on the board containing CPUs and memory modules can be relieved, the bus is replaced with point-to-point links and daisy-chained in a fully buffered DIMM [17]. However, the large energy consumption in the advanced memory buffers in a fully buffered DIMM to implement daisy-chaining limits widespread use of the interface.

Figure 4.4(a) shows a block diagram of a memory controller that is located between computation cores and memory modules. A canonical memory controller consists of channel buffers, a scheduler, and a physical interface to the memory modules. Channel buffers hold memory requests from the computation cores until the requests are served by the corresponding DRAM ranks. The channel buffers can perform the functionality of miss status holding registers (MSHRs), so that requests to the same memory address can be combined in the buffers. The scheduler checks the pending requests in the channel buffers and decides on the most appropriate commands to one of the memory ranks attached to it under timing and structural constraints of the DRAM chips. The mechanism to decide the commands in the scheduler is called a memory access scheduling policy. The simplest policy is to serve the requests in the order in which they arrive at the channel buffer without considering the requests behind the oldest request. This is called the first-come-first-serve (FCFS) policy. Since the FCFS policy does not utilize modern DRAM chips effectively, many advanced memory access scheduling policies have been proposed, which is the subject of Sect. 4.3. The physical interface in the memory controller is the counterpart of the I/O pins and corresponding circuitry in the DRAM chips, and consists of unidirectional control I/Os that send address and command signals to the DRAM ranks and bidirectional data I/Os.

## 4.3 Energy-Aware Memory Scheduling

Rixner et al. [7] introduced a technique called memory access scheduling that improves the performance of memory systems by considering the non-oldest memory requests in addition to the oldest request in the channel buffer of the memory controller. If the memory command to serve the oldest memory request cannot be scheduled due to timing or structural constraints and the commands for other requests are ready to be scheduled, an available command from the non-oldest memory requests is served so that the requests are effectively reordered. This is similar to the out-of-order execution concept in modern high-performance cores. The paper [7] proposed multiple scheduling policies. The first-ready first-come-first-serve (FR-FCFS) policy gives higher priorities to requests that stay longer in the channel buffer. Closed-row policies precharge a row in a DRAM bank as soon as none of the pending requests in the buffer have the address targeting the row, whereas open-row policies precharge a row in a DRAM bank only if there is no pending request in the buffer that has the address targeting the row and there is at least one pending request that has the address targeting a different row but in the same DRAM bank. These policies can be combined with the FR-FCFS policy to further reduce the average access latencies since they "close" unused rows and precharge the bitlines of the corresponding banks earlier on certain conditions, so that the bank can activate the subsequent row faster unless the "closed" row is used consecutively, which is rare with proper channel buffer sizes.

Ahn et al. [18] explored the design space of memory systems of data parallel processors. Their results reaffirmed that reordering memory requests in memory controllers improves the average memory access latency and hence the throughput of the memory system. However, insofar as both row-level commands and column-level commands to serve the memory requests can be reordered, a specific reordering policy does not affect its throughput significantly when the memory controllers have enough channel buffers for reordering. Another important factor to affect the memory system performance discussed in their study is the interference among memory access sequences from different request sources. Sequences of memory requests from different sources (typically computation cores) often have different types of accesses (e.g., one is read requests while the other is write requests) or cause bank conflicts (both request sequences head to the same bank of the same DRAM rank but to different rows), leading to performance degradation due to timing constraints, as explained in Sect. 4.2. In order to minimize the negative impacts of this interference, [18] suggested grouping requests from the same source as much as possible or making the requesting sources cooperate so that only a single source generates memory requests to a memory controller at a time.

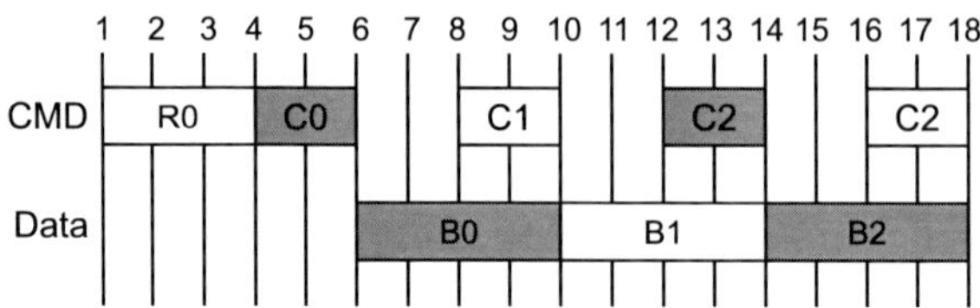



**Fig. 4.7** An example of burst scheduling. Accesses to the same row and the same bank are scheduled in an access burst [19]

Burst scheduling [19] is an access reordering technique to exploit the row-level locality and bank-level parallelism of modern DRAM chips. A burst is a set of memory accesses to the same row and the same bank. In the burst scheduling scheme, ongoing accesses are reordered, and the reordered accesses are partitioned into bursts. Burst scheduling schedules the bursts in a bank interleaved fashion, therefore exploiting the bank-level parallelism. In addition, since a single burst requires only one row activation, the overall latency for the row activation can be lowered. Therefore, with burst scheduling, the data bus utilization is improved (see Fig. 4.7). The reordering process to group accesses into bursts incurs overhead so that single access bursts or short bursts can be starved. However, this overhead can be amortized because the reordering process increases the overall utilization of the data bus.

Prefetch-aware DRAM controllers (PADC) [20] aim to maximize the benefits of useful prefetches and minimize the harmful effects from useless prefetches. To improve the performance of memory systems, a PADC prioritizes prefetch accesses and reorders ongoing accesses. Conventional memory controllers treat prefetch requests and demand requests equally, or prioritize demand requests over prefetch requests. Since the usefulness of prefetching depends on the characteristics of applications running in the computer system, system throughput cannot be maximized with a single, nonadaptive prefetch scheduling. A PADC prioritizes prefetching accesses with processor-aided prefetch accuracy data. The PADC contains two modules for prefetch prioritization: an adaptive prefetch scheduling (APS) unit and an adaptive prefetch dropping (APD) unit. The APS prioritizes the prefetch accesses. If the prefetch accuracy of a core is greater than or equal to a certain threshold, all of the prefetch requests from that core are treated the same as demand requests. As a result, useless prefetches are not prioritized. The APD removes useless prefetch requests from the request buffer if they have been placed for a long time. The PADC reduces the waste of bandwidth caused by useless prefetching; however, it requires additional hardware for caching and processing prefetch accuracy.

ATLAS [21] reduces the bandwidth contention caused by multiple memory controllers in a multicore system with global memory access reordering. Compared to previous researches that only considered the performance of a single memory controller, ATLAS focuses on systems with multiple memory controllers where the memory accesses of each computation core can be assigned to multiple memory controllers, and independent access optimization of each memory controller induces bandwidth contention and degrades system throughput. A key reliever of

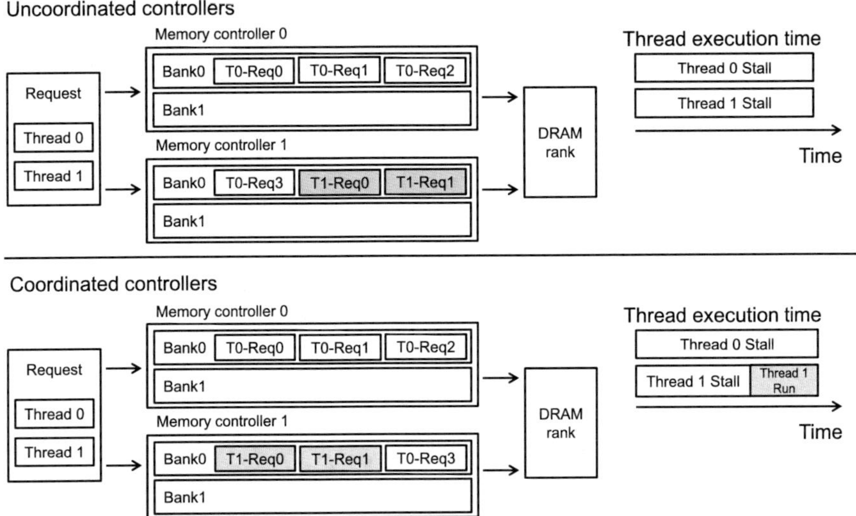


**Fig. 4.8** Conceptual example showing the importance of coordinating the action of multiple memory controllers. In both cases, memory controllers schedule shorter jobs first [21]

this problem is to allow each memory controller to exchange its access information with other controllers. With this access information coordination, each memory controller schedules ongoing accesses, thereby reducing bandwidth contention, as shown in Fig. 4.8. However, frequent coordination costs overhead, and it becomes more serious as more memory controllers communicate. In ATLAS, execution time is divided into long time intervals, called quanta, and coordination is done only when each quantum begins. In addition, ATLAS controllers utilize thread-level access scheduling to reduce the size of coordinated information; therefore, coordination overhead drops. However, the thread-level scheduling of ATLAS limits the exploitation of row locality and bank parallelism of DRAM chips.

The Reinforcement Learning (RL)-based controller [22] dynamically modifies access scheduling policies to improve the memory system performance. The controller learns to optimize its scheduling policy "on the fly" to maximize its long-term performance, while the fixed access scheduling policies only focus on hardware-level optimizations, such as row locality or bank-level parallelism for average-case applications. Since each application has different memory access patterns, a single fixed access scheduling policy cannot provide optimal performance on every application. The RL-based memory controller utilizes machine learning and selects the next scheduled access using a decision-making process based on the concept of Markov decision processes. The scheduler estimates the long-term performance impact on all the accesses in transaction queues with given system states and data bus utilization. After that, the scheduler fetches an access that has the highest long-term performance impact. In this scheme, system states can include the number of reads, writes, or load misses in the transaction, criticality of each request, row buffer hit, or the total number of reads or writes pending for each row and bank. Since system states cover a variety of attributes, trade-offs between performance and complexity can be exploited.


## 4.4 Explicitly Managing DRAM Power Consumption

Liu et al. [23] proposed a write buffer between cache and DRAM, called TAP (Throughput-Aware Page-hit-aware write buffer), to reduce the active power consumption and operation temperature of DRAM chips. The write operations that incur page misses consume much higher power than the ones without page misses. This increases the operation temperature of the DRAM chips. With TAP, write operations that are not targeting an activated row are buffered. Buffered write is scheduled when a targeting row is activated. To reduce the power consumption of TAP, unused entries of TAP are turned off through power gating. TAP leads to fewer row activations for write requests, and also decreases the page miss rate. Therefore, it reduces the power consumption and operation temperature of DRAM chips. TAP is useful in a small memory system which experiences frequent page faults.

Hur et al. [24] researched memory controller designs to improve DRAM energy efficiency and manage DRAM power. Since a memory system is operated through the interaction between DRAM ranks and its controller, a comprehensive approach is needed for DRAM energy efficiency. Contemporary DRAM chips already have power-saving mechanisms which put idle memory devices into one of the low-power modes. Hur et al. proposed specific policies to exploit these low-power modes of modern DRAM chips for commercially available server-class systems. To utilize the intrinsic power-saving mechanisms of DRAM efficiently, they proposed three key mechanisms for the memory controller: a centralized arbiter queue (CAQ), a power-aware memory scheduler, and adaptive memory throttling. CAQ monitors each rank power mode status and controls the power mode of each rank with its power down/up mechanism. Each rank is turned into power-up mode only when there is an access to it. The power-aware scheduler is an extended version of a conventional adaptive history-based scheduler. It has an additional finite state machine (FSM) to improve power efficiency. The goal of the new FSM is to reduce the total number of power-down operations through grouping of same-rank commands. With this approach, ranks having no command can remain in a power-down mode longer. Adaptive memory throttling blocks issuance of commands for some fixed periods of time, thereby protecting the system from the hazard of worst-case power consumption. In this approach, to reduce DRAM power consumption to a target level, a delay estimator produces an estimated throttling delay with power thresholds and states of DRAM chips. Since throttling forces all the ranks to turn into power-down modes, accurate delay estimation is critical to achieve high system performance. Another example of a memory controller design to manage DRAM power is the smart refresh [25] suggested by Ghosh and Lee, which targets saving refresh power, which could be a concern on servers with large main-main capacity.

Aggarwal et al. [26] explored the unnecessary power consumption of speculative DRAM accesses at a broadcast-based shared-memory multiprocessor system. They proposed the Power Efficient DRAM Speculation (PEDS), which prevents unnecessary speculative accesses to DRAM chips and thereby reduces the power consumption of the memory systems. In a shared-memory multiprocessor system, speculative access has a potential performance advantage, because it can hide DRAM latency. However, if data obtained with a speculation already exists in a cache of another processor, the speculative access wastes power. PEDS estimates unnecessary speculative access with the information provided by region coherence arrays (RCAs). RCAs provide information on memory regions that are not shared by other processor and memory regions from which the processor is caching lines. Based on given information, PEDS estimates whether a request is necessary or not. An estimated unnecessary request is buffered in a memory controller. The memory controller compares it with the snoop response. If the snoop response indicates that no processor has the data, the prediction was incorrect and the access will be scheduled. In this scheme, wrong prediction causes a latency penalty. Since a smaller memory region provides more accurate information, the estimation accuracy largely depends on the size of the RCAs.


## 4.5  Exploiting Emerging Technologies

We have experienced unprecedented device scaling in the past few decades, and the fabrication is still developing. However, the device size is approaching its physical limit. As shown in the ITRS roadmap [27], maximum die size grows very slowly and the yield and power-consumption issues limit the practical die sizes of commonly used integrated circuits. To place more transistors in the same planar area, 3D die stacking and interconnect technologies are rapidly emerging.

3D die stacking increases transistor density by vertically integrating two or more silicon dies. Furthermore, 3D stacking also improves interconnect latency by eliminating some of the off-chip wires and decreasing the length of on-chip global wires. A shorter signal path also enables smaller system form factor, lowers crosstalk, and reduces power dissipation.

The most typical type of interconnect in the early era of 3D die stacking was wire bonding. Obviously, this method requires aggressive design rules and has quality or reliability issues. To connect each die directly and in the shortest path, an electrical path that passes through the silicon is needed. This idea is realized by the through-silicon via (TSV) technology [28]. TSV connects two adjunct dies electrically. This approach is highly feasible, but difficult to utilize because it requires grinding wafers and causes miscellaneous mechanical stresses. There is another approach that makes indirect connection between multiple dies, called "indirect couplings." This approach isolates each die electrically using capacitive [29] or inductive coupling [30] between levels, but these indirect coupling technologies are not suitable to deliver power to dies. Indirect coupling continuously gains interest but is not as popular as the TSV technology as of 2010.

3D die stacking can be used to improve the latency and bandwidth between computation cores and DRAM banks. Figure 4.9 [31] shows one of the high-performance 3D integrated architectures that are claimed to achieve up to three times

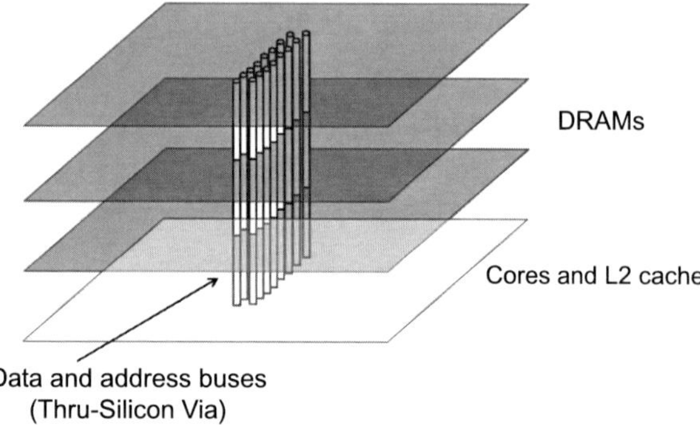

**Fig. 4.9** 3D-stacked DRAM on a chip multiprocessor with one rank per layer



the speedup of conventional multi-socket systems since off-socket interconnects are completely eliminated while accessing main memory data and on-chip global interconnects are partially removed due to vertical aligning between the processor die and DRAM dies. This approach is attractive for memory systems with relatively small storage capacity such as hand-held devices. Increasing storage capacity using multiple silicon layers is possible as well. In 2009, a single DRAM socket that implements 3D stacking technology that can store 8 Gb was presented [32].

While 3D die stacking technology is developed to increase capacity and lower latency, another emerging technology, called phase change memory (PCM), is showing its possibilities as an alternative to DRAM for main memory and last-level caches. As process technology improves, it becomes more difficult to design a DRAM cell capacitor compared to an access transistor because the cell capacitance must be sufficiently large to retain data for a certain amount of time and provide enough of a voltage change for reliable sensing on the sense amplifier. This threshold drops more slowly than the rate at which the cell capacitance drops due to cell area shrinking, and conventional techniques such as trenching and stacking to obtain higher capacitance are approaching their limits [33]. By contrast, PCM can store data in phase change material placed between the intersections of bitlines and wordlines; thus, its cell size could be smaller than that of a DRAM cell and it could be fabricated in finer process technologies more easily.

Flash memory [36] is another memory technology that can provide smaller cell size compared to DRAM, but it is too slow to replace DRAM as main memory. Table 4.1 compares multiple high-density memory technologies and shows that flash memory is orders of magnitude slower than DRAM, especially for write operations. Flash memory is designed to access data at a block granularity, which is much larger than a typical cache size, which leads to a poor performance on random accesses.

PCM is a nonvolatile memory. During writes, an access transistor injects current into phase change material that has multiple stable phases; high temperature induces phase change. The intensity of the current and its duration can be used to determine the specific phase of the material, which effectively writes one or more bits of data if we can reliably detect more than two phases of the material. This transition

**Table 4.1** Comparison of high-density memory technologies [34, 35]

|              | SRAM          | DRAM      | Flash          | PCM           |
|--------------|---------------|-----------|----------------|---------------|
| Size         | 100–200 $F^2$ | 6–8 $F^2$ | $\sim$4 $F^2$  | 4–20 $F^2$    |
| MLC supported| No            | No        | Yes            | Yes           |
| Volatile     | Yes           | Yes       | No             | No            |
| Read time    | <1 ns         | 5–100 ns  | 10–100 ns      | 10–100 ns     |
| Write time   | <1 ns         | 5–100 ns  | >1 ms          | >100 ns       |
| Write energy | 0.1 nJ/b      | 0.1 nJ/b  | 0.1–1 nJ/b     | <1 nJ/b       |
| Endurance    | $\infty$      | $\infty$  | $10^4$–$10^5$  | $10^8$        |

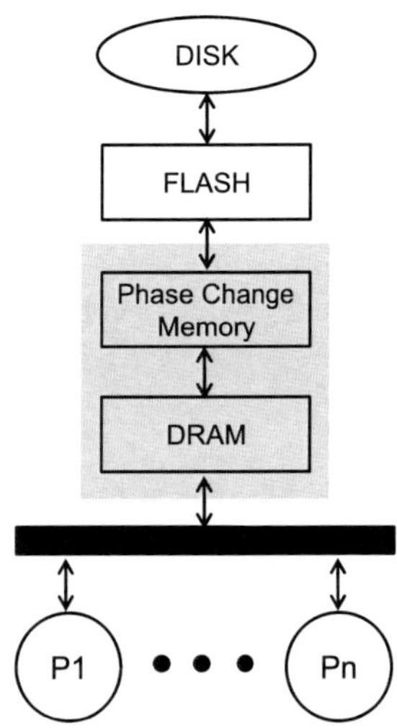


**Fig. 4.10** System with hybrid memory system configured with DRAM, PCM, flash, and magnetic disks

affects the resistivity of the material; the difference in resistivity can be detected during reads by measuring the current over the resistive material. PCM is projected to scale up to 9 nm process technology [27]. In 2010, 1 Gb PCM with a 45 nm process was realized [37]. It has a relatively slow access time and low throughput. Although PCM is a highly emerging technology, it is several times slower in access latency and has higher write energy compared to DRAM. To overcome these drawbacks, alternative architectures and design methodologies for using PCM as main memory have been proposed [33]. Since PCM has a limited write cycle count, a multi-level memory structure has also been proposed [38] to extend the lifecycle of endurance of PCM. Inserting DRAM memory between PCM main memory and last-level DRAM or SRAM cache can boost the PCM lifetime from 3 years to 9.7 years, which is comparable to the endurance of DRAM (Fig. 4.10). Compared to DRAM, PCM still has endurance and speed problems as main memory, but it is more attractive than other alternatives (as of 2010).

## 4.6 Effective Data Placements

As each chip integrates more transistors, additional performance advantages by further increasing the complexity of already complicated out-of-order speculative cores diminish, and energy efficiency gains more interest, multicore architectures or chip-multiprocessors have become popular and various manycore processors have been proposed as well. A commercial version of the MIT raw processor [39] called Tilera64 [40] integrated 64 computation cores and 4 memory controllers, while Intel showed that 80 cores could be integrated in a single die on their Polaris prototype [41]. To take full advantage of many computation cores, abundant memory bandwidth is needed for each core. For example, Sun's Niagara processor [42], which was announced as UltraSPARC T1 in 2005, contains 4 memory controllers and L2 caches to feed 32 concurrent hardware threads on 8 computation cores. However, the number of I/O pins per processor chip increases at a rate slower than that of the number of transistors that can be fabricated per chip due to packaging constraints, and increasing data transfer rates per I/O pin incurs power and area overheads. As a result, the ratio of memory controllers to computation cores decreases as the number of cores per processor increases; this lowers memory bandwidth dedicated per core and increases average access latencies.

Both Polaris and Tilera64 processors placed the memory controllers at the edges of the chips. As the number of computation cores increases, placing all the memory controllers at the edges is not practical, but placing them at each core is not practical either due to the I/O pin count constraint. Two reasonable assumptions on future manycore processors are that the ratio of the computation cores to the memory controllers is from a few to several, and the latency between a core and a memory controller varies heavily according to the topological distance between the two. These assumptions mean that careful placement of memory controllers is essential to lower average memory access latencies. For example, when 64 cores are connected through an $8 \times 8$ 2D mesh network-on-a-chip [43], placing memory controllers in a diamond shape lowers the maximum channel load by 33% compared to a configuration that places memory controllers at the top and bottom rows of the computation cores as presented by Abts et al. [44] (Fig. 4.11).

## 4.7 New Memory Module or Component Architectures

A large body of research is working to improve the performance, energy efficiency, and reliability of main memory systems by either modifying the structure of conventional memory modules or suggesting new DRAM chip organizations. Many of the proposals reviewed below are highly promising, but we note that compatibility with existing main memory standards typically limits their effectiveness and applicability.

There have been multiple proposals that advocate rank subsetting, which divides a DRAM rank into multiple rank subsets and guides a memory request to one of the rank subsets, not to an entire DRAM rank [9, 45, 46]. The rank subsetting increases

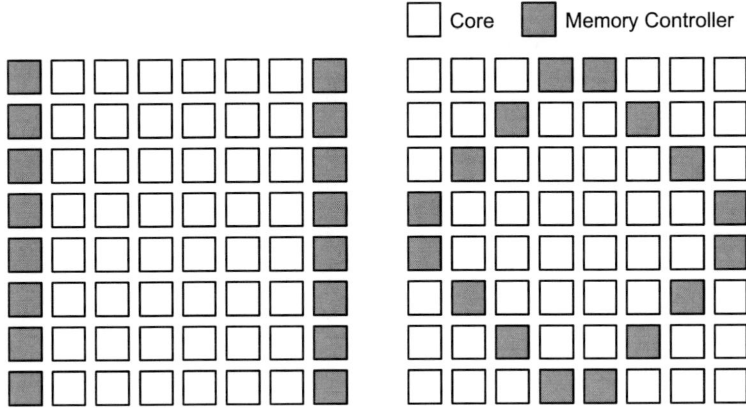


**Fig. 4.11** Different memory controller configurations. *The shaded tiles* represent memory controllers that are co-located with computation cores. All memory controllers are co-located *in the top and bottom row on the left configuration*, while memory controllers are co-located *in a diamond shape on the right configuration*

the serialization latency to transfer a cache line through a memory channel but improves the effective throughput of memory controllers, so that the average memory access latency could be lowered and less energy could be consumed per access. Among the proposals that embody rank subsetting, module threading [45] relies on high speed signaling. The memory controller of the module threading scheme selects a subset of devices by feeding separate chip-select signals to them. Figure 4.12 shows a memory channel with two Multicore DIMMs [9] that replaces a register per memory rank with a demux register, which routes or demultiplexes address and command signals to the selected subset. Zheng et al. called a subset a mini-rank [46] and proposed a design in which all mini-ranks in a memory rank communicate data with a memory controller through a mini-rank buffer. The key difference between Multicore DIMM and mini-rank is how they multiplex and demultiplex data, address, and command signals. For data signals, mini-rank has a demux per memory rank, whereas Multicore DIMM has one per memory channel. As a result, mini-rank consumes more energy and needs more components. Multicore DIMM has one demux for address and command signals per memory rank; mini-rank does not have any. Both proposals require chip-select signals per rank subset. Ahn et al. [47] further assessed the impact of rank subsetting for system-wide performance, energy efficiency, and reliability perspectives, and proposed an extension of Multicore DIMM that supports chipkill-level reliability [48]. By simulating a multicore system using multithreaded and consolidated workloads, [47] showed that rank subsetting and utilizing DRAM power-down modes are largely complementary because rank subsetting targets saving dynamic energy on memory systems, while the latter is more effective when most of the DRAM chips are idle.

Zheng et al. [49] proposed an architectural approach to utilize a memory bus that operates at a higher data rate than that of DRAM devices. Their proposed architecture, called decoupled DIMM, synchronizes a high data rate bus and low data rate

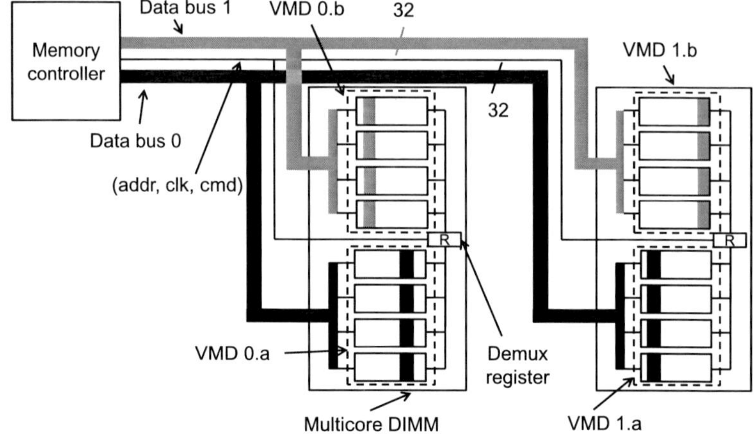



**Fig. 4.12** A memory channel with two Multicore DIMMs (MCDIMMs), each divided into two subsets called virtual memory devices (VMDs). Each MCDIMM has a demux register instead of a normal register routing control signals to each VMD [47]

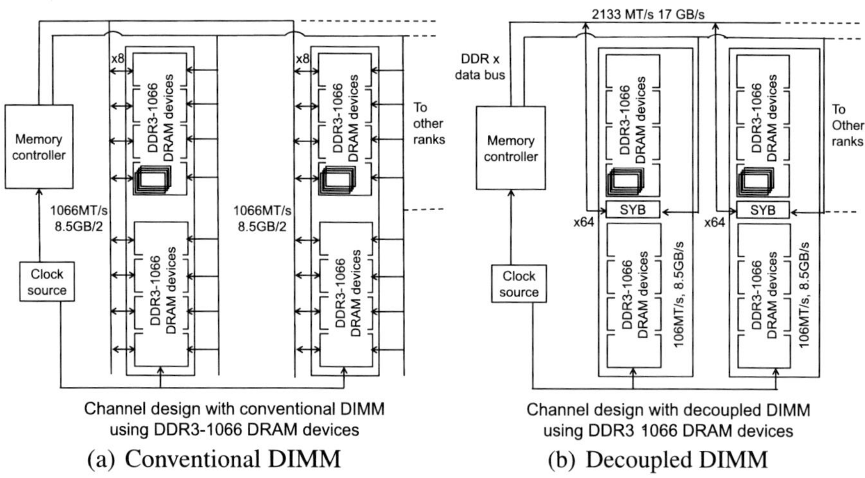



Channel design with conventional DIMM using DDR3-1066 DRAM devices

Channel design with decoupled DIMM using DDR3 1066 DRAM devices

(a) Conventional DIMM      (b) Decoupled DIMM

**Fig. 4.13** Conventional DIMM organization vs. decoupled DIMM organization with DDR3-1066, ×8 devices as an example [49]

DRAM devices with synchronization buffers, as shown in Fig. 4.13. The random access time of DRAM has not been improved as fast as the data rate improvement of the bus. In this situation, some DDR3 devices are forced to operate faster under higher supply voltage than the JEDEC recommendation [50], which makes them consume substantially more power and degrades their lifetime reliability. A synchronization buffer is connected to each rank of DIMMs, and it arbitrates the data rate gap between the bus and the DRAM devices. Since each rank can operate independently, the bandwidth of the data bus can be exploited with rank interleaving

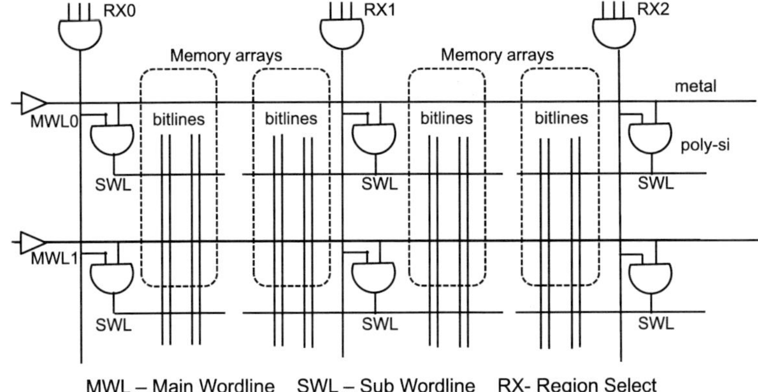



**Fig. 4.14** Selective Bitline Activation (SBA) DRAM architecture [51]

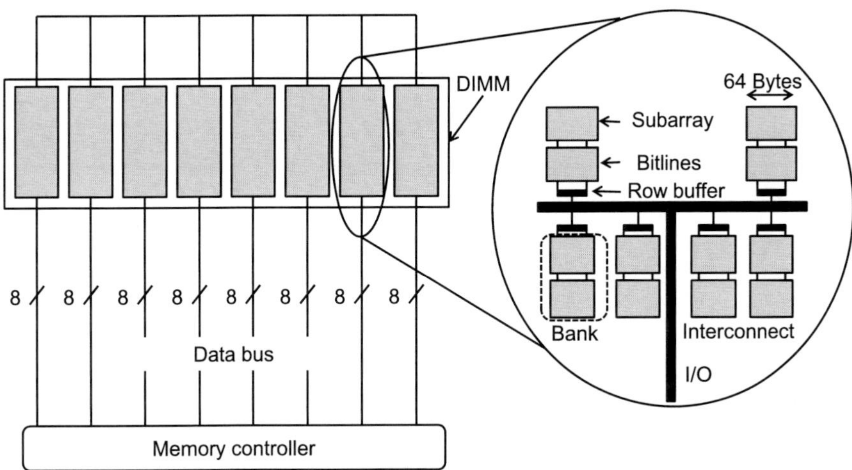



**Fig. 4.15** Selective Subarray Access (SSA) DRAM architecture [51]

memory accesses. This architecture uses low speed DRAM devices, which have less power per access than that of high speed devices; therefore, it is energy efficient. The performance of this architecture is sensitive to a scheduling policy and workload characteristics. For example, consecutive memory accesses to the same rank decrease the data bus utilization by less than half, because there is a delay from a synchronization buffer. In addition, the power consumption of the synchronization buffer makes the system consume more energy than the conventional memory system using the same data rate DRAM devices.

Udipi et al. [51] attempted to mitigate the unnecessary power consumption of DRAM chips with their overfetch-free DRAM architectures. The first new DRAM architecture they propose is Selective Bitline Activation (SBA), which activates a small segment of wordline, therefore decreasing the size of an activated row. Figure 4.14 shows its corresponding memory array design. In this design, a RAS (row access strobe) cannot activate a subrow before its corresponding CAS (column access strobe) arrives. Although the posted-RAS nature incurs an additional delay, substantial power can be saved. The second architecture is Selective Subarray Access (SSA). SSA aggressively utilizes the rank subsetting. Figure 4.15 shows the SSA DRAM architecture. In this design, each DRAM chip acts as a single rank. A bank of a DRAM chip is divided into subarrays. The row size of a subarray is the same as that of a cache line. Since each DRAM chip can service a different cache line, this design has much higher concurrency. However, the data rate of each sub-bus is clearly reduced; therefore, it incurs a higher DRAM latency. SSA also supports a chipkill-level reliability using a RAID-5 like scheme. The RAID-5 is well suited to this design, because a single memory access gets an entire cache line from a single subarray of an SSA DRAM chip. Both SBA and SSA require more area of the DRAM chips and modifications of memory controllers. However, these designs can eliminate the relatively large power consumption caused by the over-fetch with the moderate power consumed by additional components and wires in their designs.

# References


1. Jacob, B., Ng, S.W., Wang, T.D.: Memory Systems: Cache, DRAM, Disk. Morgan Kaufmann, San Mateo (2008)
2. Ho, R., Mai, K., Horowitz, M.A.: The future of wires. Proc. IEEE **89**(4) (2001)
3. Micron Technology Inc.: SDRAM datasheet. http://www.micron.com/products/ (2010). Rev. N 1/10
4. Micron Technology Inc.: DDR SDRAM datasheet. http://www.micron.com/products/ (2009). Rev. B 2/09
5. Micron Technology Inc.: DDR2 SDRAM datasheet. http://www.micron.com/products/ (2010). Rev. E 06/10
6. Samsung semiconductor: DDR3 SDRAM datasheet. http://www.samsung.com/global/business/semiconductor/ (2010). Rev. 1.0
7. Rixner, S., Dally, W.J., Kapasi, U.J., Mattson, P.R., Owens, J.D.: Memory access scheduling. In: ISCA (2000)
8. Micron Technology Inc.: DDR3 SDRAM datasheet. http://www.micron.com/products/ (2010). Rev. J 05/10
9. Ahn, J., Leverich, J., Schreiber, R.S., Jouppi, N.P.: Multicore DIMM: an energy efficient memory module with independently controlled DRAMs. Comput. Archit. Lett. **7**(1) (2008)
10. Yun, W.J., Lee, H.W., Shin, D., Kang, S.D., Yang, J.Y., Lee, H.O., Lee, D.U., Sim, S., Kim, Y.J., Choi, W.J., Song, K.S., Shin, S.H., Choi, H.H., Moon, H.W., Kwack, S.W., Lee, J.W., Choi, Y.K., Park, N.K., Kim, K.W., Choi, Y.J., Ahn, J.H., Yang, Y.S.: A 0.1-to-1.5 GHz 4.2 mW all-digital DLL with dual duty-cycle correction circuit and update gear circuit for DRAM in 66 nm CMOS technology. In: ISSCC (2008)
11. Rambus: RDRAM. http://www.rambus.com (1999)
12. Kho, R., Boursin, D., Brox, M., Gregorius, P., Hoenigschmid, H., Kho, B., Kieser, S., Kehrer, D., Kuzmenka, M., Moeller, U., Petkov, P., Plan, M., Richter, M., Russell, I., Schiller, K., Schneider, R., Swaminathan, K., Weber, B., Weber, J., Bormann, I., Funfrock, F., Gjukic, M., Spirkl, W., Steffens, H., Weller, J., Hein, T.: 75 nm 7 Gb/s/pin 1 GB GDDR5 graphics memory device with bandwidth-improvement techniques. In: ISSCC (2009)

13. Oh, T.Y., Sohn, Y.S., Bae, S.J., Park, M.S., Lim, J.H., Cho, Y.K., Kim, D.H., Kim, D.M., Kim, H.R., Kim, H.J., Kim, J.H., Kim, J.K., Kim, Y.S., Kim, B.C., Kwak, S.H., Lee, J.H., Lee, J.Y., Shin, C.H., Yang, Y.S., Cho, B.S., Bang, S.Y., Yang, H.J., Choi, Y.R., Moon, G.S., Park, C.G., Hwang, S.W., Lim, J.D., Park, K.I., Choi, J.S., Jun, Y.H.: A 7 gb/s/pin GDDR5 SDRAM with 2.5 ns bank-to-bank active time and no bank-group restriction. In: ISSCC (2010)
14. Rambus: XDR DRAM. http://www.rambus.com (2005)
15. Hennessy, J., Patterson, D.A.: Computer Architecture: A Quantitative Approach, 4th edn. Morgan Kaufmann, San Mateo (2006)
16. Smith, J., Nair, R.: Virtual Machines: Versatile Platforms for Systems and Processors. Morgan Kaufmann, San Mateo (2005)
17. Haas, J., Vogt, P.: Fully-buffered DIMM technology moves enterprise platforms to the next level. Technol. Intel Mag. (2005)
18. Ahn, J., Erez, M., Dally, W.J.: The design space of data-parallel memory systems. In: SC'06 (2006)
19. Shao, J., Davis, B.: A burst scheduling access reordering mechanism. In: HPCA(2007)
20. Lee, C.J., Mutlu, O., Narasiman, V., Patt, Y.: Prefetch-aware DRAM controllers. In: MICRO (2008)
21. Kim, Y., Han, D., Mutlu, O., Harchol-Balter, M.: Atlas: a scalable and high-performance scheduling algorithm for multiple memory controllers. In: HPCA (2010)
22. Ipek, E., Mutlu, O., Martinez, J., Caruana, R.: Self-optimizing memory controllers: a reinforcement learning approach. In: ISCA (2008)
23. Liu, S., Zhang, Y., Memik, S., Memik, G.: An approach for adaptive DRAM temperature and power management. IEEE Trans. Very Large Scale Integr. **18** (2010)
24. Hur, I., Lin, C.: A comprehensive approach to DRAM power management. In: HPCA (2008)
25. Ghosh, M., Lee, H.H.S.: Smart refresh: an enhanced memory controller design for reducing energy in conventional and 3D die stacked DRAMs. In: MICRO (2007)
26. Aggarwal, N., Cantin, J., Lipasti, M., Smith, J.: Power-efficient DRAM speculation. In: HPCA (2008)
27. Semiconductor Industries Association: International technology roadmap for semiconductors. Tech. rep. (2009)
28. Ramm, P., Wolf, M., Klumpp, A., Wieland, R., Wunderle, B., Michel, B., Reichl, H.: Through silicon via technology: processes and reliability for wafer-level 3D system integration. In: Electronic Components and Technology Conference, pp. 841–846 (2008)
29. Cardu, R., Scandiuzzo, M., Cani, S., Perugini, L., Franchi, E., Canegallo, R., Guerrieri, R.: Chip-to-chip communication based on capacitive coupling. In: IEEE International Conference on 3D System Integration, 3DIC 2009, pp. 1–6 (2009)
30. Niitsu, K., Shimazaki, Y., Sugimori, Y., Kohama, Y., Kasuga, K., Nonomura, I., Saen, M., Komatsu, S., Osada, K., Irie, N., Hattori, T., Hasegawa, A., Kuroda, T.: An inductive-coupling link for 3D integration of a 90 nm CMOS processor and a 65 nm CMOS SRAM. In: ISSCC, pp. 480–481, 481a (2009)
31. Loh, G.H.: 3D-stacked memory architectures for multi-core processors. In: ISCA (2008)
32. Kang, U., Chung, H.J., Heo, S., Ahn, S.H., Lee, H., Cha, S.H., Ahn, J., Kwon, D., Kim, J.H., Lee, J.W., Joo, H.S., Kim, W.S., Kim, H.K., Lee, E.M., Kim, S.R., Ma, K.H., Jang, D.H., Kim, N.S., Choi, M.S., Oh, S.J., Lee, J.B., Jung, T.K., Yoo, J.H., Kim, C.: 8 Gb 3D DDR3 DRAM using through-silicon-via technology. In: ISSCC (2009)
33. Lee, B.C., Ipek, E., Mutlu, O., Burger, D.: Architecting phase change memory as a scalable DRAM alternative. In: ISCA (2009)
34. Hwang, C.G.: Semiconductor memories for IT era. In: ISSCC (2002)
35. Sasago, Y., Kinoshita, M., Morikawa, T., Kurotsuchi, K., Hanzawa, S., Mine, T., Shima, A., Fujisaki, Y., Kume, H., Moriya, H., Takaura, N., Torii, K.: Cross-point phase change memory with 4F2 cell size driven by low-contact-resistivity poly-Si diode. In: VLSI, pp. 24–25 (2009)
36. Futatsuyama, T., Fujita, N., Tokiwa, N., Shindo, Y., Edahiro, T., Kamei, T., Nasu, H., Iwai, M., Kato, K., Fukuda, Y., Kanagawa, N., Abiko, N., Matsumoto, M., Himeno, T., Hashimoto, T., Liu, Y.C., Chibvongodze, H., Hori, T., Sakai, M., Ding, H., Takeuchi, Y., Shiga, H., Kajimura,

N., Kajitani, Y., Sakurai, K., Yanagidaira, K., Suzuki, T., Namiki, Y., Fujimura, T., Mui, M., Nguyen, H., Lee, S., Mak, A., Lutze, J., Maruyama, T., Watanabe, T., Hara, T., Ohshima, S.: A 113 mm$^2$ 32 Gb 3b/cell NAND flash memory. In: ISSCC (2009)

37. Villa, C., Mills, D., Barkley, G., Giduturi, H., Schippers, S., Vimercati, D.: A 45 nm 1 Gb 1.8 V phase-change memory. In: ISSCC (2010)

38. Qureshi, M.K., Srinivasan, V., Rivers, J.A.: Scalable high performance main memory system using phase-change memory technology. In: ISCA, pp. 24–33 (2009)

39. Taylor, M., Kim, J., Miller, J., Wentzlaff, D., Ghodrat, F., Greenwald, B., Hoffman, H., Johnson, P., Lee, J.W., Lee, W., Ma, A., Saraf, A., Seneski, M., Shnidman, N., Strumpen, V., Frank, M., Amarasinghe, S., Agarwal, A.: The raw microprocessor: a computational fabric for software circuits and general-purpose programs. IEEE MICRO **22**(2) (2002)

40. Wentzlaff, D., Griffin, P., Hoffmann, H., Bao, L., Edwards, B., Ramey, C., Mattina, M., Miao, C.C., Brown, J., Agarwal, A.: On-chip interconnection architecture of the tile processor. IEEE MICRO **27**(5), 15–31 (2007)

41. Vangal, S., Howard, J., Ruhl, G., Dighe, S., Wilson, H., Tschanz, J., Finan, D., Iyer, P., Singh, A., Jacob, T., Jain, S., Venkataraman, S., Hoskote, Y., Borkar, N.: An 80-tile 1.28 tflops network-on-chip in 65 nm CMOS. In: ISSCC, pp. 98–589 (2007)

42. Kongetira, P., Aingaran, K., Olukotun, K.: Niagara: a 32-way multithreaded Sparc processor. IEEE MICRO **25**(2), 21–29 (2005)

43. Dally, W.J., Towles, B.: Route packets, not wires: on-chip interconnection networks. In: DAC (2001)

44. Abts, D., Enright Jerger, N.D., Kim, J., Gibson, D., Lipasti, M.H.: Achieving predictable performance through better memory controller placement in many-core CMPs. In: ISCA(2009)

45. Ware, F.A., Hampel, C.: Improving power and data efficiency with threaded memory modules. In: ICCD(2006)

46. Zheng, H., Lin, J., Zhang, Z., Gorbatov, E., David, H., Zhu, Z.: Mini-rank: adaptive DRAM architecture for improving memory power efficiency. In: MICRO (2008)

47. Ahn, J., Jouppi, N.P., Kozyrakis, C., Leverich, J., Schreiber, R.S.: Future scaling of processor-memory interfaces. In: Proceedings of the Conference on High Performance Computing Networking, Storage and Analysis, SC'09 (2009)

48. Dell, T.J.: A white paper on the benefits of chipkill-correct ECC for PC server main memory. IBM Microelectron. Div. (1997)

49. Zheng, H., Lin, J., Zhang, Z., Zhu, Z.: Decoupled DIMM: building high-bandwidth memory system using low-speed DRAM devices. In: ISCA (2009)

50. JEDEC: DDR3 SDRAM specification. JESD79-3B. http://www.jedec.org/download/search/JESD79-3B.pdf (2007)

51. Udipi, A.N., Muralimanohar, N., Chatterjee, N., Balasubramonian, R., Davis, A., Jouppi, N.P.: Rethinking DRAM design and organization for energy-constrained multi-cores. In: ISCA (2010)

# Chapter 5
# Energy-Aware On-Chip Networks

John Kim

**Abstract** As technology continues to evolve, communication is becoming the bottleneck of future systems as it significantly impacts overall performance and cost. The energy consumed in the communication of future many-core processors will be critical in achieving a scalable many-core system. In this chapter, we present energy-aware on-chip network architectures that attempt to achieve *ideal* on-chip network behavior by approaching the latency and energy consumed in the wires in transmitting data from source to destination. We present different techniques, including topology, flow control, and router microarchitecture, that attempt to achieve this ideal on-chip network. These approaches minimize the energy and latency overhead of intermediate routers as packets traverse the network. In addition to these approaches, we describe an alternative approach, i.e., bufferless on-chip networks, which minimize the amount of network buffers to reduce energy consumption.

## 5.1 Introduction

With the increasing number of transistors in modern VLSI technology, the number of cores and components on a single chip continues to increase in order to efficiently utilize the transistors. Examples of multi-core and many-core processors include traditional latency-sensitive processors, such as Intel [1] processors, as well as throughput-oriented processors, such as the Niagara processor [2] or the NVidia general-purpose graphics processing unit (GPU) processors [3]. An efficient on-chip network or network-on-chip (NoC) is required in these architectures to connect these components. It is projected that the on-chip network will be the critical bottleneck of future many-core processors—in terms of both performance and power [4].

An on-chip network can be characterized by the following components: topology, routing, flow control, and router microarchitecture [5]. The topology defines how the routers and channels are interconnected. Once the topology is defined, the routing determines which path a packet takes from source to destination. The flow

J. Kim (✉)
KAIST, Daejeon, Republic of Korea
e-mail: jjk12@kaist.edu

C.-M. Kyung, S. Yoo (eds.), *Energy-Aware System Design*,
DOI 10.1007/978-94-007-1679-7_5, © Springer Science+Business Media B.V. 2011

control aspect of a network determines the allocation of resources, particularly, how resources are allocated when contention occurs. The router microarchitecture determines how the building blocks of the network are organized, including the switch organization and allocators. These components are common in all types of interconnection networks, and on-chip networks are no exception. On-chip networks have also evolved from large-scale off-chip networks and, as a result, have adopted many of the characteristics of large-scale networks. In this chapter, we focus on these different characteristics of multi- (many-)core on-chip networks and describe how energy-aware on-chip networks can be achieved with different techniques. In particular, we focus on how *ideal* on-chip networks [6] can be designed to minimize energy consumption, and we approach the energy consumption of the wires or the channels themselves. Various recent works have tried to achieve this goal through different techniques, including topology, flow control, and router microarchitecture.

The remainder of this chapter is organized as follows. In Sect. 5.2, we provide a background of the different components in a conventional on-chip network router. We discuss an *ideal* on-chip network in Sect. 5.3 and then present three different approaches which attempt to achieve this goal. Section 5.4 presents how the flattened butterfly topology can be mapped to on-chip networks to reduce the number of intermediate routers. Section 5.5 presents the express virtual channel flow control, which bypasses intermediate routers to improve efficiency; Sect. 5.6 presents an alternative router microarchitecture to reduce the cost and complexity of an on-chip network. In addition to these three approaches, an alternative on-chip network based on bufferless router microarchitecture which trades off performance for lower energy is presented in Sect. 5.7. We conclude this chapter in Sect. 5.8.

## 5.2 Conventional On-Chip Network Router Organization

In this section, we provide a background discussion on the main components of an on-chip network router. The block diagram of a conventional router microarchitecture is shown in Fig. 5.1. The main components include the input buffers, the crossbar switch, and the control logic, which includes the switch and the virtual channel allocators. We briefly discuss the impact of each component on overall performance and cost, including the area and energy consumption of an on-chip network.

### 5.2.1 Buffers

Unlike off-chip networks, where bandwidth is expensive and buffers are relatively cheap, the constraints for an on-chip network are different—wires (bandwidth) are relatively cheap, while buffers are expensive [7]. Buffers are used to decouple the allocation of resources in interconnection networks and simplify the flow control by using buffered flow control such as virtual cut-through or wormhole flow control.

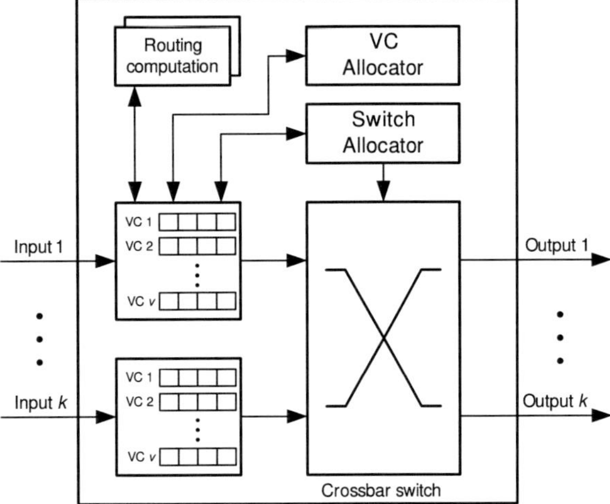



**Fig. 5.1** Block diagram of a conventional router microarchitecture [16]. © 2005 IEEE

However, input buffers represent a significant portion of the area and power in an on-chip network router. For a static random access memory (SRAM) buffer implementation, the input buffers can consume 46% of the total on-chip network power while occupying 17% of the total area [8]. SRAM is preferred over a register-based first-in first-out (FIFO) memory because of its area efficiency, as flipflop-based input buffers can occupy up to 51% of the total area [9]. However, SRAM incurs latency overhead in terms of accessing the buffers. For example, two of the five router pipeline stages in the Intel TeraFlop are dedicated to accessing the buffers (buffer write and buffer read stages) [10]. To efficiently utilize buffers in on-chip networks, dynamic buffer management schemes have been proposed to dynamically share the buffers among different virtual channels [6, 9, 11]. However, these microarchitectures can add additional design and verification complexity and impact the router pipeline latency.

The different microarchitectures described in this chapter attempt to reduce the amount of buffers through reduced number of routers (flattened butterfly topology [12]), reduced per-hop buffers (low-cost router microarchitecture [13]), or complete removal of buffers through bufferless flow control [14, 15].

### 5.2.2 Switch

The area of a crossbar switch is often the dominant area component of an on-chip router, as the area is proportional to $O(p^2 w^2)$, where $p$ is the number of router ports and $w$ is the datapath width. Compared to the datapath width ($w$), the number of ports ($p$) for on-chip routers is relatively small, e.g., $p = 5$ for a two-dimensional (2D) mesh network and $p = 10$ for high-radix on-chip network routers [12] while $w = 128$ to $w = 256$ because of the abundant on-chip bandwidth. The wire dominated crossbar area can occupy up to 64% of the total router area [8]. As a result, to minimize the area impact of on-chip routers, the crossbar area must be minimized.

The low-cost router microarchitecture reduces the switch area by using dimension-sliced routers with minimal performance loss. The flattened butterfly topology can actually increase the area of a single switch since it requires a high-radix router [16], but it reduces the number of switches required because the network diameter is reduced.

### 5.2.3 Arbitration

The power consumption or the area from the arbitration logic is very minimal [17]. However, poor arbitration can limit the throughput of the router and reduce the overall performance of on-chip networks. The latency of the arbitration logic also often determines the router cycle time. Separable allocators have been proposed for on-chip networks to separate the allocation into two stages: input and output arbitration. These allocators require an efficient matching algorithm, and novel switch allocation has been proposed to increase the matching efficiency for on-chip networks [8]. However, arbitration is still often in the critical path. Arbitration is needed since resources (such as channel bandwidth) are shared, but if they are reserved ahead of time, the arbitration complexity can be reduced or removed completely.

Express virtual channel (EVC) flow control [6] is an example that exploits arbitrations and bandwidth reservation through the use of a dedicated virtual channel and enables bypassing of the pipeline stages of intermediate routers. The EVC approach was also shown to increase the network throughput as well, in addition to reducing network latency. In comparison, the low-cost router microarchitecture takes an alternative approach; the arbitration is greatly simplified by giving priority to packets already in the network that continue to travel in the same dimension—thus removing the switch arbitration from the critical path.

### 5.2.4 Latency in On-Chip Networks

Latency is a critical performance metric for on-chip networks. The latency of a packet through an interconnection network can be expressed as the sum of the header latency ($T_h$), the serialization latency ($T_s$), and the time of flight on the wires ($T_w$),

$$T = T_h + T_s + T_w = H t_r + L/b + T_w$$

where $t_r$ is the router delay, $H$ is the hop count, $L$ is the packet size, and $b$ is the channel bandwidth.

Minimizing latency requires establishing a careful balance between $T_h$ and $T_s$. For on-chip networks, wires are abundant and on-chip bandwidth plentiful; consequently, $T_s$ can be reduced significantly by providing very wide channels. However,

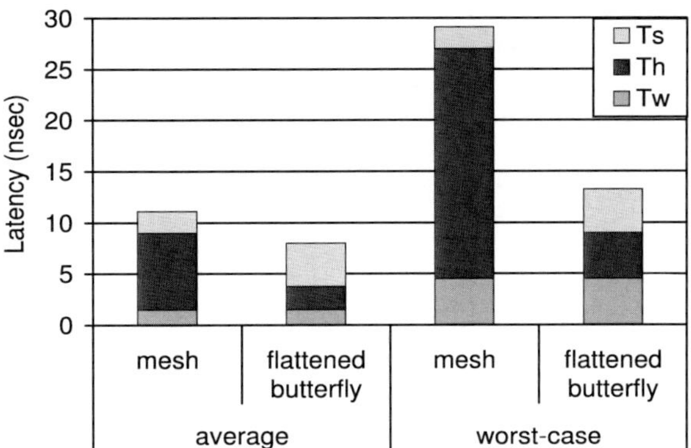

**Fig. 5.2** Latency of a packet in on-chip networks [12]. © 2007 IEEE

traditional 2D mesh networks tend to establish $T_s$ and $T_h$ values that are unbalanced, with wide channels providing a low $T_s$ while $T_h$ remains high due to the high hop count. Consequently, these networks fail to minimize latency. For example, in the 2D mesh network used in the Intel TeraFlop [18], with uniform random traffic, $T_h$ is approximately 3 times $T_s$ and for worst case traffic, there is approximately a $10\times$ difference.[1]

In order to reduce on-chip network latency and approach an *ideal* on-chip network latency (see Sect. 5.3), the number of intermediate routers must be reduced. One such example is the use of the flattened butterfly topology. By using high-radix routers and the flattened butterfly topology, the hop count can be reduced at the expense of increasing the serialization latency, assuming that the bisection bandwidth is held constant, as is shown in Fig. 5.2. As a result, the flattened butterfly achieves a lower overall latency. The wire delay ($T_w$) associated with the Manhattan distance between the source and the destination nodes generally corresponds to the minimum packet latency in an on-chip network [6].

## 5.3 Approaching *Ideal* On-Chip Network

Many recent works have focused on reducing energy consumption in the on-chip network. In this chapter, we discuss how some of these works have attempted to reduce the energy consumption to approach an *ideal* on-chip network. An on-chip network consists of two primary components: routers, described in the previous section, and channels. In an *ideal* on-chip network [6] the impact of the routers on performance and energy is minimized, and the performance/energy of communicating data between source and destination approaches that of the channels. In terms of performance, an ideal on-chip network performance can be described as a network where the latency approaches that of the wire latency from source to destination (Fig. 5.3(a)). With long wires, the channels will likely include one or more

---

[1] The latency calculations were based on Intel TeraFlop [18] parameters ($t_r = 1.25$ ns, $b = 16$ GB/s, $L = 320$ bits) and an estimated value of wire delay for 65 nm ($t_w = 250$ ps/mm).

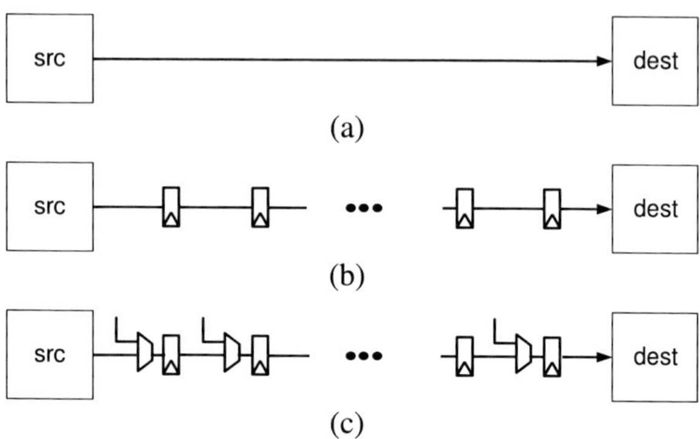


**Fig. 5.3** (**a**) Ideal on-chip network with only wire delay, (**b**) wire delay pipelined with registers, and (**c**) muxes inserted to share the wire resources with other nodes [13]. © 2009 IEEE

pipeline registers and create a multicycle channel (Fig. 5.3(b)). This ideal on-chip network also minimizes the energy consumption of an on-chip network, as the energy overhead of the router components is nearly negligible. Different approaches to on-chip networks have tried to achieve this on-chip network behavior. In this section, we describe three different approaches that try to achieve this ideal behavior by reducing or removing the intermediate routers through which a packet traverses. Each of these techniques approaches this problem from a different aspect of on-chip networks including topology, flow control, and router microarchitecture:

1. Topology—reduce the number of intermediate routers by reducing the network diameter [12].
2. Flow control—avoid intermediate routers through bypassing of internal router with flow control [6].
3. Router microarchitecture—simplify the router microarchitecture such that minimal overhead is required in traversing each router [13].

In addition to these approaches, we also describe recently proposed bufferless routers [14, 15]. These approaches require the support of router microarchitecture as well as routing to minimize the cost of on-chip networks while removing the need for input buffers. The removal of buffers trades off reduction of network cost and energy for reduced network throughput when the network load is increased.

## 5.4 Flattened Butterfly Topology

In on-chip networks, 2D mesh topology is commonly assumed because of its simplicity and because it maps well to the 2D VLSI planar layout. However, alternative topologies for on-chip networks have been proposed, not only to increase performance, but also to reduce the network power consumption. Recently proposed topologies include concentrated mesh [19], hierarchical organization [20], multidrop express channels (MECs) [21], and flattened butterfly [12]. In this section, we describe the impact of concentration which is leveraged in most of these new topologies and describe one of these recently proposed topologies (flattened butterfly) in detail.

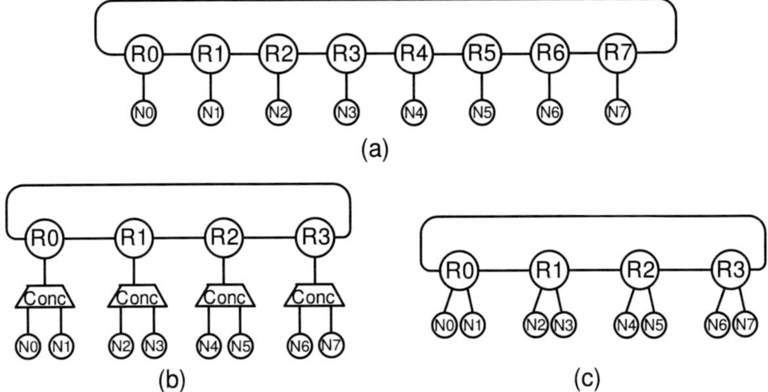

**Fig. 5.4** The use of concentration in interconnection networks. (**a**) 8-node (N0–N7) ring with 8 routers (R0–R7) without concentration, (**b**) 4-node ring with 2-way concentrator (external concentration), and (**c**) the same topology as (**b**) with the 2-way concentrator integrated into the router (integrated concentration) [12]. © 2007 IEEE

### 5.4.1 Concentration

The cost of an off-chip interconnection network typically increases with the channel count. As the channel count increases with the hop count, reducing the diameter of the network reduces the cost of the network [16]. Similar benefits can exist for on-chip networks as well since a lower diameter reduces the number of intermediate routers and, thus, reduces the overall power consumption in the network. Furthermore, since buffered flow control is often used in on-chip networks, the aggregate area allocated to the input buffers in the routers decreases as the number of channels is reduced. Topologies such as the flattened butterfly can achieve this goal of lower network diameter and lead to lower cost by reducing the number of channels and the amount of buffers required in the network. However, reducing network diameter can significantly increase network cost. For example, the generalized hypercube structure [22] or the torus with express channels [23] can reduce network diameter; however, the cost becomes prohibitively high with the large port count required at all route nodes. Thus, to reduce network diameter without increasing network cost, concentration [5] must be exploited.

The use of *concentration*, where network resources are shared among different processor nodes, can improve the efficiency of the network while reducing network cost. An example of concentration is shown in Fig. 5.4 using a ring topology. Eight nodes can be connected with 8 routers, as shown in Fig. 5.4(a), without any concentration. By using a concentration factor of 2, the 8 nodes can be connected in a 4-*node* ring, where each *node* consists of two terminal nodes and a router, as shown in Fig. 5.4(b). The use of concentration aggregates traffic from different nodes into a single network interface. This reduces both the number of resources allocated to the network routers and the average hop count, which can improve latency. Thus, while providing the same bisection bandwidth, concentration can reduce the cost of the network by reducing the network size. The concentrator can also be integrated into the router by increasing the radix of the router, as shown in Fig. 5.4(c). This allows all of the terminal nodes to access the network concurrently, instead of allowing only one terminal node associated with a router to access the network during any one cycle. The use of concentration in on-chip networks also reduces the wiring complexity. The concentration implementation in Fig. 5.4(b) can be referred to as *external* concentration; the implementation in Fig. 5.4(c) can be referred as *integrated* concentration [24]. In this example, an external multiplexer or an integrated crossbar switch is used to implement concentration, but there can be other implementations; e.g., the hierarchical topology [20] uses a local bus as a local network and implements a different form of concentration.

Concentration is also practical for on-chip networks because the probability that more than one of the processors attached to a single router will attempt to access the network on a given cycle is relatively low. For example, in a chip multiprocessor (CMP) architecture with a shared L2 cache distributed across the chip, the L1 miss rate is often under 10% [25], which results in relatively low traffic injection rates at the processors. Consequently, using concentration to share the network resources is an effective technique for CMP traffic.

### 5.4.2 Topology Description

The flattened butterfly topology [26] is a cost-efficient topology for use with high-radix routers, originally proposed for large-scale networks. The flattened butterfly is derived by combining (or *flattening*) the routers in each row of a conventional butterfly topology while preserving the inter-router connections. The flattened butterfly is similar to a generalized hypercube [22]; however, by leveraging concentration in the routers, the flattened butterfly significantly reduces the wiring complexity of the topology, allowing it to scale more efficiently.

An example of a 64-node flattened butterfly is shown in Fig. 5.5. To derive the flattened butterfly topology, we collapse a 3-stage radix-4 butterfly network (4-ary 3-fly) to produce the flattened butterfly shown in Fig. 5.5(a). The resulting flattened butterfly has two dimensions and uses radix-10 routers. With four processor nodes attached to each router, the routers have a concentration factor of 4. The remaining 6 router ports are used for inter-router connections: 3 ports are used for the dimension 1 connections, and 3 ports are used for the dimension 2 connections. To map the flattened butterfly topology to a 64-node on-chip network, routers are placed as shown in Fig. 5.5(b) to embed the topology in a planar VLSI layout with each router placed in the middle of the 4 processing nodes. Routers connected in dimension 1 are aligned horizontally, while routers connected in dimension 2 are aligned vertically; thus, the routers within a row are fully connected, as are the routers within a column.

The wire delay associated with the Manhattan distance between a packet's source and its destination provides a lower bound on the latency required to traverse an

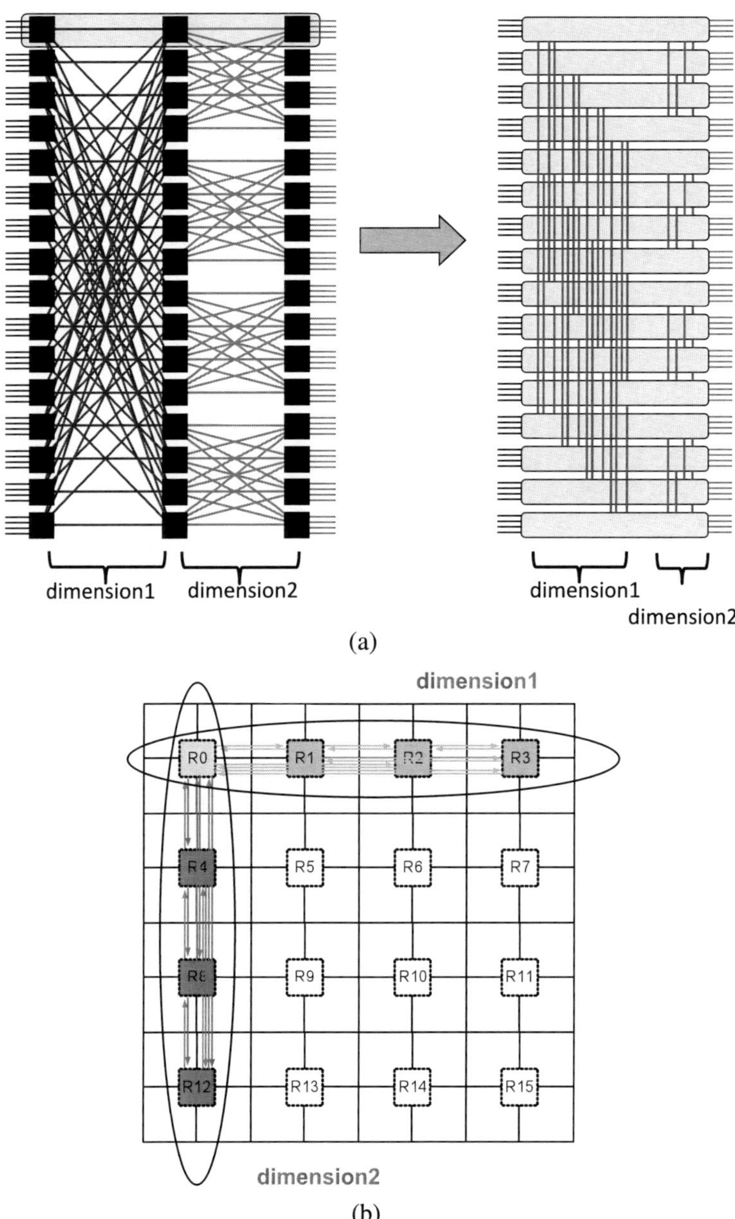



**Fig. 5.5** (**a**) Block diagram of a 2D flattened butterfly consisting of 64 nodes and (**b**) the corresponding layout of the flattened butterfly where dimension 1 routers are horizontally placed and dimension 2 routers are vertically placed [12]. © 2007 IEEE

on-chip network. When minimal routing is used, processors in this flattened butter-fly network are separated by only two hops (or two-channel traversal), which is a significant improvement over the hop count of a 2D mesh. The flattened butterfly attempts to approach the wire delay bound by reducing the number of intermediate routers—resulting in not only lower latency but also lower energy consumption. However, the wires connecting distant routers in the flattened butterfly network are necessarily longer than those found in the mesh. The adverse impact of long wires on performance is readily reduced by optimally inserting repeaters and pipeline registers to preserve the channel bandwidth while tolerating channel traversal times that may be several cycles. The longer channels also require deeper buffer sizes to cover the credit round trip latency in order to maintain full throughput. However, this increase in buffers is offset by the reduction in the number of buffers through reduction in the number of routers.

### 5.4.3 Routing and Deadlock

Both minimal and nonminimal routing algorithms can be implemented on the flattened butterfly topology. Depending on the routing algorithm, a limited number of virtual channels (VCs) [27] is needed to prevent routing deadlock within the network. Additional VCs may be required for other purposes such as separating traffic into different classes or avoiding deadlock in the client protocols.

Dimension-order routing (DOR) can be used as a minimal routing algorithm for the flattened butterfly (e.g., route in dimension 1, then route in dimension 2); in this case, the routing algorithm itself is restrictive enough to prevent deadlock, but path diversity is not exploited. Other minimal routing algorithms such as randomized DOR [28] can be implemented on the flattened butterfly topology as well—route in dimension 1 and then dimension 2, or route first in dimension 2 and then dimension 1. This routing algorithm would require two VCs for a 2D flattened butterfly to avoid routing deadlock.

Nonminimal routing in the flattened butterfly topology allows the path diversity in the network to be exploited to better load balance the channels and improve performance. For nonminimal routing, Valiant's routing [29] can be leveraged, where the packet is first routed to a randomly selected intermediate router before being routed to the destination. An example of nonminimal routing is shown in Fig. 5.6; one minimal path and two different nonminimal paths are shown. However, always using nonminimal routing removes any traffic locality or load-balancing that might already exist in the traffic pattern. Thus, adaptive routing such as the global adaptive routing (UGAL) [30] algorithm is needed on the flattened butterfly topology. UGAL load balances by determining whether it is beneficial to route minimally or nonminimally. If it selects nonminimal routing, UGAL routes minimally to an intermediate router in the first phase, and then routes minimally to the destination in the second phase—this is identical to Valiant's routing. Otherwise, the packet is routed minimally. The number of VCs needed to avoid deadlock can be proportional to the

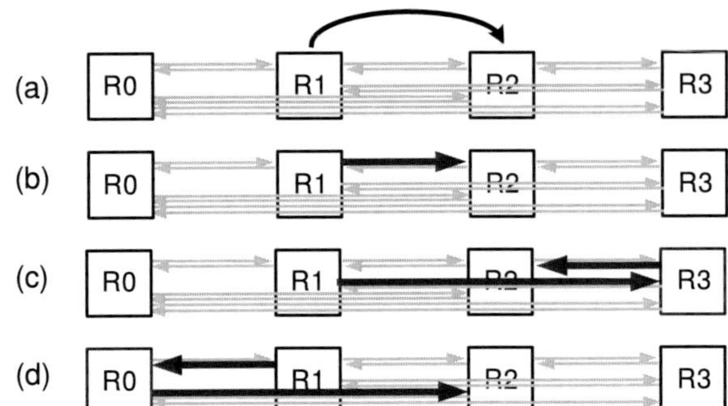

**Fig. 5.6** For traffic from R1 to R2 (**a**), routing path using minimal routing (**b**), and two different nonminimal routing paths (**c**, **d**). The intermediate router is R3 in (**c**) and R1 in (**d**). For simplicity, the processing nodes attached to these routers are not shown, and only the first row of routers is shown [12]. © 2007 IEEE

diameter of the flattened butterfly. However, to reduce the number of VCs that are needed, DOR can be used within each phase of UGAL routing. Thus, only two VCs are needed—one VC for minimal routing to the intermediate router and another VC to route to the destination.

## 5.5 Express Virtual Channel

While the flattened butterfly provides a topology with *physical* express channels, express channels or bypassing intermediate routers can also be implemented with flow control to create a *virtual* express topology [31]. The network diameter and the number of hop counts is reduced with a physical express topology; however, in a virtual express topology, the number of hop counts does not change, but the internal router pipeline of intermediate routers is bypassed, as logically shown in Fig. 5.7. The difference between physical and virtual express topologies is illustrated in Fig. 5.8. While the physical express topology requires an alternative topology with different channel organization, a virtual express topology can leverage a baseline topology, such as a 2D mesh network, and provide the ability to bypass intermediate routers. The use of express virtual channels (EVCs) [6] is an example of a flow control mechanism that *virtually* bypasses intermediate routers without requiring dedicated physical channels; it will be discussed in this section.

### 5.5.1 Router Microarchitecture

There are two types of virtual channels in EVC flow control: normal virtual channels (NVCs) and express virtual channels (EVCs). NVCs are allocated similar to conventional VC-based flow control [27] when a packet is not bypassing any intermediate router but traversing through each router. In comparison, EVCs are only allocated when the packet will be bypassing intermediate routers. In addition to the different type of VCs, there are also two different types of router nodes in EVC flow control: bypass nodes and EVC source/sink nodes. The bypass nodes are the virtually bypassed, intermediate router nodes. They do not serve as either a source

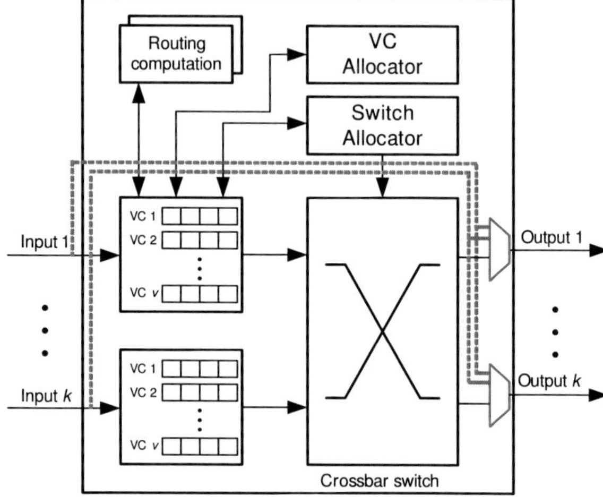

**Fig. 5.7** Bypass path shown in a conventional router microarchitecture [13]. © 2009 IEEE

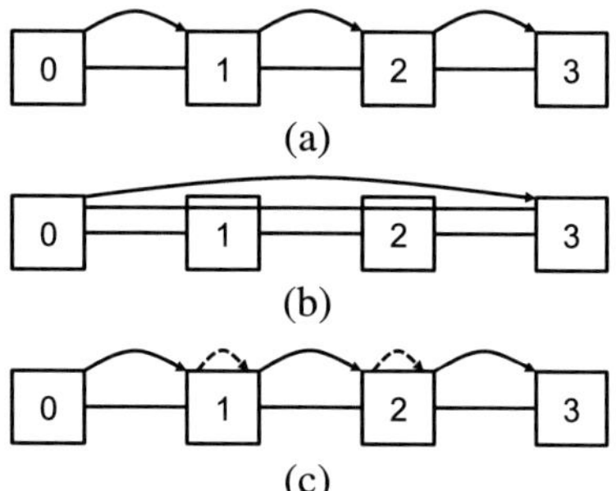

**Fig. 5.8** Comparison of (**a**) a baseline mesh topology, (**b**) a physical express topology where mesh is extended with physical channels, and (**c**) a virtual express topology. The *dotted line* shows the internal pipeline of a router being bypassed in a virtual express topology

or a sink node for an EVC packet and can only allocate NVCs. In comparison, the EVC source node is where an EVC can be allocated, and the EVC is released at the EVC sink node. An EVC source node also serves as an EVC sink node, and NVCs can also be allocated at the EVC source/sink nodes. Each EVC is a $k$-hop EVC, as it is able to bypass $k$ intermediate routers. As a result, the buffer organization of the routers must be modified to support EVC. For the bypass router nodes, only a single entry EVC (or an EVC latch) is needed, whereas for the source/sink nodes, a separate EVC buffer is needed, as shown in Fig. 5.9.

To illustrate the behavior of EVC, an example is shown in Fig. 5.10 for a 64-node 2D mesh network, with a packet being sent from node 1 to node 46, assuming $k = 3$ and $xy$ dimension-order routing (DOR). The shaded router nodes represent EVC source/sink nodes; the other router nodes are bypass nodes. Since node 1 is not an EVC source node, NVC is used to route the packet to node 3, which is an EVC source node. At node 3, an EVC is used to route the packet to node 6, which serves as a sink node, thus bypassing the intermediate router, node 4 and node 5. At node 6, EVC is used again to traverse to node 30 and bypass node 14 and node 22. The final two hops of the routing are done by using NVC, since using EVC again would result in the packet bypassing its destination of node 46. Thus, with conventional

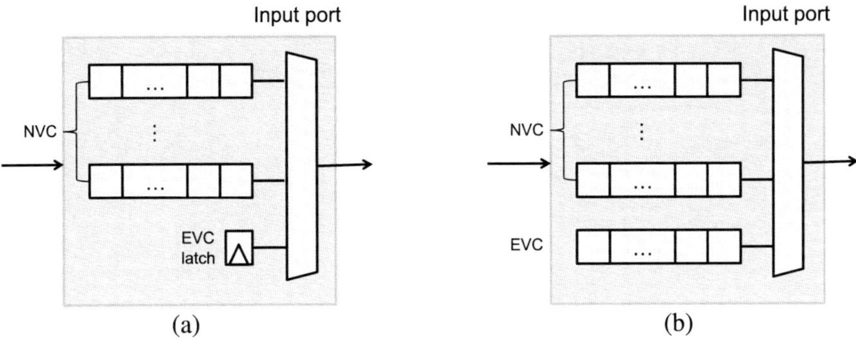



**Fig. 5.9** EVC buffer organization at (**a**) bypass nodes and (**b**) EVC source/sink nodes

NVC, 10 intermediate routers must be traversed, but with EVC, only 5 intermediate routers must be traversed, while the internal pipeline of 5 intermediate routers can be bypassed.

### 5.5.2 Flow Control

Since buffered flow control is used in EVC, proper buffer management is needed to ensure packets are not dropped. Credit-based flow control is commonly used to manage downstream buffers. With NVC, credits for only neighboring downstream

**Fig. 5.10** Example of EVC when sending a packet from node 1 to node 46

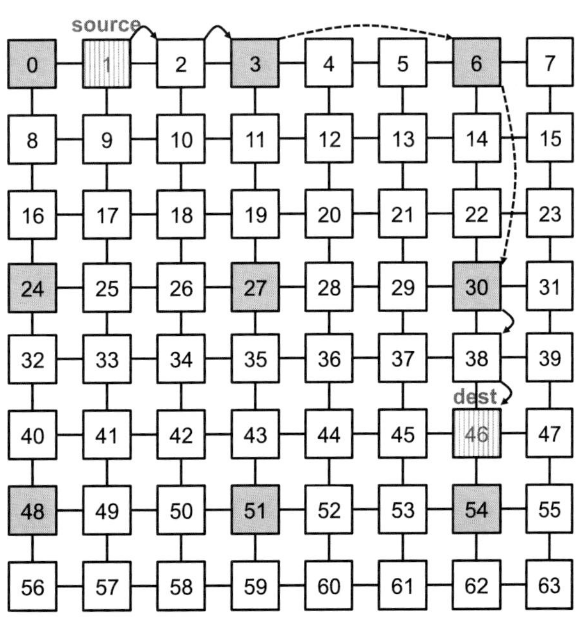

**Table 5.1** Qualitative comparison between physical and virtual express topologies

|  | Physical express topology (flattened butterfly [12]) | Virtual express topology (EVC [6]) |
| --- | --- | --- |
| Intermediate router | Bypass intermediate routers using dedicated channels | Virtually bypass intermediate routers using dedicated VCs |
| Routing | Minimal routing can be used, but nonminimal, adaptive routing is needed to exploit the path diversity | Minimal (DOR) routing |
| Router microarchitecture | Increase in port count (high-radix routers) | Increase in VCs |
| Additional complexity | Nonuniform channels (longer channels) | Starvation avoidance mechanism needed |

nodes must be maintained. However, to support EVC, the credit (or buffer) information of $k$-hop away routers must be maintained as well. Since the credit round-trip latency is proportional to the latency between the nodes, the EVC buffers need to be deeper to cover the credit round-trip latency. With static buffer assignment, EVC buffers can become underutilized where significant traffic locality exists. To overcome this limitation, dynamic buffer management [6] can be used, where the buffer is shared between NVCs and EVCs.

Since EVC is essentially reserving channel bandwidth between the EVC source and sink nodes, EVC is given priority in arbitration over NVC nodes in the intermediate router nodes. As a result, starvation can be problematic with EVC. For example, in the example shown earlier in Fig. 5.10, if node 1 continues to send traffic to node 64, the NVCs in node 7 can be indefinitely starved as EVC continues to have priority. There can be other traffic patterns which cause node 7 to be starved; this is just one example. Thus, a starvation avoidance mechanism is needed to avoid NVCs in bypass nodes from being starved. To avoid starvation, a starvation token can be transmitted upstream after a node has been starved for $n$ cycles and prevent EVCs from being allocated for $p$ cycles, where $n$ and $p$ are parameters that can be adjusted appropriately. Once the upstream node stops allocating EVC, the bypass nodes can use the channel bandwidth and avoid being starved.

In this section, we described static EVC, in which each EVC enables bypassing of $k$ consecutive nodes in the network where $k$ is a fixed value in the network. The use of static EVC can create asymmetry in the network as EVC source nodes can leverage EVC much more efficiently than non-EVC source nodes. To overcome this limitation, dynamic EVC can be used [6], which can increase performance but also at an increase in cost as all router nodes need to be a bypass node and an EVC source/sink node.

A qualitative comparison between flattened butterfly and EVC is shown in Table 5.1. A detailed comparison of physical and virtual express topologies is also given in [31]. However, their analysis does not include the benefits of concentration in an on-chip network topology. It is clear that concentration can help reduce the complexity of physical express topologies (such as the flattened butterfly topology described earlier in Sect. 5.4), but the impact of concentration in virtual express topologies is not clear.

## 5.6 Low-Cost Router Microarchitecture

Both the flattened butterfly topology and the express virtual channel flow control assume the *conventional*, baseline router microarchitecture shown earlier in Fig. 5.1. However, this router microarchitecture is adapted from the router microarchitecture of a large-scale system network such as the routers found in a supercomputer network. On-chip networks provide very different constraints compared with off-chip networks. As a result, the low-cost router microarchitecture [13] described in this section takes a different approach in the design of router microarchitecture—using the implementation shown earlier in Fig. 5.3(c) as the starting point. Along with this datapath, prioritized switch allocation and a partitioned crossbar are used to minimize the complexity of a router microarchitecture for a 2D mesh network.

### 5.6.1 Switch Organization

The low-cost router microarchitecture block diagram for the 2D mesh network is shown in Fig. 5.11. Instead of a 5-port router used in a conventional 2D mesh topology, the router is partitioned or *sliced* into two separate routers—one for each dimension of the network—to create a *dimension-sliced* router. The dimension-sliced router was used for the Cray T3D router [32] as the router was partitioned into three separate router chips, one for each dimension of the 3D torus networks. Technology constraints prevented the router from fitting on a single chip, and it was necessary to partition the router across multiple chips. However, we leverage the same microarchitectural technique to reduce the cost of on-chip networks and simplify the router microarchitecture. The number of router ports is reduced from a single router with 5 ports to two routers with 3 ports.

A dimension-sliced router partitions the crossbar switch into two smaller crossbar switches: the $x$ router ($R_x$) and the $y$ router ($R_y$). $R_x$ ($R_y$) is used to route packets that continue to traverse in the $x$ ($y$) dimension, respectively. Since we assume dimension-ordered ($x-y$) routing, a packet that needs to traverse both dimensions to reach its destination will need to change dimension once (i.e., switch routers from $R_x$ to $R_y$ once). However, even if the source and the destination share the same row or column, the packet will still need to be routed from $R_x$ to $R_y$ because the local injection port is only connected to $R_x$, while the local ejection port is connected to $R_y$ (see Fig. 5.12). Using the dimension-sliced router enables us to create *overpass* channels in the on-chip network which are similar to overpasses in highways and roads. The dimension-sliced router does not require packets that travel in different dimensions to stop at the current router.

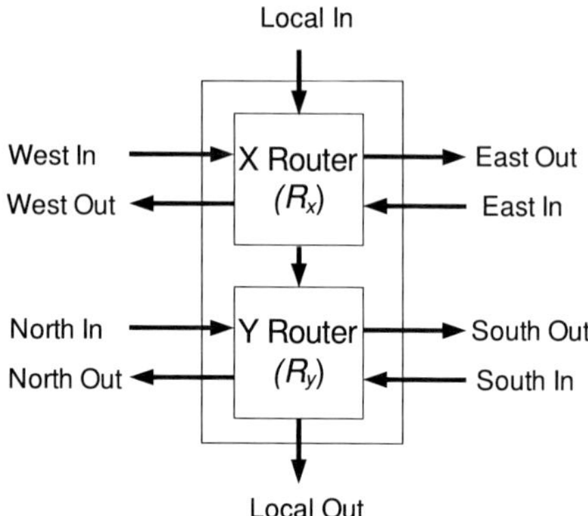

**Fig. 5.11** High-level block diagram of a dimension-sliced router in a 2D mesh network [13]. © 2009 IEEE



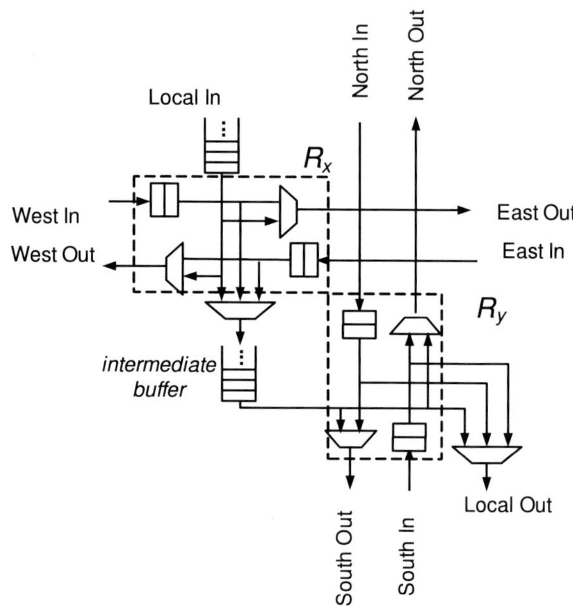

**Fig. 5.12** Detailed logic diagram of the proposed router microarchitecture using (**a**) a shared, intermediate buffer and (**b**) a dedicated, separate intermediate buffer [13]. © 2009 IEEE



## 5.6.2 Buffer Organization

With a single-cycle router, only two buffer entries are needed to cover the credit round-trip latency. The buffer management provides backpressure and avoids the need to drop packets or misroute them as in a bufferless router [15]. However, intermediate buffers are placed between $R_x$ and $R_y$ to decouple the flow control and routing of $R_x$ from $R_y$. If a packet traversing the $x$ dimension switches to the $y$ dimension, the packet is buffered in this intermediate buffer before traversing in the $y$ dimension. Thus, a packet will encounter an additional cycle when switching from $R_x$ to $R_y$.

With a shared intermediate buffer organization (Fig. 5.12), switch arbitration (discussed in Sect. 5.6.3) is needed before the packet is buffered in the intermediate buffer. The intermediate buffer removes packets that are changing dimensions and allows other packets in the $x$ dimension to continue traversing the network. As the intermediate buffers are located locally, no complex buffer management flow control is needed. For example, the east/west input units need to share the credit count for the intermediate buffer. If the intermediate buffer is full, the input buffers will hold the packets until an intermediate buffer slot becomes available.

## *5.6.3 Arbitration*

To reduce the complexity of switch arbitration, simple priority arbitration is used, where packets in flight that continue to travel in the same direction have priority over other packets. For example, if a packet arriving from the west port in $R_x$ must be routed through the east port, it is given priority over packets injected from the local port that need to be routed through the east port. Similarly, a packet arriving from the north port in $R_y$ that must be routed to the south port will have priority over packets being injected from the intermediate buffer. Thus, a packet continuing to travel in one dimension will encounter delay very similar to that in Fig. 5.3(c) shown earlier, with additional delay encountered only when the packet "turns" into the intermediate buffer. For packets injected into the network from the injection port or the intermediate buffer, if the router output port is not used by packets in flight, the packet is injected.

Switch arbitration is needed for intermediate buffers, as multiple requests can be added with a shared intermediate buffer (Fig. 5.12(a)). With a dedicated intermediate buffer (Fig. 5.12(b)), switch arbitration is required after the packets are buffered in the intermediate buffer. However, switch arbitration can be done in parallel with writing a packet in the buffer—with the result of switch arbitration used in the following cycle. Thus, switch arbitration is not on the critical path of the microarchitecture.

The pipeline diagram of a conventional baseline router microarchitecture is shown in Fig. 5.13(a) based on a 3-cycle router [33] and compared with the proposed router microarchitecture in Fig. 5.13(a). The dimension-sliced router microarchitecture results in the reduction of router latency, as the router pipeline assumed for the conventional router microarchitecture can be avoided, and both the router delay and link delay can be combined into a single cycle, resulting in a single pipeline stage router. An extra pipeline stage is needed when the packet must change dimensions as shown in Fig. 5.13(b) because of the use of intermediate buffers. Switch arbitration is removed for packets continuing to travel in the same dimension, as packets in the network have priority over those that have not been injected.

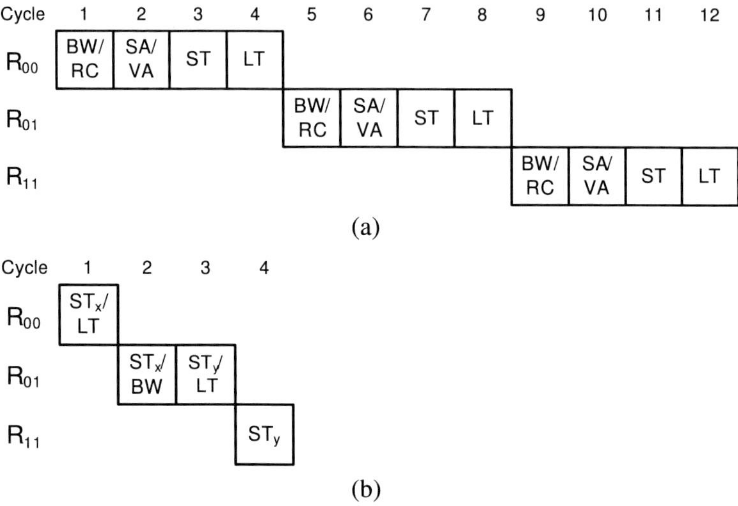

Fig. 5.13 Pipeline diagram of a single-flit packet in the (**a**) baseline, conventional router and the (**b**) proposed low-cost router in a 2D mesh network, routing from R00 to R11 in the network shown in Fig. 5.14. $ST_x$ is the switch traversal of $R_x$, and $ST_y$ is the switch traversal $R_y$ [13]. © 2009 IEEE

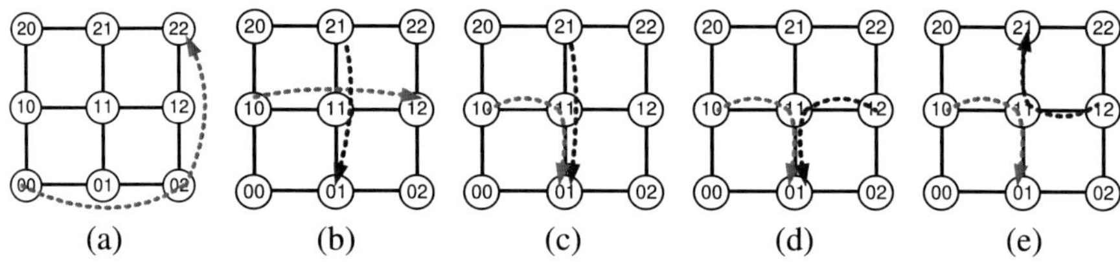

Fig. 5.14 Example flow control in the low-cost, router microarchitecture when there is no contention (**a**, **b**) and when packets contend for the same resource (**c**, **d**, **e**) on a $3\times3$ mesh network. *The different arrows* represent different packets in the network [13]. © 2009 IEEE

### 5.6.4 Routing/Flow Control Examples

To illustrate the behavior of the low-cost router microarchitecture, examples of routing and flow control on a $3\times3$ mesh are shown in Fig. 5.14. As shown in Fig. 5.14(a), without any congestion in the network, it can approach *ideal* latency using the proposed low-cost router microarchitecture. If traffic from the two different dimensions cross over a single router (Fig. 5.14(b)), each packet does not affect the other packets because of the dimension-sliced architecture, and minimal latency is achieved as well.

When one packet continues to travel in a dimension while another packet turns from a different dimension into the same dimension (Fig. 5.14(c)), contention will occur for the same channel resource. By prioritizing packets continuing in the same dimension, the packet that is "turning" will be buffered in the intermediate buffer

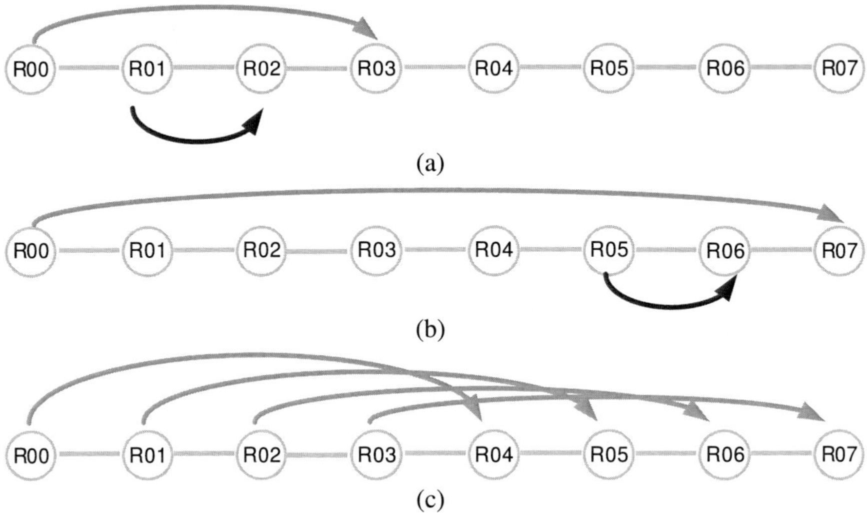



**Fig. 5.15** Illustration of (**a, b**) local starvation and (**c**) global starvation in the proposed router microarchitecture [13]. © 2009 IEEE

until the output channel becomes available. When two packets arrive at the same router node from the same dimension and want to turn to the *same* direction of a different dimension (Fig. 5.14(d)), arbitration is needed to access the shared intermediate buffer before traversing in the new dimension. The winner accesses the intermediate buffers, while the packet that does not have its access granted remains buffered in the input buffer.

The contention of resource shown in Fig. 5.14(c, d) is similar to that observed in a conventional microarchitecture; the only difference is how the contention is resolved. However, Fig. 5.14(e) is a contention that is unique to our proposed architecture with a shared intermediate buffer. Two packets arrive at a router from the same dimension and want to turn to a *different* direction of the new dimension. Because of the limited connections in a dimension-sliced crossbar, the intermediate buffer becomes a shared resource and creates a bottleneck. Thus, in the worst-case scenario, the throughput of the network can be degraded by one-half compared with a network using a conventional on-chip network router.

### 5.6.5 *Fairness/Starvation*

By simplifying the arbitration in the proposed lightweight router microarchitecture, fairness can become an issue. Fairness is not an issue for uniform random traffic near zero-load, but as the load increases and approaches saturation, fairness will become a problem as the packets being injected from the edge of the network will always have priority. Nonuniform traffic patterns will also cause starvation. For example, with the traffic pattern shown in Fig. 5.15(a), R01 can be starved indefinitely if

R00 continues to inject packets into the network. Similarly, R05 can be starved in Fig. 5.15(b), as well as R01, R02, and R03 in Fig. 5.15(c).

To overcome this limitation, we can send an explicit control signal upstream to prevent starvation. For example, R01 in Fig. 5.15(a) can wait $n$ cycles and if it is still starved, it can send a control signal to R00 to halt its transmission of packets. Once R00 stops transmission and R01 does not see any more packets, it can inject its packet. A similar scheme was proposed in the EVC flow control [6] for starvation avoidance, as express VCs can starve normal VCs. Thus, *tokens* were proposed in the EVC to prevent starvation. However, using control signals can be very complex and add latency and complexity. For the traffic pattern shown in Fig. 5.15(a), control signals can be sent relatively quickly because the injecting node is only 1 hop away. However, in Fig. 5.15(b), sending an explicit control signal or token can be very time consuming, as illustrated in the time diagram shown in Fig. 5.16(a). If R05 decides to send an explicit control signal at $t_0$, in the worst-case scenario (when R00 continues to inject packets), R05 would not be able to inject packets into the network until $t_1$; thus, $(t_1 - t_0)$ corresponds to the round-trip delay from R05 to R00. For traffic patterns such as the one shown in Fig. 5.15(c), where R01, R02, and R03 are all starved from R00 traffic, each node sending a control signal (or tokens) will complicate the starvation avoidance scheme.

To prevent starvation, we propose a simplified distributed approach where starvation is prevented by manipulating local router credits. A credit is decremented when a flit is sent downstream. Once the flit departs from the downstream node, a credit is sent back upstream and the appropriate credit count is incremented. However, by stalling the return of credits, an artificial backpressure can be created which prevents upstream nodes from injecting packets into the network, and allowing the current node to inject packets into the network.

In this scheme, each router maintains a starvation count $n$; once it reaches the maximum value ($n_{max}$), the router stops the transmission of credits upstream. By delaying the credit upstream, the backpressure ensures that upstream nodes do not inject any more packets into the network. Once the source router injects its packet, $n$ is reset to 0. Each router must maintain four separate values of $n$: $n_E$ and $n_W$ for the east and the west port of $R_x$, respectively, and $n_N$ and $n_S$ for the north and the south port of $R_y$, respectively. In each cycle, if the injection port of the router has a packet that it is trying to inject into the network, the appropriate $n$ value is incremented each cycle if the packet is unable to inject the packet into the network. When $n$ reaches $n_{max}$, it means that the $n_{max}$ continuous stream of flits has flowed through the current router node in one particular direction while the router waited, and, thus, credits are not immediately returned.

Figure 5.16(b) shows the time diagram of the starvation avoidance scheme. Assume R00 continues to inject packets into the network destined for R07 (Fig. 5.15(b)), and, for simplicity, assume that $n_{max} = 1$. Initially, $n_E = 0$ at R05 and at $t_0$, R05 wants to inject a packet to R06. In $t_0$, flit 0 also arrives from R04. Due to the prioritized allocation, flit 0 is granted access to the output port and transmitted to R06 in the next cycle while $n_E$ is incremented to 1. Since there is a packet waiting at R05 to be injected into the network and $n_E = n_{max}$, instead

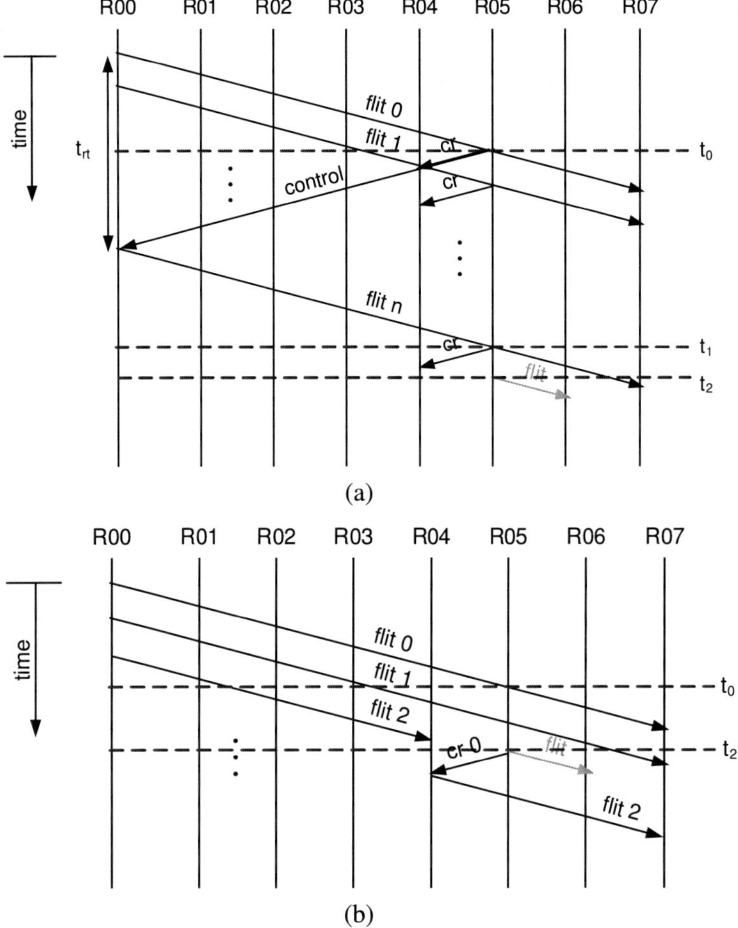



**Fig. 5.16** Time diagram of using (**a**) explicit control signals to avoid starvation and (**b**) stalling credits to provide local starvation avoidance scheme for the traffic illustrated in Fig. 5.15(b) [13]. © 2009 IEEE

of immediately sending a credit back upstream at $t_0$, the credit return is stopped. As another packet can be in flight, R05 will have to wait another cycle for flit 1 to pass through. At this point, R04 does not have any credits and is thus required to stop sending packets to R05—allowing R05 to inject flits into the network at $t_2$ and also start returning credits back upstream. As long packets in the network require multiple flits, we prevent the stalling of the credits until the tail flit is received at the current node. This ensures that packets will not be interleaved at the flit granularity and allows the body and tail flits to follow the head flit to its destination.

## 5.7 Bufferless Flow Control

As mentioned in Sect. 5.2.1, buffers can represent a significant portion of over-all power and area. To reduce the energy consumption and provide an energy/performance trade-off, an alternative approach to an on-chip network is buffer-less flow control where the input buffer is essentially removed.[2] At low traffic, the bufferless approach results in minimal loss of performance since very few packets are in the network. However, at high traffic, the lack of input buffers can lower performance with reduced throughput. This trade-off can be acceptable in many-core processors if the traffic in the network is often low.

By removing the input buffers from the routers, the allocation of resources can be problematic where there is contention for resources. This contention can be resolved via two different methods:

1. Misrouting (or deflection routing)—misroute one of the packets to a different direction.
2. Packet dropping—drop the packet and retransmit.

In this section, we describe two different implementations of bufferless flow control for on-chip networks: BLESS (bufferless routing algorithms) [15], which is based on deflection routing, and SCARAB [14], which is based on dropping packets and retransmitting.

### *5.7.1 BLESS*

With deflection routing, when two or more packets need to be routed through the same output port, only one packet can use this resource in that particular cycle. Buffered flow control is able to decouple this dependency in time by delaying the allocation of the output port resource for other packets to a subsequent cycle. Thus, in order for deflection routing to operate properly, the number of input ports and the number of output ports must be identical. This property ensures that there is at least an available output port where a packet can be misrouted or deflected when contention for the output port occurs.

Thus, the use of deflection routing can be used with the bufferless router mi-croarchitecture. Without contention, a packet is routed using a conventional routing algorithm, such as dimension-order routing (DOR). This baseline routing is the *preferred* route a packet takes, but that packet can be deflected when contention occurs. Once a packet is deflected, the packet will attempt to continue to use its baseline routing to make progress toward the destination, but it can also be deflected again.

Thus, the following problems can occur with deflection routing:

- Livelock—forward progress must be guaranteed since packets can continue to be deflected.

---

[2]Some amount of storage is still needed; for example, pipeline registers are still needed and, if the router microarchitecture requires multiple cycles, additional internal storage would be required.

- Packet re-ordering—since packets are often partitioned into multiple flits and each flit is routed separately, they can arrive out of order at the destination.

To prevent livelock and guarantee progress, age-based arbitration or oldest-first packet policy can be implemented. This priority in the arbitration ensures that a packet will eventually become the oldest packet at some point and guarantee progress toward the destination. To avoid the problem of packet re-ordering, a re-order buffer is needed at the destination such that flits within a packet are properly ordered before sending the packet to destination.

The primary advantage of BLESS is the simplified router microarchitecture, as the input buffers are removed, and the area and energy consumed by the buffers are also reduced. However, the disadvantages, in addition to the additional buffering at the receivers, is the higher latency and reduced bandwidth with deflection routing. At near zero-load, there is minimal impact, but as the traffic increases, more contention occurs and additional hops taken by the packet not only consume additional bandwidth but also increase latency. Thus, at high traffic load, the bufferless approach can result in lower performance, such as reduced throughput, but also with reduced energy consumption.

## 5.7.2 SCARAB

In contrast to the deflection routing used in BLESS, the single-cycle minimally adaptive routing and bufferless router (SCARAB) [14] handles contention differently by dropping packets that are not granted the resource it needs, e.g., output channel. However, since the on-chip network needs to be lossless and cannot drop packets, the dropped packet must be retransmitted. The source needs to be notified of the dropped packet so that the packet can be retransmitted.

To support packet retransmission for dropped packets, SCARAB provides a NACK network. In sending a packet from source to destination, a reverse path NACK network is reserved at each hop. If the packet is dropped, the reserve NACK path is used to send a NACK back to the source. With this approach, the source must store the packet until the packet is received at the destination without being dropped in the network. With a NACK network, the ACK is implicit, because if the packet is not dropped, the maximum delay in receiving a NACK is deterministically bounded and is proportional to the hop count between source and destination. The NACK network optimizes the network retransmission, but if the NACK network cannot be reserved, the packet also must be dropped as well.

For multiflit packets, the resource allocation is held such that body and tail flits will follow the head flit and packets are guaranteed to arrive in order. In the SCARAB architecture, there are three parallel networks: the data network for the payload (or the data), the allocation network, which allocates the output channels based on packet priority, and the NACK network used to communicate a NACK for when a packet is dropped. Communication between these three networks is necessary to ensure proper management of the entire bufferless network.

**Table 5.2** Qualitative comparison of bufferless flow control

|  | BLESS [15] | SCARAB [14] |
|---|---|---|
| Contention resolution | Deflection routing | Packet dropping and retransmission |
| Additional increase in buffers | Receiver side for packet re-ordering | Source (processor) side for packet re-transmission |
| Routing | Deflection routing—can result in a nonminimal path | Minimal, adaptive routing |
| Additional complexity | Oldest-first arbitration flow control | Dedicated NACK network |

In addition to the NACK network, SCARAB introduces opportunistic buffering to optimize the performance and cost of bufferless flow control. Instead of retransmitting packets from the source for dropped packets, the packet can be opportunistically stored in intermediate node. This can be accomplished by leveraging the unused MSHR buffer space entry in the intermediate routers. Thus, instead of sending a NACK back to the source, if the ejection port of the current intermediate router is full and the miss status holding register (MSHR) buffer is available, the packet is opportunistically buffered in the intermediate router. This can help reduce the latency and power consumption of having to retransmit from the source node. A qualitative comparison of the two bufferless approach is shown in Table 5.2.

## 5.8 Conclusion

As technology evolves and transistors shrink, the number of components integrated into a single chip will continue to increase. As a result, the energy to move data across the chip will be higher than the actual energy cost of computation. In this new era, it is critical that the communication aspect of future multi-core and many-core processors implement an energy-aware on-chip network to reduce energy consumption while minimizing the impact on performance. In this chapter, we described three different approaches that attempt to achieve the characteristics of an *ideal* on-chip network (in terms of both performance and area) through various techniques, including topology, flow control, and router microarchitecture. In addition to these methods, another energy-aware on-chip network design was presented in this chapter: a bufferless on-chip network approach that removes input buffers in the router microarchitecture.

## References


1. Intel: From a few cores to many: a tera-scale computing research overview (2006)
2. Kongetira, P., Aingaran, K., Olukotun, K.: Niagara: a 32-way multithreaded Sparc processor. IEEE MICRO **25**, 21–29 (2005). 10.1109/MM.2005.35. http://portal.acm.org/citation.cfm?id=1069597.1069758

3. Nickolls, J., Dally, W.J.: The GPU computing era. IEEE MICRO **30**, 56–69 (2010). http://doi.ieeecomputersociety.org/10.1109/MM.2010.41
4. Owens, J.D., Dally, W.J., Ho, R., Jayasimha, D.N., Keckler, S.W., Peh, L.S.: Research challenges for on-chip interconnection networks. IEEE MICRO 96–108 (2007)
5. Dally, W.J., Towles, B.: Principles and Practices of Interconnection Networks. Morgan Kaufmann, San Francisco (2004)
6. Kumar, A., Peh, L.S., Kundu, P., Jhay, N.K.: Express virtual channels: towards the ideal interconnection fabric. In: Proc. of the International Symposium on Computer Architecture, ISCA, San Diego, CA (2007)
7. Dally, W.J., Towles, B.: Route packets, not wires: on-chip interconnection networks. In: Proc. of the 38th Conference on Design Automation, DAC, pp. 684–689 (2001)
8. Kumar, A., Kundu, P., Singh, A., Peh, L.S., Jha, N.: A 4.6 Tbits/s 3.6 GHz single-cycle NOC router with a novel switch allocator in 65 nm CMOS. In: International Conference on Computer Design, ICCD (2007). http://www.gigascale.org/pubs/1218.html
9. Nicopoulos, C.A., Park, D., Kim, J., Vijaykrishnan, N., Yousif, M.S., Das, C.R.: Vichar: a dynamic virtual channel regulator for network-on-chip routers. In: Annual IEEE/ACM International Symposium on Microarchitecture, MICRO, Orlando, FL (2006)
10. Hoskote, Y., Vangal, S., Singh, A., Borkar, N., Borkar, S.: A 5-GHz mesh interconnect for a teraflops processor. IEEE MICRO **27**(5), 51–61 (2007)
11. Kodi, A.K., Sarathy, A., Louri, A.: iDEAL: inter-router dual-function energy and area-efficient links for network-on-chip. In: Proc. of the International Symposium on Computer Architecture, ISCA, Beijing, China (2008)
12. Kim, J., Balfour, J., Dally, W.J.: Flattened butterfly for on-chip networks. In: Proc. of the 40th Annual IEEE International Symposium on Microarchitecture, MICRO, Chicago, IL, Dec. 2007, pp. 172–182 (2007)
13. Kim, J.: Low-cost router microarchitecture for on-chip networks. In: Proc. of the 42nd Annual IEEE International Symposium on Microarchitecture, MICRO 42, New York, NY, Dec. 2009, pp. 255–266 (2009)
14. Hayenga, M., Jerger, N.E., Lipasti, M.: Scarab: a single cycle adaptive routing and bufferless network. In: Proc. of the 42nd Annual IEEE/ACM International Symposium on Microarchitecture, MICRO 42, pp. 244–254 (2009)
15. Moscibroda, T., Mutlu, O.: A case for bufferless routing in on-chip networks. In: Proc. of the 36th Annual International Symposium on Computer Architecture, pp. 196–207 (2009)
16. Kim, J., Dally, W.J., Towles, B., Gupta, A.K.: Microarchitecture of a high-radix router. In: Proc. of the 32nd IEEE International Symposium on Computer Architecture, ISCA, Madison, WI, pp. 420–431 (2005)
17. Wang, H., Peh, L.S., Malik, S.: Power-driven design of router microarchitectures in on-chip networks. In: Proc. of the 36th Annual IEEE/ACM International Symposium on Microarchitecture, pp. 105–116 (2003)
18. Vangal, S., Howard, J., Ruhl, G., Dighe, S., Wilson, H., Tschanz, J., Finan, D., Singh, A., Jacob, T., Jain, S., Erraguntla, V., Roberts, C., Hoskote, Y., Borkar, N., Borkar, S.: An 80-tile sub-100-W TeraFLOPS processor in 65-nm CMOS. IEEE J. Solid-State Circuits **43**(1), 29–41 (2008)
19. Balfour, J., Dally, W.J.: Design tradeoffs for tiled CMP on-chip networks. In: ICS'06: Proc. of the 20th Annual International Conference on Supercomputing, pp. 187–198 (2006)
20. Das, R., Eachempati, S., Mishra, A.K., Vijaykrishnan, N., Das, C.R.: Design and evaluation of a hierarchical on-chip interconnect for next-generation CMPs. In: International Symposium on High-Performance Computer Architecture, HPCA, Raleigh, NC, pp. 175–186 (2009)
21. Grot, B., Hestness, J., Keckler, S.W., Mutlu, O.: Express cube topologies for on-chip interconnects. In: International Symposium on High-Performance Computer Architecture, HPCA, Raleigh, NC, pp. 163–174 (2009)
22. Bhuyan, L.N., Agrawal, D.P.: Generalized hypercube and hyperbus structures for a computer network. IEEE Trans. Comput. **33**(4), 323–333 (1984)
23. Dally, W.J.: Express cubes: improving the performance of $k$-ary $n$-cube interconnection networks. IEEE Trans. Comput. **40**, 1016–1023 (1991)

24. Kumar, P., Pan, Y., Kim, J., Memik, G., Choudhary, A.N.: Exploring concentration and channel slicing in on-chip network router. In: International Symposium on Networks-on-Chips, NOCs, La Jolla, CA, pp. 276–285 (2009)
25. Cho, S., Jin, L.: Managing distributed, shared l2 caches through OS-level page allocation. In: Annual IEEE/ACM International Symposium on Microarchitecture, MICRO, Orlando, FL, pp. 455–468 (2006)
26. Kim, J., Dally, W.J., Abts, D.: Flattened butterfly: a cost-efficient topology for high-radix networks. In: Proc. of the International Symposium on Computer Architecture, ISCA, San Diego, CA (2007)
27. Dally, W.J.: Virtual-channel flow control. IEEE Trans. Parallel Distrib. Syst. 3(2), 194–205 (1992)
28. Seo, D., Ali, A., Lim, W.T., Rafique, N., Thottethodi, M.: Near-optimal worst-case throughput routing for two-dimensional mesh networks. In: Proc. of the International Symposium on Computer Architecture, ISCA, pp. 432–443 (2005)
29. Valiant, L.G.: A scheme for fast parallel communication. SIAM J. Comput. 11(2), 350–361 (1982)
30. Singh, A.: Load-balanced routing in interconnection networks. Ph.D. thesis, Stanford University, Palo Alto, CA (2005)
31. Chen, C.H.O., Agarwal, N., Krishna, T., Koo, K.H., Peh, L.S., Saraswat, K.C.: Physical vs. virtual express topologies with low-swing links for future many-core NoCs. In: Proc. of the 2010 4th ACM/IEEE International Symposium on Networks-on-Chip, NOCS'10, pp. 173–180. IEEE Comput. Soc., Washington (2010)
32. Kessler, R., Schwarzmeier, J.: Cray T3D: a new dimension for Cray research. Compcon Spring'93, Digest of Papers, pp. 176–182 (22–26 Feb. 1993)
33. Kumar, A., Peh, L.S., Jha, N.K.: Token flow control. In: Annual IEEE/ACM International Symposium on Microarchitecture, MICRO, Lake Como, Italy, pp. 342–353 (2008)

# Chapter 6
# Energy Awareness in Video Codec Design

Jaemoon Kim, Giwon Kim, and Chong-Min Kyung

**Abstract** In portable multimedia devices, one of the most critical issues is to minimize the energy consumption and thereby prolong the operational lifetime of the system while maintaining the required video quality. In this chapter, we discuss several methods for minimizing the energy consumption of video codec. First, we review the H.264/AVC codec and analyze the computational complexity of the H.264/AVC codec functional blocks. Second, we describe the method of low power integer and fractional motion estimation which occupies a significant part of the total computational complexity. We also explain embedded compression to reduce the power consumed by memory access. Finally, we introduce a power–rate–distortion ($P$–$R$–$D$) model for a video coding system to maximize its lifetime. The $P$–$R$–$D$ video encoder model is generated in two steps. The first step is modeling the relationship between the power consumption and the distortion of video encoder based power-scalable architecture of the H.264/AVC encoder using the power consumption data of each functional module. The second step is generating the unified $P$–$R$–$D$ model based on the $P$–$D$ model and the conventional rate–distortion ($R$–$D$) model.

## 6.1 Introduction

Video capturing has become a key feature of wireless sensor networks and portable multimedia devices. Since the raw format of the video data is voluminous, efficient video compression is necessary to reduce the amount of transmission bandwidth

J. Kim (✉)
Samsung Electronics, Seoul, Republic of Korea
e-mail: jaemoon.kim@gmail.com

G. Kim · C.-M. Kyung
KAIST, Daejeon, Republic of Korea

G. Kim
e-mail: gwkim@vslab.kaist.ac.kr

C.-M. Kyung
e-mail: kyung@ee.kaist.ac.kr

C.-M. Kyung, S. Yoo (eds.), *Energy-Aware System Design*,
DOI 10.1007/978-94-007-1679-7_6, © Springer Science+Business Media B.V. 2011

or required storage space. However, video compression is often computationally intensive and energy consuming. Considering that the portable devices or sensor networks are powered by batteries, energy minimization of the video codec is one of the challenging issues involved in prolonging the operational lifetime. In this chapter, we discuss the relations among power consumption, bit rate, and distortion of the video encoding process to reduce the power consumption of video codec while maintaining the video quality.

## 6.2 Overview of H.264/AVC

H.264/AVC [1] was finalized by the Joint Video Team (JVT) from ISO/IEC MPEG and ITU-T VCEG in 2003. Compared with the previous video coding standards such as MPEG-2, MPEG-4, and H.263, H.264/AVC provides much better coding efficiency as shown in Fig. 6.1, which represents the rate–distortion curve for different coding standards for two video sequences, *Tempete* and *Paris* [2]. Table 6.1 shows the relative bit-rate savings for equivalent perceptual image quality for video streaming and conferencing applications [3]. The table shows that, compared to MPEG-2, MPEG-4, and H.263, H.264/AVC can achieve a high bit-rate savings for equivalent perceptual image quality. The high coding efficiency of H.264/AVC comes from new techniques such as variable block size motion estimation (VB-SME), quarter-pixel fractional motion estimation, multiple reference frame (MRF), context-adaptive binary arithmetic coding (CABAC), and in-loop deblocking filter. However, these advanced coding techniques have introduced a much higher computational complexity than that of previous video coding standards, e.g., 10 times than that of MPEG-2.

## 6.3 H.264/AVC Video Codec

In this section, we explain the H.264/AVC codec, which has a conventional four-stage macroblock (MB) pipeline architecture, and its process of pipeline operation. We then analyze the computational complexity of the main functional blocks of the H.264/AVC codec. The analysis helps in understanding the power scalability of video codec as well as the design of low power video codec.

### *6.3.1 H.264/AVC Video Codec Architecture*

Figure 6.2 shows the block diagram of the H.264/AVC codec architecture, which adopts a conventional four-stage MB pipeline architecture. The architecture consists of multiple functional blocks: the integer motion estimation (IME) unit, fractional

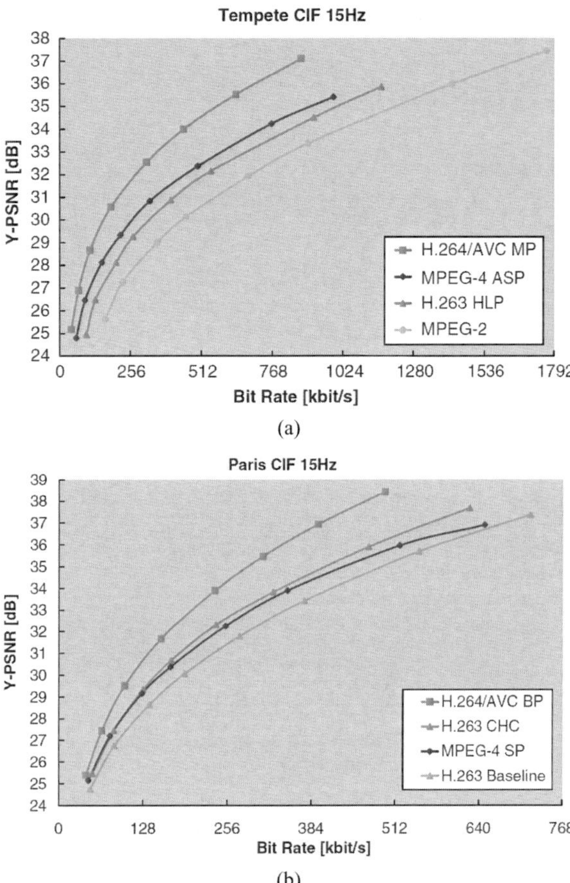

**Fig. 6.1** Luminance PSNR versus average bit rate for different coding standards for (**a**) *Tempete* and (**b**) *Paris* [2]. © 2002 IEEE



**Table 6.1** Average bit-rate savings for video streaming applications [3]. © 2003 IEEE

| Coder | MPEG-4 ASP | Bit-rate savings relative to: | |
| --- | --- | --- | --- |
| | | H.263 HLP | MPEG-2 |
| H.264/AVC MP | 37.44% | 47.58% | 63.57% |
| MPEG-4 ASP | – | 16.65% | 42.95% |
| H.263 HLP | – | – | 30.61% |

motion estimation (FME) unit, chroma motion compensation (CMC) unit, intra prediction (Intra) unit, reconstruction (Rec) unit, deblocking filtering (DF) unit, entropy coding (EC) unit, and frame memory controller (FMC) unit. In the video codec architecture, each pipeline stage is connected by buffers, which are implemented by single-port SRAMs (light-gray colored), except for the "Luma ref." buffer (dual-port SRAM, dark-gray colored) located between the IME and FME modules. The

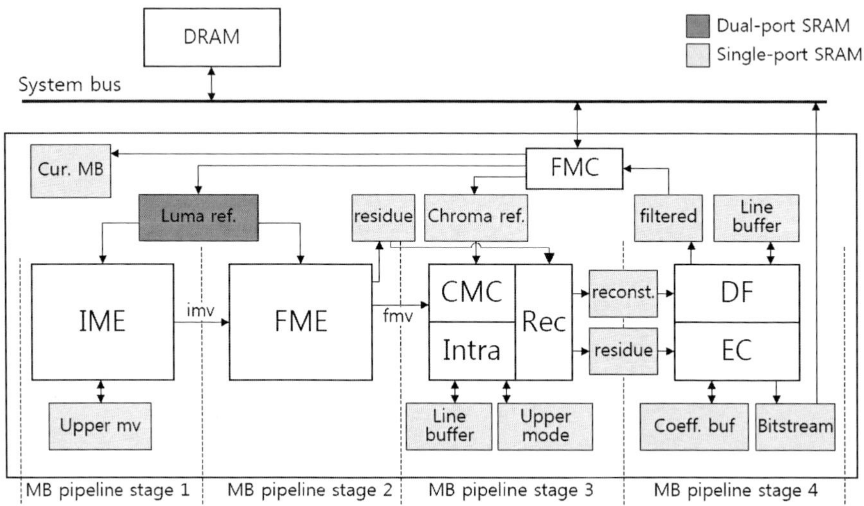



**Fig. 6.2** The entire H.264/AVC codec architecture. *The dark-gray* and *light-gray colors* represent dual-port and single-port RAM units, respectively. The video codec is connected to the system bus to access external DRAM components used as the reference frame memory

video codec is connected to the system bus to access the external DRAM, which is used as the reference frame memory.

In the first stage, the IME module searches the best block matching position within a given search range. The reference frame data, used in block matching processing, are loaded from the external DRAM through the FMC module. The fetched data are stored in the Luma ref. buffer and are referenced during IME. Since the Luma ref. buffer consists of dual-port SRAM, simultaneous accesses from the IME module and FME module are possible. The output of the IME module is the integer motion vector, "imv", which represents the best block matching position in terms of integer pixel resolution. This imv is used as the initial search point of FME.

In the second stage, the FME module searches the best fractional motion vector, "fmv", to reduce additional temporal redundancy by using the interpolation of neighboring pixel values. In H.264/AVC, there are four block modes: $16 \times 16$, $16 \times 8$, $8 \times 16$, and $8 \times 8$. If $8 \times 8$ is chosen as a candidate block mode, the $8 \times 8$ block can be further divided into four block types: $8 \times 8$, $8 \times 4$, $4 \times 8$, and $4 \times 4$. Thus, the FME module selects the block mode with the best fmv among seven candidate block modes.

In the third stage, the CMC module generates the motion compensated pixel values of the chrominance components based on fmv. While CMC is performed, the Intra module predicts the pixel values of the current MB using the neighboring pixels. When the motion compensated pixel values and intra predicted pixel values are ready, the Rec module compares each rate–distortion cost and decides which one is better. It then reconstructs the pixel values using the predicted pixel values of CMC or Intra. These reconstructed pixel values are transferred to DF through the "reconst." buffer while the residue pixel values, i.e., the subtractions of the current

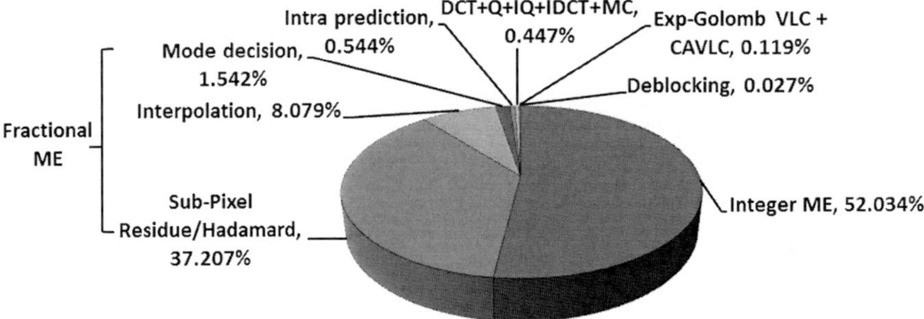


**Fig. 6.3** Computational complexity of main blocks in H.264/AVC codec [4]. © 2004 IEEE

pixel values and the reconstructed pixel values, are stored into the "residue" buffer for compression by EC.

Finally, in the fourth stage, deblocking filtering and entropy coding are performed in parallel. DF is one of the tools used to improve the image quality during video coding; it removes blocking artifacts caused by the block-based transform. The filtered data, which are used as reference frame data in the next frame, are transferred to the external DRAM through the FMC module. The residue pixel values are entropy coded by the EC module. In H.264/AVC, context-adaptive variable length coding (CAVLC) and context-adaptive binary arithmetic coding (CABAC) are defined as entropy coding methods. In the architecture presented in Fig. 6.2, we adopt CAVLC as an entropy coding method to reduce the power consumption of the video codec.

### 6.3.2 Computational Complexity of H.264/AVC Codec

Based on the H.264/AVC codec architecture discussed in Sect. 6.3.1, in this section we analyze the computational complexity of the main functional blocks in the video codec for the design of a low power video codec. In the H.264/AVC codec, the main blocks are IME, FME, Intra, Rec, DF, and EC as shown in Fig. 6.2. Figure 6.3 shows the computational complexity of the main functional blocks in the H.264/AVC encoding process [4]. As shown in the figure, motion estimation (ME) occupies about 98% of the total computational complexity of the encoding process. IME occupies about 52% of the total computational complexity due to the huge search range for best block matching and increased DRAM access for loading multiple reference frames. FME occupies about 47% of the total computational complexity due to refinement of quarter-pixel accuracy and the variable block size modes. Therefore, reduction of computational complexity of ME is essential to reduce the total power consumption of the H.264/AVC codec.

## 6.4  Design of Low Power H.264/AVC Codec

Our analysis of the computational complexity of the H.264/AVC codec has shown that it is important to reduce the computational complexity in ME to reduce the total power consumption of the H.264/AVC codec. In this section, we describe two methods to reduce the computational complexity in IME and FME: bit truncated IME and single-pass FME. Moreover, we explain the embedded compression to reduce the DRAM access.

### *6.4.1  Bit Truncated Integer Motion Estimation*

In typical video coding systems, ME is widely used to improve the coding efficiency by removing the temporal redundancy. However, as shown in Fig. 6.3, IME has a high computational complexity, leading to a high computational power consumption. To resolve this problem, a hardware implementation is necessary, and many IME hardware architectures have been proposed. Most IME hardware has parallel processing units, which calculate the sum of absolute differences (SAD) in parallel to support real-time video coding. However, these SAD units require a lot of hardware resources. One SAD unit for an $n \times n$ block has $n \times n$ processing elements, which calculate the absolute difference between the value of a current pixel and the value of a reference pixel. To reduce the hardware cost and the long critical path of the architecture, bit truncation algorithms are used.

Figure 6.4 shows the probability distribution of the absolute value of the residue at the best block matching, which is produced by subtracting the value of the reference pixel from that of the current pixel. In Fig. 6.4, most of the absolute residues (ARs) are concentrated at zero and 99% of the ARs are placed under the value of 32. Hence, most of the ARs can be expressed with reduced bit width by truncating the most significant bits (MSBs) of the AR.

Figure 6.5 shows the bit truncation algorithm. The least significant bits (LSBs) of the value of the current pixel and that of the reference pixel are truncated first according to the value of the number of truncated LSBs (NTL). The AR is produced by subtracting the value of the truncated reference pixel from that of the truncated current pixel. To reduce the bit width of the AR further, the MSBs of the AR are truncated according to the value of the number of truncated MSBs (NTM). If the AR exceed the maximum value of the reduced bit width before the MSB truncation, then the AR is replaced by the clipped value, i.e., $2^{8-\text{NTBL}-\text{NTBM}}$. As a result, the bit width of the AR is reduced to $(8 - \text{NTL} - \text{NTM})$ bit. By using this clipping, severe errors due to just truncating the MSBs of the AR are avoided.

Figure 6.6 shows the overflow probability at the best block matching, which occurs when the AR is larger than the clipped value. The overflow probability is lower than 0.4% when NTM and NTL are 3 and 2, respectively. This means that the overflow occurs by just one pixel among the 256 pixels in a $16 \times 16$ block. Therefore, the best matched block can be found, even though the MSBs of the AR are truncated.

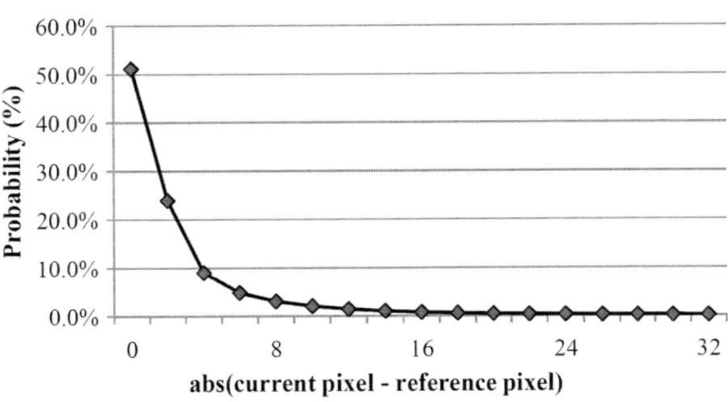

**Fig. 6.4** Distribution of absolute value of the residues at the best block matching

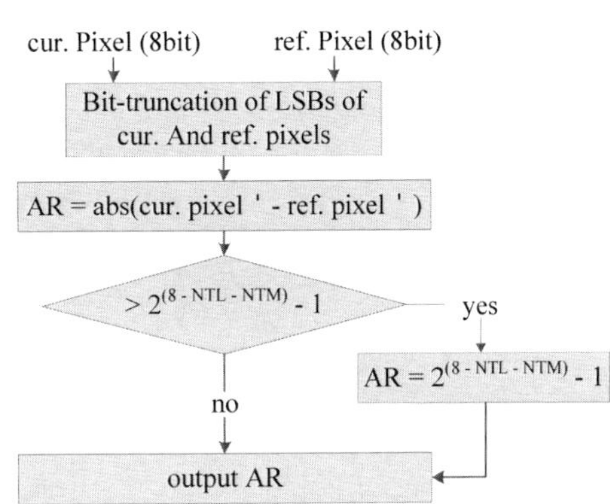

**Fig. 6.5** Bit truncation algorithm on both ends (MSB and LSB)



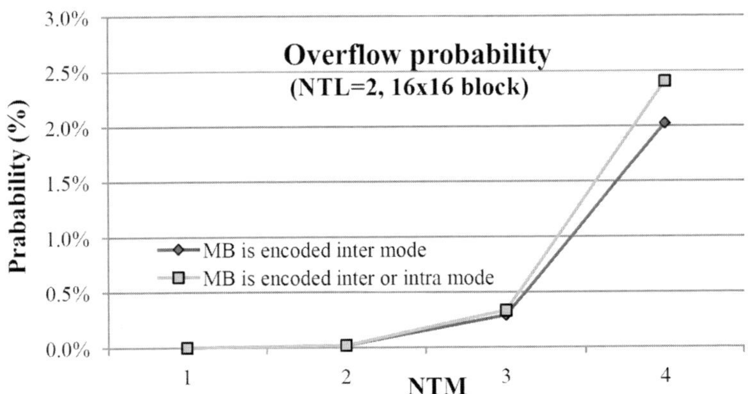

**Fig. 6.6** The overflow probability according to NTM

## *6.4.2 Single-Pass Fractional Motion Estimation*

Figure 6.7 shows the full-search FME method defined in the H.264/AVC reference software [5]. Since the FME has quarter-pixel accuracy in H.264/AVC, FME is divided into two search passes, as shown in Fig. 6.7: fractional motion vector search with half-pixel accuracy among eight candidates with half-pixel accuracy includ-

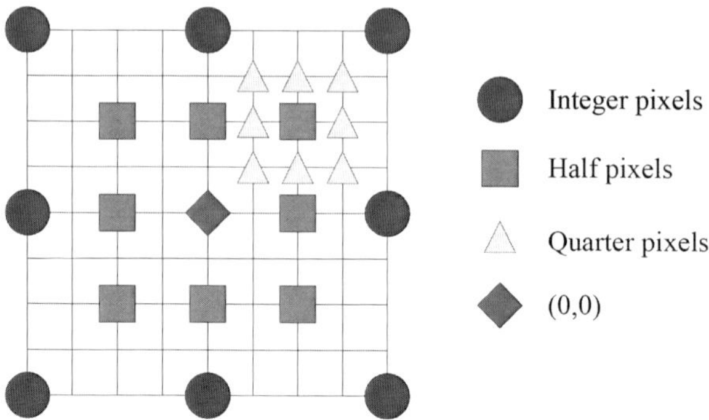

**Fig. 6.7** Full-search FME in H.264/AVC reference software [5]. *Diamond shapes* denote the search center of the first pass, i.e., the best integer motion vector



ing best integer motion vector (first pass), and fractional motion vector search with quarter-pixel accuracy among nine candidates with quarter-pixel including best half-pixel fractional motion vector (second pass). In addition, H.264/AVC adopts seven block modes for each macroblock, namely $16 \times 16$, $16 \times 8$, $8 \times 16$, $8 \times 8$, $8 \times 4$, $4 \times 8$, and $4 \times 4$. Since H.264/AVC has two search passes in each block mode, two search passes in the variable block size modes contribute to the high computational complexity of FME as shown by Fig. 6.3.

To reduce the high computational complexity of full-search FME, single-pass fractional motion estimation (SPFME) algorithms have recently been proposed. SPFME has one search pass which directly searches the best fractional motion vector among various candidates with quarter-pixel accuracy. To maintain low computational complexity, while satisfying the required compression efficiency, SPFME bypasses the search pass with half-pixel accuracy. Among the proposed SPFME methods, we have introduced one that has a low hardware cost and negligible loss of image quality; see [6]. Compared to the full-search FME [5] whose number of candidates for full-search FME is seventeen, the SPFME presented in [6] has only ten quarter-pixel candidates, thereby reducing the computational complexity and hardware cost.

In [6], SPFME has two search centers: $(0, 0)$ and the predicted fractional motion vector (*pred_frac_mv*). The predicted motion vector (*pred_mv*) is defined as the median of three neighboring motion vectors in H.264/AVC. The *pred_frac_mv* [7] is extracted from *pred_mv* and the best integer motion vector (*imv*),

$$pred\_frac\_mv = (pred\_mv - mv) \quad \text{modulo } 4 \tag{6.1}$$

where modulo 4 operation is applied to obtain the fractional component by removing the integer part. The basic idea of obtaining the *pred_frac_mv* according to (6.1) is based on the assumption that most of the best fractional motion vectors (*best_frac_mv*) lie on either *pred_frac_mv* or its four neighbors (top, down, left, and right).

Figure 6.8 shows the probability distribution of the *best_frac_mv* when it is located at neither *pred_frac_mv* nor its four neighbors (top, down, left, right) for three HD 1080p video sequences classified into three groups. The first group includes *sunflower* with low motion. The second group includes *station2* with moderate motion

**Fig. 6.8** The probability distribution of the location of best fractional motion vector when it is located at neither predicted fractional motion vector nor its four neighbors (*top*, *bottom*, *left*, and *right*) for three video sequences: (**a**) *sunflower*, (**b**) *station2*, and (**c**) *tractor*

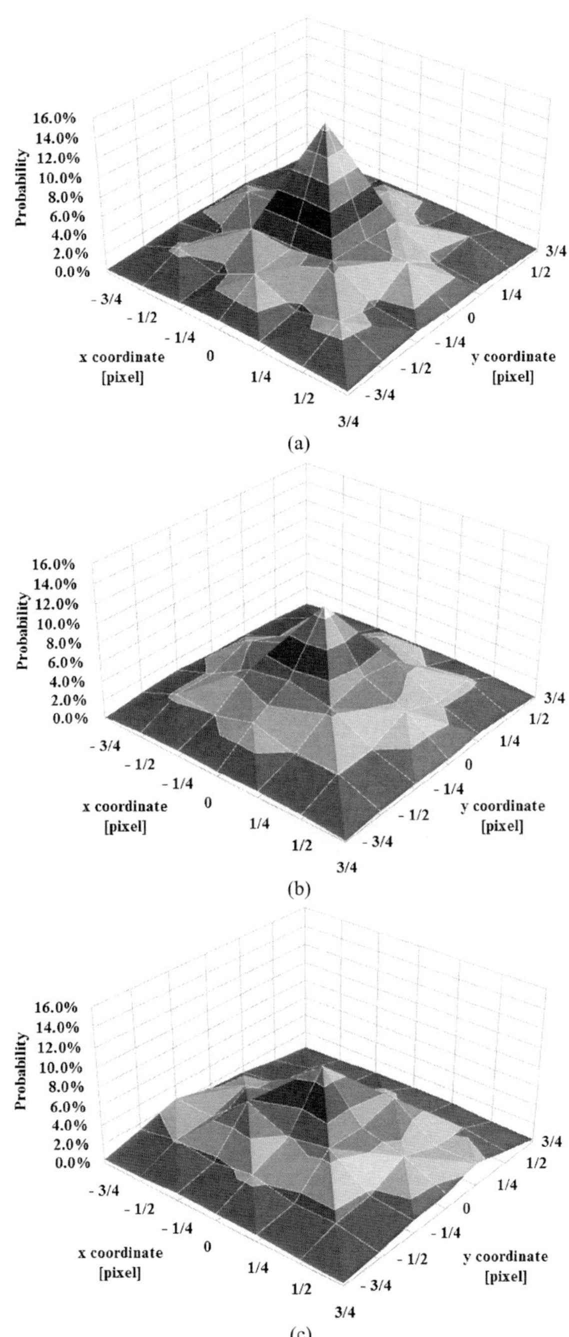

**Fig. 6.9** Two SPFME algorithms with different candidate locations for the fractional motion vector are shown. (**a**) Has candidates as predicted fractional motion vector and its four surroundings (*top, down, left,* and *right*). (**b**) Has fractional predicted motion vector, its four surroundings (*top, down, left,* and *right*), (0, 0), and four diagonal surroundings [6]. © 2010 IEEE

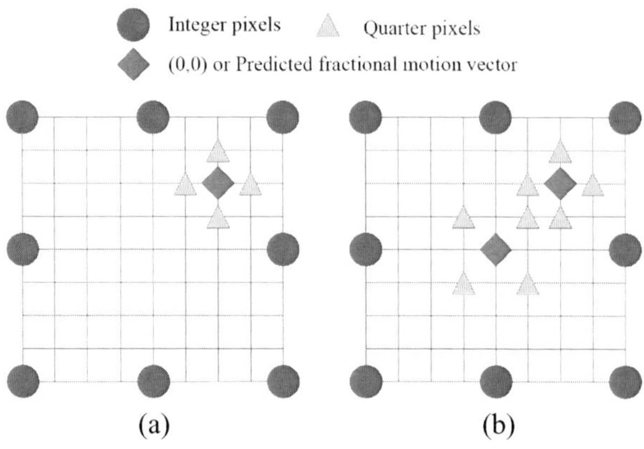


**Table 6.2** Prediction ratio and $\Delta$PSNR in Figs. 6.9(a) and (b) [6]. © 2010 IEEE

| Sequences | Fig. 6.9(a) | | Fig. 6.9(b) | |
|---|---|---|---|---|
| | Prediction ratio (%) | $\Delta$PSNR | Prediction ratio (%) | $\Delta$PSNR |
| Sunflower | 78.0 | −0.153 | 83.4 | −0.019 |
| Tractor | 59.6 | −0.142 | 68.0 | −0.049 |
| Average | 68.8 | −0.148 | 75.7 | −0.034 |

contents. The third group includes *tractor* with high motion contents. In Fig. 6.8, the $x$-axis and $y$-axis are in quarter-pixel resolution. It shows that the distribution of *best_frac_mv*, when it is located at neither pred_frac_mv nor its four neighbors, is concentrated at the search center, (0, 0), and its surroundings. Because contents without motion, like the background, occupy a substantial part of the total image, *best_frac_mv* can be found at either the search center, (0, 0), or its surroundings.

To observe the effect of searching (0, 0) and its neighbors in addition to *pred_frac_mv* with its four neighbors on the image quality for two HD 1080p video sequences: *sunflower* and *tractor*, two SPFME algorithms are compared. Figure 6.9 shows two SPFME algorithms with different candidate locations for the fractional motion vector. In (a), *pred_frac_mv* and its four neighbors (top, down, left, and right) are searched. In (b), (0, 0) and its four neighbors in a diagonal direction are also searched [6].

Table 6.2 shows a comparison of prediction ratio, which denotes the probability of the *best_frac_mv* being found among the candidate locations, and $\Delta$PSNR of Figs. 6.9(a) and (b) in comparison with full-search in Fig. 6.7. The average PSNR degradation of Fig. 6.9(b) compared to the full-search FME [5] is negligible, i.e., 0.034 dB, while the improvement of image quality of Fig. 6.9(b) in comparison with Fig. 6.9(a) is 0.114 dB. Because *best_frac_mv*, when it is located at neither *pred_frac_mv* nor its four neighbors, is concentrated at (0, 0) and its surroundings as shown in Fig. 6.8, including (0, 0) with its four neighboring quarter pixels has proven to be quite effective. In Fig. 6.8(b), four diagonal neighbors of (0, 0) (left-

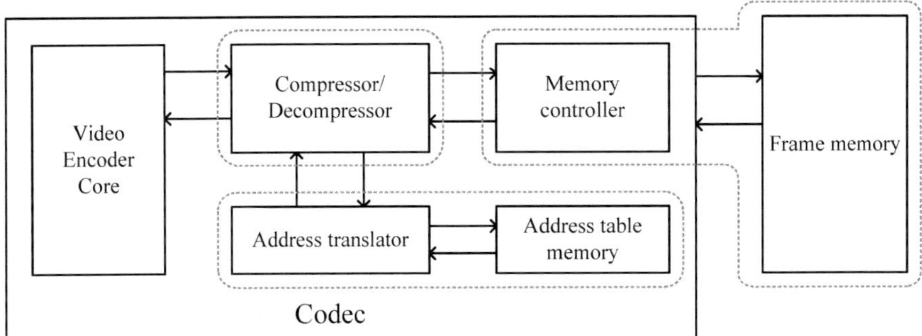


**Fig. 6.10** Block diagram of video codec with EC scheme

up, right-up, left-down, and right-down) as additional search candidates are used to avoid the duplication of search candidates when *pred_frac_mv* is $(0, 0)$.

### 6.4.3 Embedded Compression

Because the reduction of memory bandwidth leads to the reduction of memory access cycle required to encode one macroblock, it can also reduce the system clock frequency and power consumption. Embedded compression (EC) is used to compress the pixel data before they are written to a frame memory so that the bandwidth requirement of data communication to/from the frame memory decreases.

There are several important features of EC [8]. First, the compression ratio (CR) is maximized to minimize the off-chip memory access bandwidth. CR is defined as follows:

$$CR = \frac{\text{Original frame size}}{\text{Compressed frame size}} \tag{6.2}$$

Data compression is performed on each block, called a basic compression unit or accessing unit. Generally, a bigger block results in a better CR, but with a higher probability of reading unnecessary data. Therefore, an optimal block size must be determined at a trade-off point in between. It is also important to minimize the distortion or loss of data. To reduce the power consumption, design cost, and system operating cost, it is critical to minimize the hardware for compression/decompression. Finally, to reduce the effective bandwidth, the compression latency as well as the decompression latency must be maximally reduced.

To achieve successful bandwidth reduction, we should not consider the pixel data compression ratio only, but instead take the whole system into account. In this section, we explain the lossless frame memory recompression (FMR) scheme presented in [9]. The EC system [9] consists of mainly three parts: off-chip frame memory, on-chip address table memory with address translator, and compression/decompression unit. Figure 6.10 shows these system components.

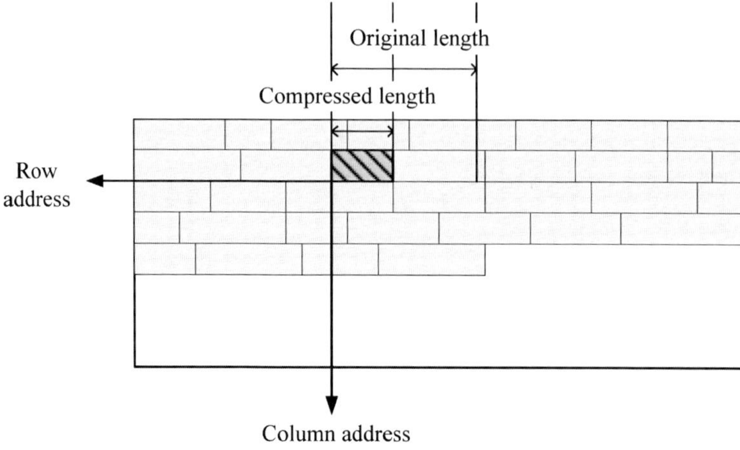


**Fig. 6.11** Address table components: starting address and length

The objective of the compressor/decompressor is to get as good a compression ratio as possible while keeping hardware resource and decompression latency in mind. The compression latency has more of a margin than the decompression latency, because the compressed data is to be used for encoding/decoding the next incoming frame.

We cannot predetermine the bit length of the compressed data by the compressor in lossless way, because it varies case by case. To access an arbitrary data item which is on demand, we need information about the starting address and the length of the data. Otherwise, we cannot determine where to read and how long to read. If we choose the regular placement scheme, then we know the starting address as the coordinates of the requested block. However, we still want to know the length of the compressed data, because we can expect the actual bandwidth reduction only when we can access the memory with its exact compressed length, not the original length. For the irregular placement scheme, both the starting address and the length of the data are required. A kind of address table is essential, and keeping it in the internal SRAM is advantageous over using external memory, in a latency sense. Figure 6.11 shows components of the address table. The address table is too large to be stored in the internal SRAM as is; it has to be compressed as well as the pixel data. This compression algorithm should also be lossless and have a high CR.

For frame memory, a high compression ratio is very important, because it directly affects the number of memory access cycles. However, there is another issue. SDRAM is usually chosen as the off-chip frame memory due to its high density. But because the memory placement scheme which allocates the compressed data in the SDRAM affects the memory access latency, it should be carefully considered. Allocating block data, i.e., macroblocks, in regular form is possible and natural with raw or uncompressed pixel data. The same regular allocation method may be used for compressed data by placing data at a fixed position, and leaving empty space between data. There is another option; the compressed data having arbitrary bit length

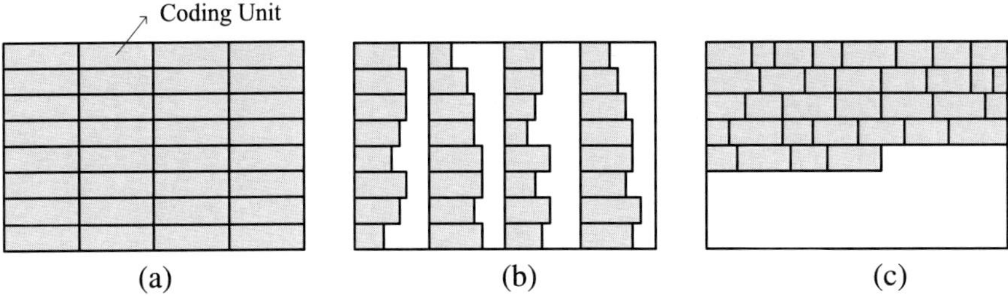



**Fig. 6.12** Memory placement schemes: (**a**) original data placement, (**b**) regular placement for compressed data, and (**c**) irregular placement for compressed data

**Fig. 6.13** The EC system pipeline consists of address translation, memory access, and decompression. $T_a$, $T_m$, and $T_d$ represent the execution time for each operation, respectively

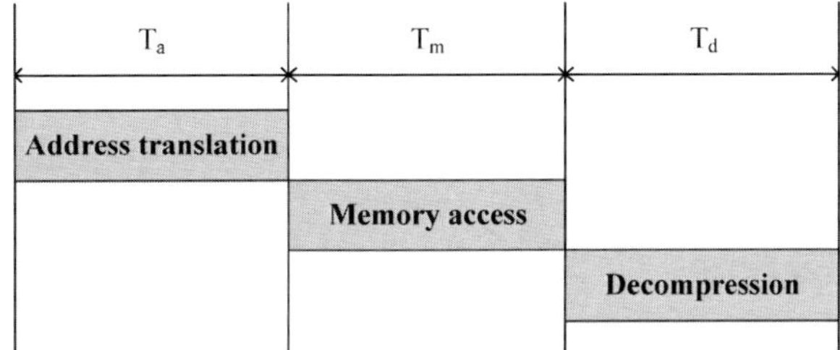



can be allocated tightly and compactly, forming irregular data placement system, as shown in Fig. 6.12. The latter has some advantages. First, we can save some free memory space, even though the amount is not guaranteed. This space may be used by other system components such as the processor core. Otherwise, we can cut down the frame memory size. In the case of a video resulting in poor CR, the lacking memory would be borrowed from another system memory temporarily. Second, power consumption and latency would be reduced as the number of row changes decreases, because a smaller number of rows is enough to store a frame. Third, when consecutive MBs are read with a burst operation, for the regular placement case the burst start address should be changed for every MB, but in the tight placement case there is no need for the SDRAM controller to drive address signals several times.

In an EC system, the three components of address translation, memory access, and decompression operate collectively, not independently. Thus, the optimization effort for each individual component affects the others, and they should be well organized and balanced. For higher throughput, the three functions may operate in a pipelined manner, as shown in Fig. 6.13. When $T_a$, $T_m$, and $T_d$ represent the execution time of each operation, respectively, the throughput is $1/\max\{T_a, T_m, T_d\}$. Thus, the goal in designing the EC system becomes the minimization of the $\max\{T_a, T_m, T_d\}$, not the minimization of individual execution times. The execution time of the three components should be well balanced.

To apply the EC algorithm to a closed-loop video coding system like H.264/AVC or MPEG-4, lossless compression is necessary to avoid a severe image quality drop. Most existing EC algorithms use variable-length codes (VLCs), e.g., Huffman [10] and Golomb coding [11], as an entropy coder. The throughput of existing EC archi-

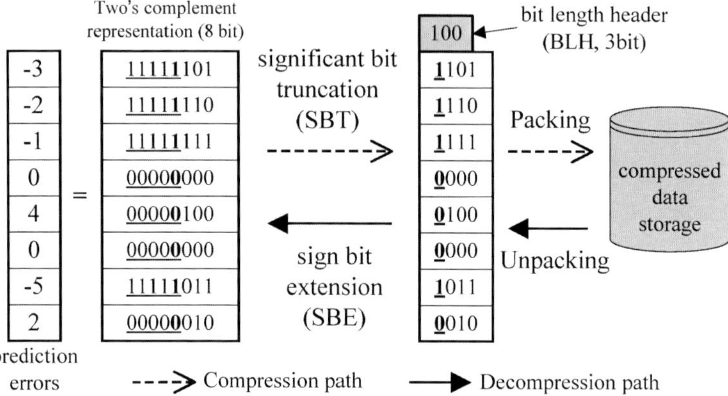



**Fig. 6.14** Basic concept of SBT coding. Each prediction error in a group has the same bit length indicated by bit length header [16]. © 2010 IEEE

tectures using VLC [4–6, 12–14] cannot be easily increased due to the VLC data dependency. In variable length decoding, there is no explicit boundary information for detecting the end or beginning of the codeword in the coded input stream. Therefore, the length of the current codeword must be known before the next codeword is decoded [15]. This data dependency between codewords complicates the architecture of the decoder so that the data dependency becomes an obstacle to implementing a parallel or pipelined architecture.

A parallel EC algorithm and its hardware architecture are proposed in [16] to overcome the data dependency problem and to achieve high CR. The proposed algorithm consists of two steps. The first is a hierarchical prediction method based on pixel averaging and copying. The second step involves significant bit truncation (SBT), which encodes prediction errors in a group with the same number of bits so that the multiple prediction errors are decoded in a clock cycle.

In the algorithm, prediction errors calculated from the hierarchical prediction method are compressed by SBT coding. The basic concept of SBT coding is to express all of the prediction errors in a group with as many significant bits truncated as necessart to preserve the information of each prediction error. In Fig. 6.14, eight prediction errors are expressed in eight bits in a two's complement representation. The underlined bits of each prediction error can be regarded as an extension of its sign bit. Therefore, four significant bits of each prediction error can be truncated without causing any loss of information in this example. Finally, the reduced prediction errors are packed with the bit length header (BLH) representing the bit length of each prediction error, which is equal to four in Fig. 6.14. As a result, 64 bits for eight prediction errors are compressed into 35 ($=4 \times 8 + 3$) bits.

The decompression process is exactly the inverse of the compression process, i.e., sign bit extension. The packed data is parsed according to the value of the BLH. Each prediction error obtained by 4-bit truncation is then recovered to an 8-bit value by extending its MSB. Here, each truncated prediction error has the same number of bits as represented in the BLH; thus, all of the prediction errors in a group can be parsed at once. In comparison with existing VLC algorithms such as Huffman or Golomb coding, the proposed algorithm is relatively simple.

The coding performance of the algorithm is explained in [16]; the SBT coding experiences nearly 9.3% deterioration of performance, i.e., a compression ratio compared to the theoretical limit of the data compression. From the experimental results of [16], the data reduction ratio of the proposed EC algorithm is approximately 60% on average, i.e., data transfers from frame memory are reduced by 60% using the proposed algorithm.

## 6.5 Power Scalability of H.264/AVC Codec

We have explained the low complexity ME (IME and FME) based on the H.264/AVC codec architecture presented in Sect. 6.3 to reduce the computational complexity of ME, since ME occupies most of computational complexity during video encoding. In this section, we explain the power-scalable H.264/AVC video codec. We control the power consumption of the video codec by selecting an appropriate operation mode for it according to a given power constraint.

### 6.5.1 GOP Structure of Video Codec

Before explaining the power scalability of the architecture, the group of pictures (GOP) and its effectiveness to the power scalability will be discussed. In Fig. 6.15, the GOP size is 5, i.e., one I-frame and four P-frames are included in a GOP. A P-frame references the previous frames (I-frame or P-frame) to find the best block matching during ME. An I-frame does not require the reference frame because intra prediction only predicts pixel values within the current frame. Therefore, in this example, there are four reference frames which are generated and referenced in a GOP. As the access to the external DRAM occupies a significant amount of power consumption in the video encoding, the number of required reference frames in a GOP affects the total power consumption of the video encoding. For example, if the GOP size is 5, then only four frames are required as the reference frame data. On the other hand, if the GOP size is 30, then 29 frames in a GOP are required as the reference frame data from the external DRAM. Hence, the power consumption of the video coding increases as the size of the GOP increases.

### 6.5.2 Power Scalability in I-Frame Coding

Figure 6.16 shows the invoked modules during I-frame coding. In I-frame coding, the Intra module is invoked to predict pixel values within the current frame. Since the I-frame is the first frame in a GOP, the reconstructed frame of the I-frame must

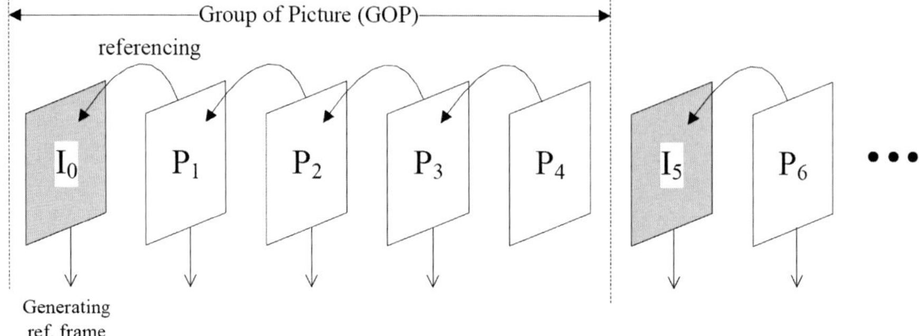



**Fig. 6.15** The group of pictures (GOP) consists of one I-frame and multiple P-frames. In this figure, the GOP size is 5, i.e., one I-frame and four P-frames are included in a GOP. The P-frame references the previous frame (I-frame or P-frame) to find the best block matching during ME. The I-frame does not require the reference frame because intra prediction only predicts pixel values within the current frame. Therefore, in this example, there are four reference frames which are generated and referenced in a GOP

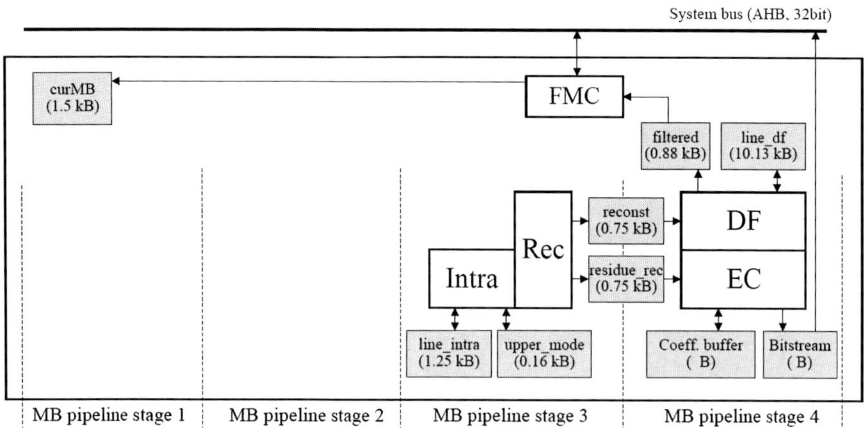



**Fig. 6.16** In I-frame coding only intra prediction is used to code the frame. To deliver the current frame and transfer the reconstructed frame, the FMC and DF modules are invoked during I-frame coding

be kept for the next P-frame coding. Therefore, the DF and FMC modules are also invoked during I-frame coding.

The power consumption of the video encoding depends on the number of accesses to the external DRAM. As discussed earlier, if the size of the GOP is small, then the number of accesses to the external DRAM is also small and vice versa. Hence, we can control the power consumption of the video encoder with the size of the GOP. The GOP size can be represented with the I-frame period, as shown in Table 6.3. In Table 6.3, we have defined eight modes according to the I-frame period, i.e., GOP size. If the "I-frame mode" is 8, then the power consumption is

**Table 6.3** Intra modes for power-scalability

| I-frame mode | I-frame period | Number of I-frame within 30 frames |
| --- | --- | --- |
| 8 | 30 | 1 |
| 7 | 15 | 2 |
| 6 | 10 | 3 |
| 5 | 8 | 3.75 |
| 4 | 6 | 5 |
| 3 | 5 | 6 |
| 2 | 4 | 7.5 |
| 1 | 3 | 10 |

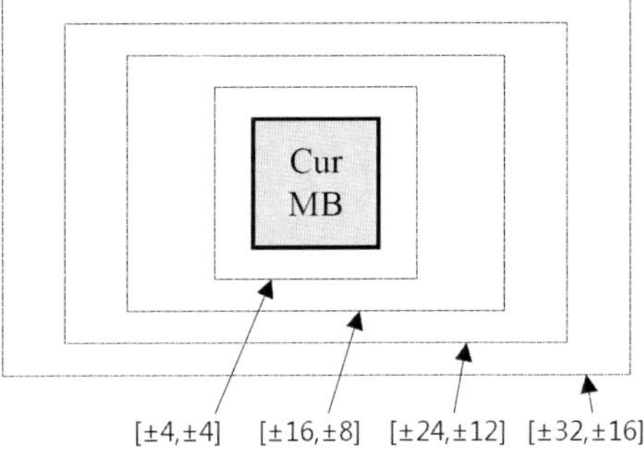


**Fig. 6.17** The search range of IME. In the IME module, IME performs with four kinds of search ranges; [±32, ±16], [±24, ±12], [±16, ±8], and [±4, ±4]. As the search range becomes small, the power consumption of the IME module decreases because of reduced logic switching power caused by reduced search area and reduced external DRAM access power

maximized due to the increased number of accesses to the external DRAM, whereas the power consumption is minimized when the "I-frame mode" is 1.

### 6.5.3 Power Scalability in P-frame Coding

P-frame coding consists of two prediction methods: ME using a block matching algorithm and intra prediction using neighboring pixels. The block matching algorithm usually consists of IME and FME. In a typical video encoding, the block matching algorithm occupies most of the computational complexity during the video encoding, as shown in Fig. 6.3. Therefore, the power scalability of the block matching algorithm is the key feature in implementing a power-scalable video codec.

In IME, the computational complexity depends on a given search range. Figure 6.17 shows the multiple search ranges for IME: [±32, ±16], [±24, ±12], [±16, ±8], and [±4, ±4]. In the encoder architecture presented in Fig. 6.2, the IME

**Table 6.4** IME modes for power scalability

| IME mode | Search range |
| --- | --- |
| 4 | $[\pm32, \pm16]$ |
| 3 | $[\pm24, \pm12]$ |
| 2 | $[\pm16, \pm8]$ |
| 1 | $[\pm4, \pm4]$ |

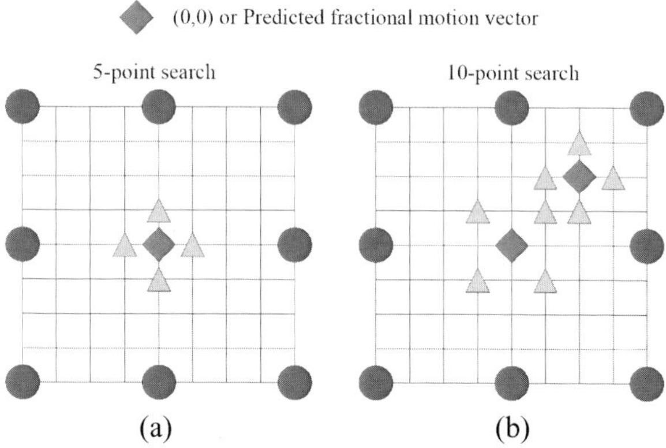


**Fig. 6.18** The number of search points of FME. Multiple fractional search points are simultaneously compared in the FME module. The power consumption of the FME module depends on the number of search points because the number of invoked processing units depends on the number of search points during the decision of the best fractional motion vector of one block mode

module uses four search ranges for the scalability. As the search range becomes small, the power consumption of the encoder decreases because of the reduced logic switching power of the IME module as well as reduced external DRAM access. Hence, we can define the IME modes to control IME power consumption according to the amount of search range, as shown in Table 6.4.

In FME, the computational complexity depends on the fractional search points and the number of block modes to be performed. Figure 6.18 shows that the multiple fractional search points are compared in parallel in the FME module. The power consumption of the FME module depends on the number of search points because the number of invoked processing units depends on the number of search points. The power consumption of the FME module also depends on the number of block modes to be performed. In the FME module of the encoder architecture, up to four candidate block modes, i.e., $16 \times 16$, the better one between $16 \times 8$ and $8 \times 16$, and $8 \times 8$ are compared to decide the best block mode. If we restrict the candidate block modes, then we can reduce the power consumption of the FME module. From these two factors, the FME mode for the power scalability can be defined as shown in Table 6.5. The table shows seven modes; the higher six modes are the combinations of the number of candidate block modes (=3) and the number of search points (=2). The "mode 1" indicates that no FME is performed in the FME module to reduce power consumption as much as possible. In this mode, the FME module performs only motion compensation to generate residue pixel values for the next MB pipeline stage.

**Table 6.5**  FME modes for power-scalability

| FME mode | Candidate block modes | Fractional search points |
| --- | --- | --- |
| 7 | $16 \times 16$, $16 \times 8$ or $8 \times 16$, $8 \times 8^a$ | 10 |
| 6 | $16 \times 16$, $16 \times 8$ or $8 \times 16$, $8 \times 8^a$ | 5 |
| 5 | $16 \times 16$, $8 \times 8^a$ | 10 |
| 4 | $16 \times 16$, $8 \times 8^a$ | 5 |
| 3 | $16 \times 16$ | 10 |
| 2 | $16 \times 16$ | 5 |
| 1 | $16 \times 16$ | 0 (no FME) |

$^a$$8 \times 8$ block can consist of $8 \times 8$, $8 \times 4$, $4 \times 8$, and $4 \times 4$

With the power-scalable ME architecture, the power consumption of P-frame coding in video codec can be controlled. Since P-frame coding uses both inter and intra prediction, all functional blocks of the encoder are invoked during P-frame coding. However, the IME and FME modules control their operation according to a given power constraint.

## 6.6  Power-Rate-Distortion Modeling of H.264/AVC Codec

In this section, we explain the rate–distortion and power-distortion relationships of a video codec. From these two relations, we extract a model of the power–rate–distortion relation of a hardware-based video codec. The power–rate–distortion $(P-R-D)$ model can be used to reduce the power consumption of the video codec while satisfying the required distortion.

### 6.6.1 Rate-Distortion Model

During the past decades, several distribution models of transformed coefficients were proposed to find the relationship between the rate and distortion in image and video coding systems based on a two-dimensional discrete cosine transform (DCT). Among various models of transformed coefficients, the Laplacian distribution has been widely used.

In [17], two formulas are used for the rate and distortion as functions of the quantization parameter $Q$. The rate term

$$R(Q) = \begin{cases} \frac{1}{2}\log_2(2e^2\frac{\sigma^2}{Q^2}), & \frac{\sigma^2}{Q^2} > \frac{1}{2e} \\ \frac{e}{\ln 2}\frac{\sigma^2}{Q^2}, & \frac{\sigma^2}{Q^2} < \frac{1}{2e} \end{cases} \tag{6.3}$$

was derived from the entropy of a Laplacian distribution random variable with variance $\sigma^2$. The distortion model is simply

$$D = \frac{1}{N} \sum_{i=1}^{N} \frac{Q_i^2}{12} \tag{6.4}$$

which is derived from the quantization error of a uniform random variable at a quantization level of $Q_i$ for the $i$th MB, where $N$ is the number of MBs.

He [18, 19] proposed a rate–distortion ($R$–$D$) model based on the fraction of zeros among the quantized DCT coefficients, denoted as $\rho$. The authors assumed that the DCT coefficients have a Laplacian distribution. Based on the observation that the bit rate and $\rho$ have a linear relationship, they proposed the rate model as

$$R = \theta(1 - \rho) \tag{6.5}$$

and the distortion model as

$$D = \sigma^2 e^{-\alpha(1-\rho)} \tag{6.6}$$

where $\theta$ and $\alpha$ are model parameters.

However, in [14], the authors claimed that the actual DCT coefficient histogram has heavy tails in several video sequences, so the Laplacian distribution model is not suitable for those cases. To overcome this problem, Kamaci et al. proposed an $R$–$D$ model [20] based on a zero-mean Cauchy distribution with parameter $\mu$, having the pdf

$$p(x) = \frac{1}{\pi} \frac{\mu}{\mu^2 + x^2} \tag{6.7}$$

where $x$ is the DCT coefficient value, and the parameter $\mu$ depends on the picture content. The authors proposed an $R$–$D$ model based on the Cauchy density distribution as follows:

$$D = cR^{-\gamma} \tag{6.8}$$

where $D$ is the distortion of the encoded video sequences in terms of mean squared error (MSE), $R$ is the bit rate in terms of bits per pixel, and $c$ and $\gamma$ are model parameters depending on the video sequence. As shown in Fig. 6.19, the Cauchy-based $R$–$D$ model shows a better fit to the DCT coefficients than the Laplacian-based model in most cases. Hence, in this book, the Cauchy-based model is adopted as the $R$–$D$ model for the video codec.

### 6.6.2 Power Modeling of Video Codec

The power consumption of the hardware-based video codec depends not only on the video configuration or input video sequence but also on the technology used to implement the video encoder. The logic power consumption can be reduced by using advanced CMOS technology. Moreover, we need to find the power model of the

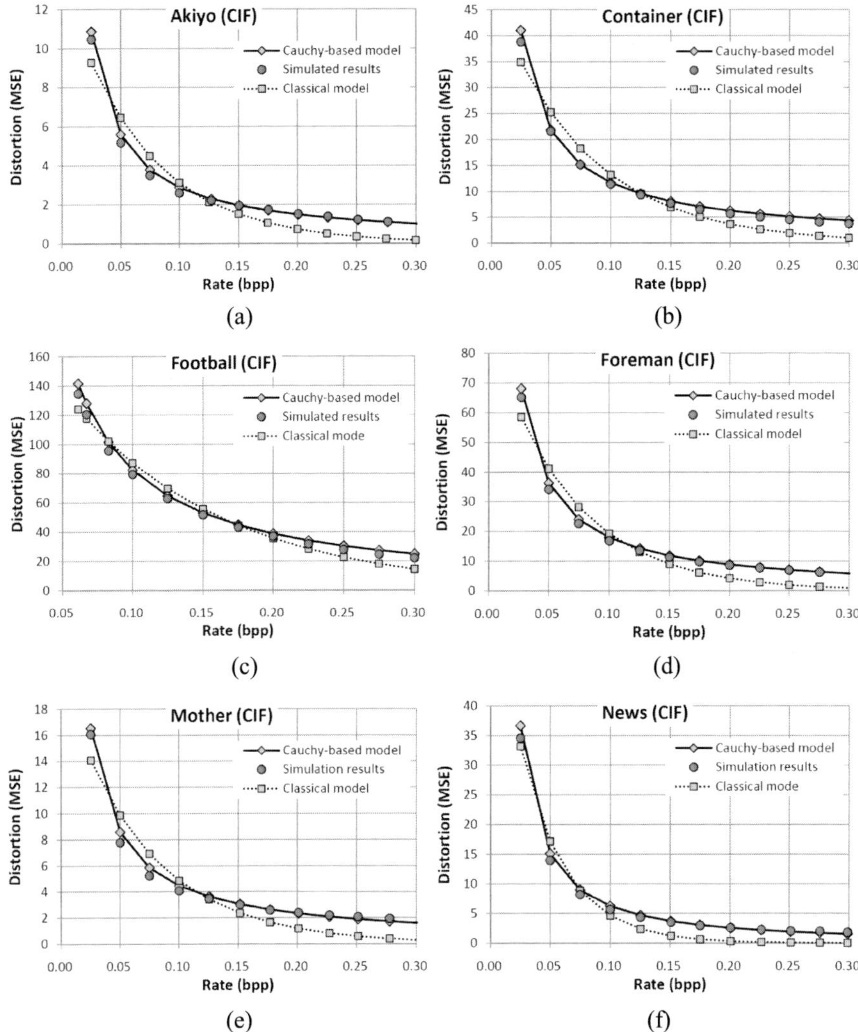

**Fig. 6.19** Rate-distortion relation of various video sequences: (**a**) *akiyo*, (**b**) *container*, (**c**) *football*, (**d**) *foreman*, (**e**) *mother*, and (**f**) *news*. The Cauchy-based model shows a better fitting than that of the classical exponential $R$–$D$ model

video encoder to optimize the entire system before the encoder is implemented as a chip. Therefore, a method for the power modeling based on gate-level simulation is required.

The gate-level simulation can report almost completely accurate power consumption, as it is implemented in a chip. Figure 6.20(a) shows the block diagram of the power modeling method. To develop a power model of the proposed video codec, all functional blocks in the proposed architecture are described with the Verilog hardware description language (HDL). The synthesizable register transfer level (RTL)

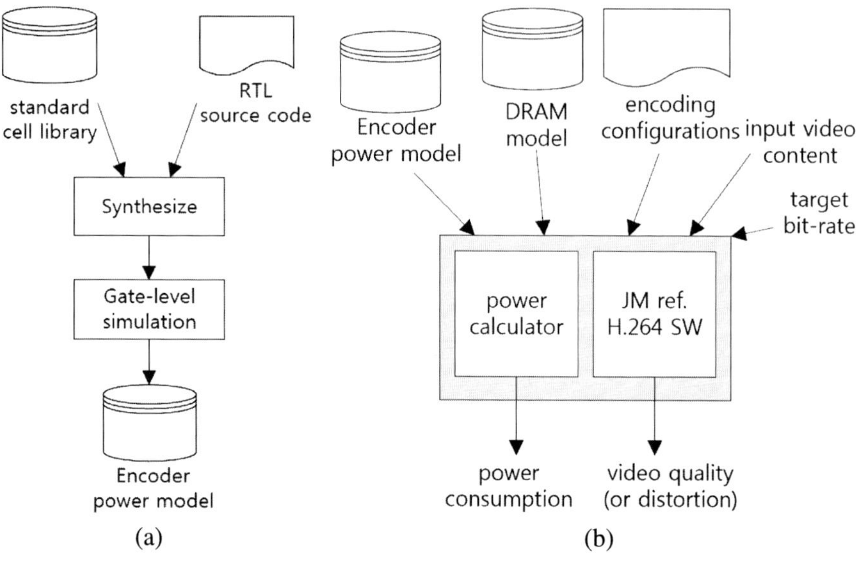



**Fig. 6.20** (**a**) The process of the power modeling of the video encoder through gate-level simulation. (**b**) The process of power–distribution ($P-D$) modeling: the inputs of the modeling process are the encoder power model, DRAM, encoding configuration, input video content, and target bit rate; the power consumption under given encoding configurations is calculated in the power calculator while its output distortion or output video quality is reported by reference video coding software, e.g., the JM reference software of H.264 [5]

source codes are synthesized with the standard cell library, which is provided from the chip manufacturing company. After the synthesizing process, the gate-level simulation is performed. In this process, the power consumption is classified according to the functional blocks, i.e., IME, FME, etc., to develop the encoder power model. The input video sequence does not significantly affect the power consumption during the gate-level simulation because the input video sequence can be treated as a random input to each functional block. Hence, the encoder power model can be represented as a function of the encoding configurations, e.g., the size of the search range, the number of candidate block modes, and the I-frame period.

### 6.6.3 Power-Distortion Model

The power–distortion ($P-D$) model of a video codec is developed based on the encoder power model. The process of the $P-D$ modeling is shown in Fig. 6.20(b). The inputs of the modeling process are the encoder power model, DRAM model, encoding configurations, input video content, and target bit rate. The encoder power consumption can be calculated by using the encoder power model. However, the entire power consumption during video encoding depends not only on the encoder power itself but also on the DRAM power for handling the reference frames. The DRAM

**Table 6.6** The experimental environment for Fig. 6.21

| Parameter | Description |
| --- | --- |
| Profile | Baseline |
| IME method | Two-level hierarchical IME |
| FME method | Single-pass FME |
| Number of reference frame | 1 |
| Intra method | Full-search, $4 \times 4$ and $16 \times 16$ |
| Entropy coder | CAVLC |
| Rate-distortion optimization | Low complexity (JM14.0) |

power consumption depends on the number of accesses to the DRAM, which varies with the encoding configurations, i.e., the size of the search range and the I-frame period. Therefore, to develop a power model of the video encoding, one must consider the power consumption of both the video encoder and the DRAM.

In Fig. 6.20(b), the output power consumption is calculated by the "power calculator" while the corresponding video quality or the distortion of encoded video sequence is reported by running the reference H.264 software [5]. In the power-scalable video codec architecture explained in the previous section, there are 224 ($=4 \times 7 \times 8$) kinds of power states, which are combinations of the IME modes ($=4$), FME modes ($=7$), and I-frame mode ($=8$).

Figure 6.21 shows the experimental results of output distortion according to the power consumption. The experimental environment is shown in Table 6.6. In these experiments, the power consumption is controlled by three parameters: IME mode (search range), FME mode (search points and block modes), and I-frame mode (I-frame period). All results are obtained at a fixed bit rate, i.e., 0.2 bits per pixel (bpp) to analyze the relationship between the power consumption and the distortion. We represented the bit rate and the power consumption with bpp and the normalized power consumption, respectively, to derive the $P$–$D$ relationship. The normalized power consumption $P$ is calculated as follows:

$$P = \frac{P_{\text{enc}}}{P_{\text{max}}} \tag{6.9}$$

where $P_{\text{enc}}$ and $P_{\text{max}}$ represent the current encoding power consumption and the maximum power consumption of the video encoding, respectively.

In Fig. 6.21, the black diamonds represent the experimental results obtained from different configurations of power control parameters, while the gray circles represent the convex hulls of the experimental results. The points on the convex hull are the *distortion–minimal encoding configuration* for a given bit rate. Hence, the $P$–$D$ model can be derived from the *distortion–minimal encoding configuration*.

Figure 6.22 shows the approximated model for the *distortion–minimal encoding configurations* and their experimental results. As shown in the figure, the distortion of each *distortion–minimal encoding configuration* shows exponential characteris-

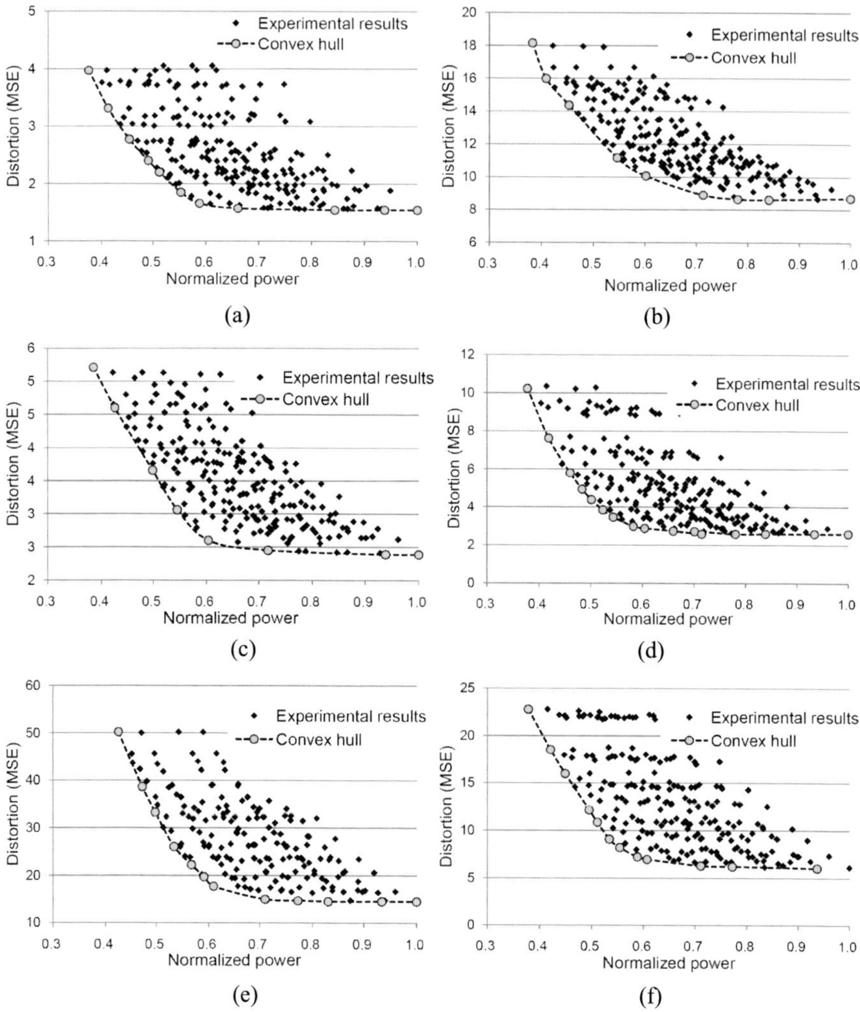

**Fig. 6.21** Power-distortion ($P-D$) relation of various video sequences at a fixed bit rate, 0.2 bpp: (**a**) *akiyo*, (**b**) *foreman*, (**c**) *mother*, (**d**) *news*, (**e**) *paris*, and (**f**) *salient*. *The black diamonds* and *the dashed line* represent the experimental results and their convex hull, respectively

tics according to its power consumption. Hence, an approximated model for the $P-D$ relationship is

$$D = d_0\left(d_1 e^{-d_2 P} + 1\right) \tag{6.10}$$

where $d_0$, $d_1$ and $d_2$ are fitting parameters and $P$ is the normalized power consumption. In Fig. 6.22, the $P-D$ model follows the actual experimental results well.

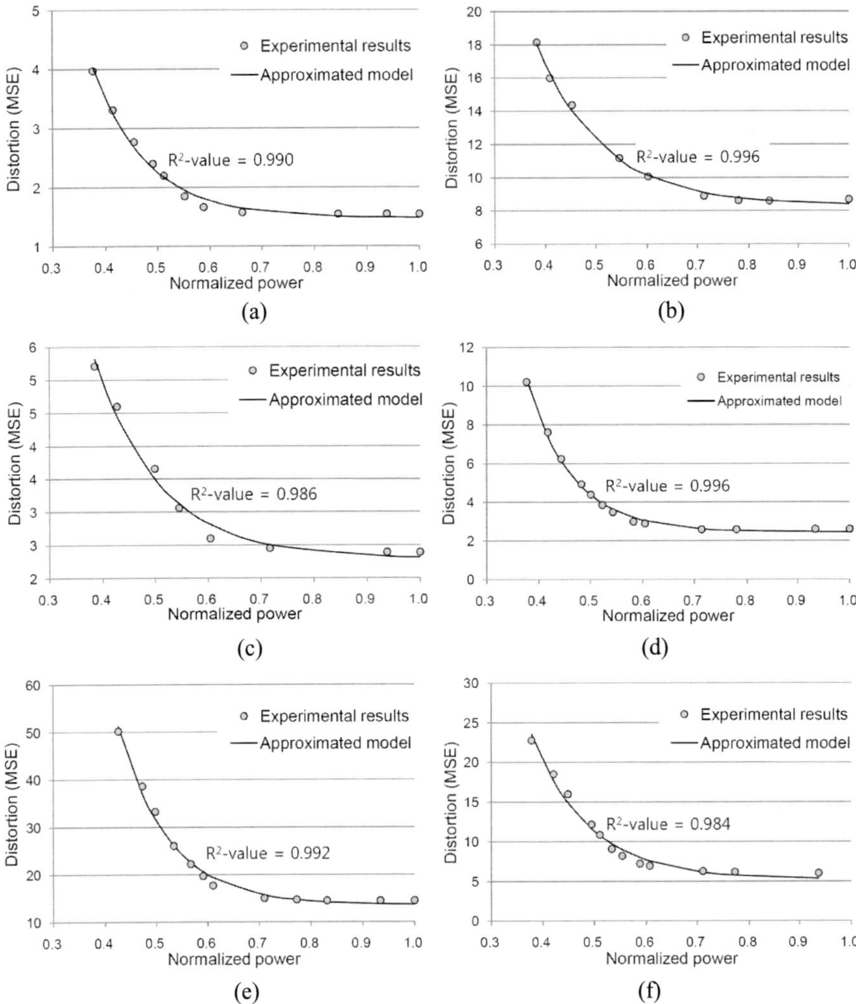

**Fig. 6.22** Approximated power-distortion ($P$–$D$) model. *Gray circles* represent experimental results, the points of the convex hulls in Fig. 6.21, and *the solid line* represents the proposed approximated model. All experimental results are obtained under a fixed bit rate, 0.2 bpp

### 6.6.4 Encoder Configuration for P–D Relationship

Although we have derived the $P$–$D$ model of the video codec, there still exists a question; How can we achieve the given power consumption with the proper video encoding configuration which minimizes the distortion? In this section, we discuss the proper encoding configuration for the $P$–$D$ relationship.

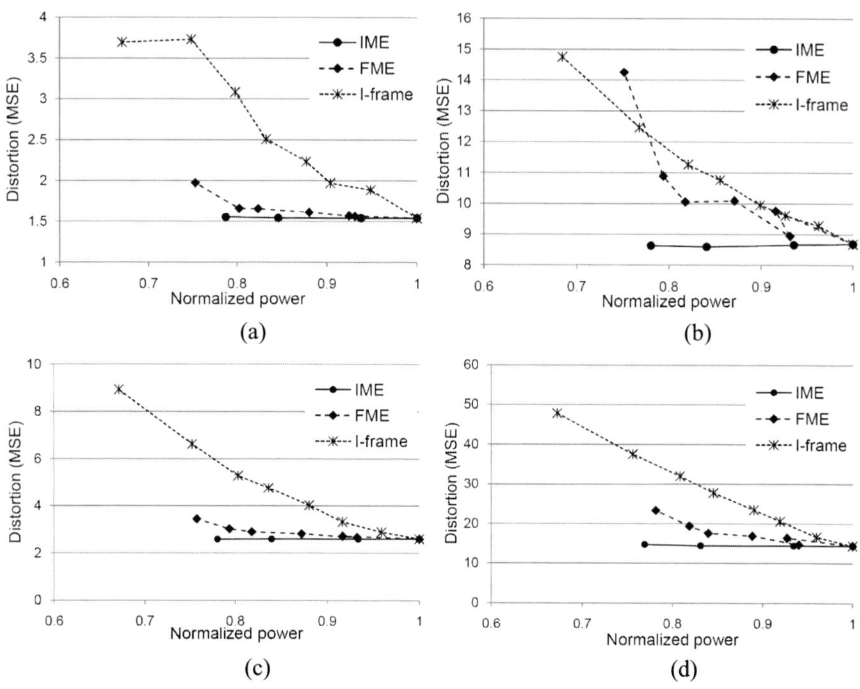

**Fig. 6.23** Distortion effectiveness of the power control parameters. All video sequences are CIF size and encoded at a 0.2 bpp bit rate: (**a**) *akiyo*, (**b**) *foreman*, (**c**) *news*, and (**d**) *paris*. The IME mode does not significantly affect the distortion during reduction of the power consumption, whereas the I-frame mode is sensitive to reduction of power consumption

The power consumption can be controlled by three control parameters: IME, FME, and I-frame period. Since we have 4, 7, and 8 modes of IME, FME, and I-frame period, respectively, there are 224 combinations of control parameters.

The optimal method of finding the proper encoding configuration is to use the convex hull points. However, the video encoding must perform at least 224 times to use convex hull as an encoding configuration, as represented in Fig. 6.21. This is not practical in a portable video coding system. R. Vanam et al. proposed a method to find distortion-complexity optimized encoding configurations using the GBFOS (generalized Breiman, Friedman, Olshen, and Stone) algorithm [14]. The number of required iterative encoding processes $N$ is given as follows:

$$N = 1 - M + \sum_{i=1}^{M} n_i \qquad (6.11)$$

where $M$ and $n_i$ represent the number of encoder parameters and the number of options corresponding to the parameters. In our architecture, $M$ is 3 and $\{n_1, n_2, n_3\} = \{4, 7, 8\}$ respectively. Therefore, the required encoding process $N$ is 17. Although 17 iterative encoding processes is smaller than that required in the convex hull based method ($=224$), the iterative encoding processing is still too high to ap-

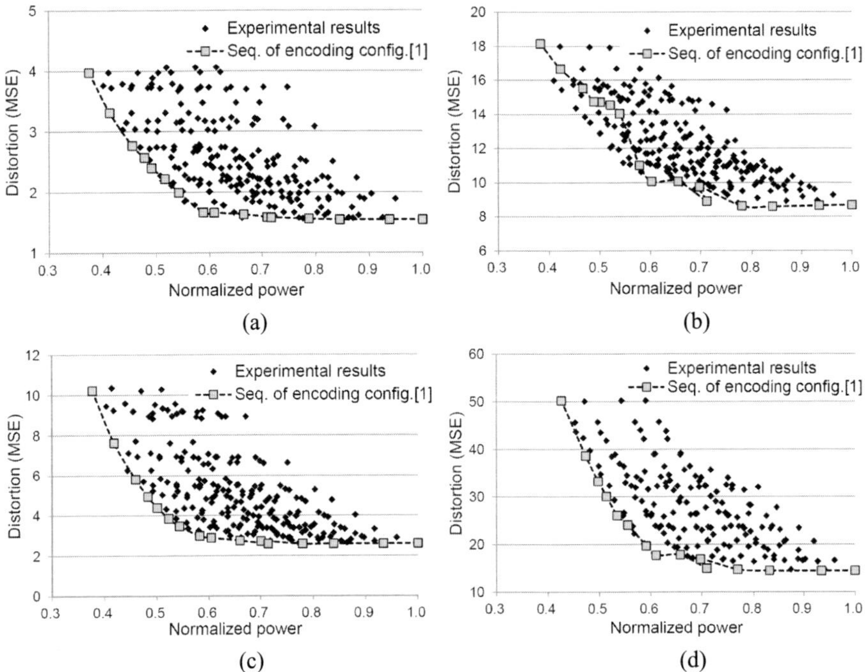

**Fig. 6.24** The experimental results of the sequences of encoding configurations: (**a**) *akiyo*, (**c**) *news*, and (**d**) *paris* video sequences show that the configuration sequence follows the convex hull of the *P–D* relationship, while (**b**) *foreman* sequence shows distorted results in some power consumption range

ply in a portable video coding system or surveillance camera system; there will be a waste of power consumption and a long delay to find the proper configurations.

To overcome this problem, one uses a sequence of encoding configurations to satisfy the given power consumption while the distortion is minimized. Figure 6.23 shows the distortion effectiveness of the encoding configurations. As shown in Fig. 6.23, the IME mode shows little effect on the distortion during the reducing the power consumption, whereas the distortion variation of the I-frame mode is sensitive to reducing the power consumption. The effectiveness can be represented with the slope of each graph according to the encoding control parameters.

From Fig. 6.23, we can find that the degrees of the slopes are different among the control parameters; the degree of the slope is increased in order of IME, FME, and I-frame mode. Using these features, we can make a sequence of encoding configuration $P$ as follows: (a) first, initialize $P$ as $P = (4, 7, 8)$, i.e., set with the highest modes; (b) second, reduce the IME mode until $P = (1, 7, 8)$; (c) third, reduce the FME mode until $P = (1, 1, 8)$; (d) fourth, reduce the I-frame mode until $P = (1, 1, 1)$. In Figs. 6.24(a), (c), and (d), the *akiyo, news,* and *paris* video se-

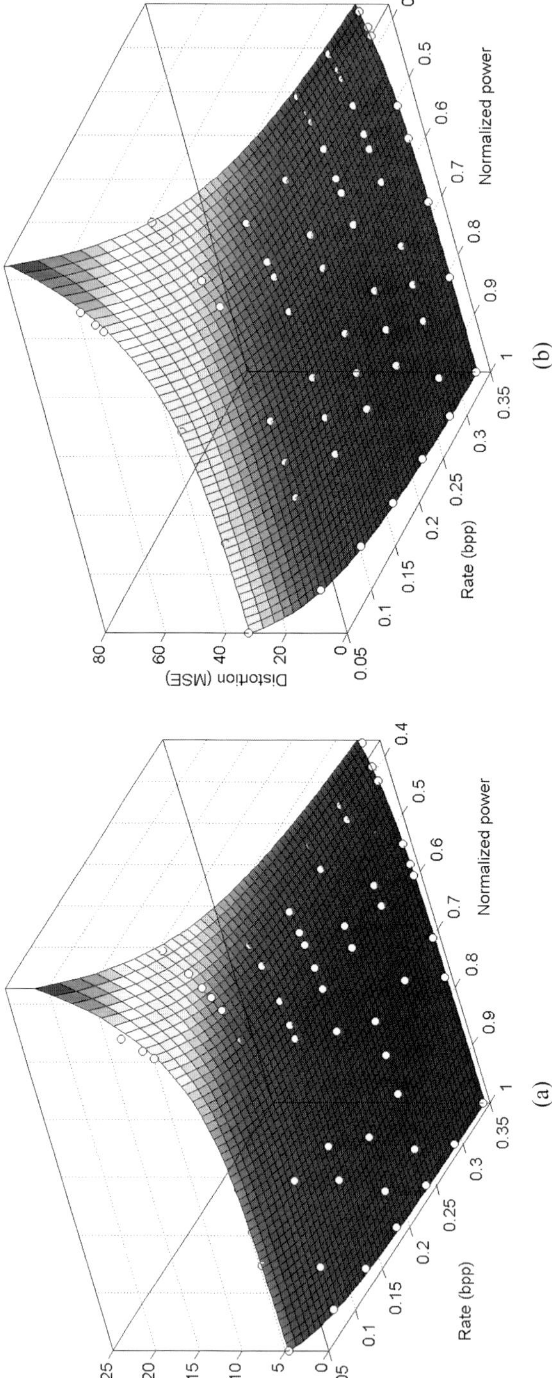

**Fig. 6.25** The power–rate–distortion ($P$–$R$–$D$) model for (**a**) *akiyo* and (**b**) *foreman* video sequences. *The white circles and the surface of the graph* represent the experimental results and the $P$–$R$–$D$ model, respectively

quences show that the sequence of the encoding configurations follows the convex hull of the $P-D$ relationship. However, Fig. 6.24(b) shows that there is a distorted region in some power consumption range. The reason for the distorted region is that, in the *foreman* sequence, the characteristics of the slopes according to the parameters, shown in Fig. 6.23(b), are different from those of the other video sequences.

The sequence of encoding configuration is not an optimal method, and it is not effective on video sequences which have various moving objects such as *foreman*. However, it is quite effective on video sequences which have fixed background images or small moving objects. Moreover, these characteristics often occur in a surveillance camera system.

### 6.6.5 Power–Rate–Distortion Model

In Sects. 6.6.1 and 6.6.3, we showed that the rate–distortion model and power-distortion model as expressed as follows:

$$D = cR^{-\gamma}$$
$$D = d_0\left(d_1 e^{-d_2 P} + 1\right)$$

(6.12)

where $c$, $d_1$, $d_2$ and $d_3$ are fitting parameters. $D$, $R$ and $P$ represent the distortion in terms of MSE, bit rate in terms of bpp, and normalized power consumption of the video encoder. To derive the $P-R-D$ model, these equations are unified as follows:

$$D = c_0\left(c_1 e^{-c_2 P} + 1\right)R^{-\gamma}$$

(6.13)

where $c_0$, $c_1$, and $c_3$ are fitting parameters. Figure 6.25 shows the experimental results and the $P-R-D$ model for the *akiyo* and *foreman* sequences. The white circles and the surface of the graph represent the experimental results and the $P-R-D$ model, respectively. As shown in the figure, the surface follows the white circles well, thus the $P-R-D$ model represents the relationship among power, rate, and distortion of the video codec. The $R$-square values of the model, a measure of how well the experimental results are likely to be predicted by the model, are 0.973 and 0.997 for *akiyo* and *foreman*, respectively.

The fitting parameters $c_0$, $c_1$, and $c_3$ can be obtained using a least-square fitting method. Note that the battery often has an operational lifetime of several weeks, or even several months. Therefore, there is no need to find fitting parameters too often because the power supply condition or input video characteristics do not change quickly, especially in a surveillance camera system or monitoring system. Therefore, the overhead for finding the fitting parameters is relatively small.

## References

1. Draft ITU-T recommendation and final draft international standard of Joint Video Specification (ITU-T Rec. H264-ISO/IEC 14496-10:2005 AVC). JVT G050 (2005)

2. Joch, A., Kossentini, F., Schwars, H., Wiegand, T., Sullivan, G.: Performance comparison of video coding standards using Lagrangian coder control. In: Proc. IEEE Intl. Conf. Image Processing, ICIP, part II, Sept. 2002, pp. 501–504 (2002)
3. Wiegand, T., Schwars, H., Joch, A., Kossentini, F.: Rate-constrained coder control and comparison of video coding standards. IEEE Trans. Circuits Syst. Video Technol. **13**(1), 688–702 (2003)
4. Chen, T.-C., Huang, Y.-W., Chen, L.-G.: Analysis and design of macroblock pipelining for H.264/AVC VLSI architecture. In: Proc. IEEE Int. Symp. Circuits Syst., ISCAS, May 2004, pp. 273–276 (2004)
5. Joint Video Team (JVT) reference software version 14.0. http://iphome.hhi.de/suehring/tml/download/old_jm/
6. Kim, G., Kim, J., Kyung, C.-M.: A low cost single-pass fractional motion estimation architecture using bit clipping for H.264 video codec. In: Proc. IEEE Intl. Conf. Multimedia and Expo, ICME, July 2010, pp. 661–666 (2010)
7. Yang, L., Yu, K., Li, J., Li, S.: Prediction-based directional fractional pixel motion estimation for H.264 video coding. In: Proc. IEEE Intl. Conf. Acoustic, Speech and Signal Processing, ICASSP, March 2005, pp. 901–904 (2005)
8. De With, P.H.N., Frencken, P.H., Scharr-Mitrea, M.: An MPEG decoder with embedded compression for memory reduction. IEEE Trans. Consum. Electron. **44**(3), 545–555 (1998)
9. Lee, S.-H., Chung, M.-K., Park, S.-M., Kyung, C.-M.: Lossless frame memory recompression for video codec preserving random accessibility of coding unit. IEEE Trans. Consum. Electron. **55**(4), 2105–2113 (2009)
10. Huffman, D.A.: A method for the construction of minimum-redundancy codes. In: Proc. Inst. Radio Engineers, vol. 40, Sep. 1952, pp. 1098–1101 (1952)
11. Golomb, S.W.: Run-length encodings. IEEE Trans. Inf. Theory **12**(3), 399–401 (1966)
12. Chang, H.-C., Chen, J.-W., Su, C.-L., Yang, Y.-C., Li, Y., Chang, C.-H., Chen, Z.-M., Yang, W.-S., Lin, C.-C., Chen, C.-W., Wang, J.-S., Quo, J.-I.: A 7 mW-to-183 mW dynamic quality-scalable H.264 video encoder chip. In: Proc. IEEE Intl. Solid-State Circuits Conf., ISSCC, Feb. 2007, pp. 280–603 (2007)
13. Chen, Y.-H., Chen, T.-C., Tsai, C.-Y., Tsai, S.-F., Chen, L.-G.: Algorithm and architecture design of power-oriented H.264/AVC baseline profile encoder for portable devices. IEEE Trans. Circuits Syst. Video Technol. **19**(8), 1118–1128 (2009)
14. Chang, H.-C., Chen, J.-W., Wu, B.-T., Su, C.-L., Wang, J.-S., Guo, J.-I.: A dynamic quality-adjustable H.264 video encoder for power-aware video applications. IEEE Trans. Circuits Syst. Video Technol. **19**(12), 1739–1754 (2009)
15. Nikara, J., Vassiliadis, S., Takala, J., Liuha, P.: Multiple-symbol parallel decoding for variable length codes. IEEE Trans. Very Large Scale Integr. Syst. **12**(7), 676–685 (2004)
16. Kim, J., Kyung, C.-M.: A lossless embedded compression using significant bit truncation for HD video coding. IEEE Trans. Circuits Syst. Video Technol. **20**(6), 848–860 (2010)
17. Lin, Y.-K., Li, D.-W., Lin, C.-C., Kuo, T.-Y., Wu, S.-J., Tai, W.-C., Chang, W.-C., Chang, T.-S.: A 242 mW 10 mm$^2$ 1080p H.264/AVC high-profile encoder chip. In: Proc. IEEE Intl. Solid-State Circuits Conf., ISSCC, Feb. 2008, pp. 314–615 (2008)
18. He, Z., Mitra, S.K.: A unified rate-distortion analysis framework for transform coding. IEEE Trans. Circuits Syst. Video Technol. **11**(12), 1221–1236 (2001)
19. He, Z., Kim, Y.K., Mitra, S.K.: Low delay rate control for DCT video coding via $\rho$-domain source modeling. IEEE Trans. Circuits Syst. Video Technol. **11**(8), 928–940 (2001)
20. Kamaci, N., Altunbasak, Y., Mersereau, R.M.: Frame bit allocation for the H.264/AVC video coder via Cauchy-density-based rate and distortion models. IEEE Trans. Circuits Syst. Video Technol. **15**(8), 994–1006 (2005)

# Chapter 7
# Energy Generation and Conversion for Portable Electronic Systems

Naehyuck Chang

**Abstract** Portable computing and communication devices such as cellular phones, PDAs, MP3 players, and laptop computers are now being used as essential devices in our daily life. The operational lifetime of these devices is determined by the capacity of the energy source, usually a battery. The capacity of a battery is proportional to its volume and weight; however, portability places very stringent constraints on its size, weight, and form factor. Unfortunately, improvements in the energy density of batteries have lagged far behind the increasing energy demand of many portable microelectronic systems. This widening gap between the capabilities of batteries (energy source) and the demands of the processor and peripherals (energy consumers) is one of the primary challenges in the design of portable systems. Obviously, this gap can be reduced by either improving the energy efficiency of the consumer (performance per watt) or by increasing the energy density of the producer.

The ultimate goal of low-power design is holistic optimization of the system-wide power consumption. Most of the low-power research literature deals with minimization of power consumption of the energy consumers, e.g., microprocessors, memory devices, buses, and peripheral devices, as the primary issue in low-power design. However, efficient power conversion and delivery is equally important for the energy efficiency of the whole system.

Currently, there exists a large body of literature on improving the efficiency of the energy consumers, that is, processors and peripherals. For processors, the basic techniques involve dynamic voltage and frequency scaling (DVFS); for subsystems such as disk drives and other peripherals, the methods involve various forms of speed control, and are generally referred to as dynamic power management. Nevertheless, power consumption still continues to plague the industry because of the continuing increase in leakage current, and dynamic power consumption is also growing as computational demand continues to increase.

Due to discrepancies in device technologies, I/O interface, nondigital elements, and so on, each device requires different supply voltages. Some analog devices require very low ripple power supplies. Consequently, many different types of voltage regulators are used in a system. DC–DC converters and linear regulators cannot

N. Chang (✉)
Seoul National University, Seoul, Republic of Korea
e-mail: naehyuck@elpl.snu.ac.kr

C.-M. Kyung, S. Yoo (eds.), *Energy-Aware System Design*,
DOI 10.1007/978-94-007-1679-7_7, © Springer Science+Business Media B.V. 2011

exhibit an acceptable conversion efficiency at all times, so enhancement of power conversion efficiency is crucial in leveraging the efficiency of the entire system.

In general, 20% to 30% power reduction of a target component is not easily achievable. In addition, even a dominant power-consuming component occupies around 10% of the whole system power consumption. As a result, a 30% power savings from a component achieves 3% extended battery life of the target system. However, power generation and conversion efficiency directly impacts the power of the whole system. Recovering 10% of the power conversion and generation provides an actual 10% battery life extension.

This chapter introduces power conversion subsystems and their efficiency characteristics followed by system-level solution to leverage the power conversion efficiency. The subtopics include:

- Power sources and energy storage devices
- DC–DC conversion and efficiency
- Applications of power source-aware power consumption

## 7.1 Energy Storage Devices as Power Sources

### 7.1.1 Battery Technologies

Battery cells convert chemically stored energy into electrical energy through electrochemical reaction. Figure 7.1 shows a schematic picture of an electrochemical cell [1]. A cell consists of an anode, a cathode, and electrolyte, which separates the two electrodes. Discharge of a battery incurs an oxidation reaction at the anode. The reductants, $R_1$, donate $m$ electrons, which are released into the connected circuit. The cathode experiences a reduction reaction, where $m$ electrons are accepted by an oxidant, $O_2$:

$$\begin{cases} R_1 \rightarrow O_1 + me^-, & \text{at the anode,} \\ O_2 + ne^- \rightarrow R_2, & \text{at the cathode.} \end{cases} \tag{7.1}$$

Modeling the behavior of batteries is complicated because of nonlinear effects during discharge. In the ideal case, the voltage stays constant during discharge, with an instantaneous drop to zero when the battery is empty. The ideal capacity would be constant regardless of the magnitude of the discharge current, and all the energy stored in the battery could be used without loss. However, in a real battery, the voltage slowly drops during discharge, and the effective capacity becomes lower for high discharge currents. This effect is termed the *rate capacity effect*. Besides this, there is the *recovery effect*: during a period of no or very low current discharge, the battery can recover the capacity which has been lost during the previous high current discharge period. Of course, the recovery process cannot perfectly restore all the lost charge, but the recovery process increases the effective capacity and lifetime of the battery. The rate capacity and recovery effects exist in all types of

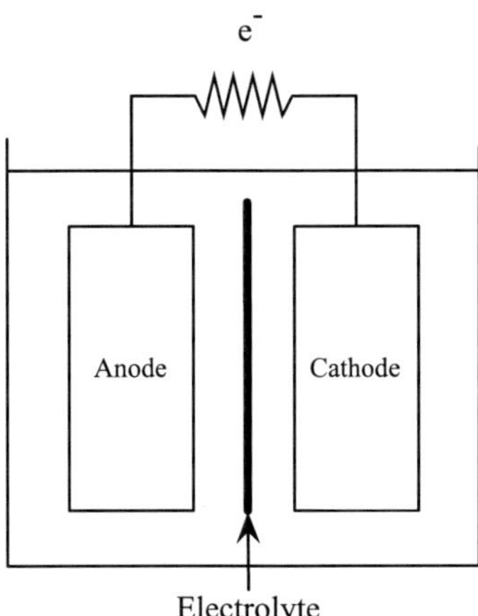


**Fig. 7.1** Conceptual diagram of electrochemical cell

batteries. However, the amount of charge loss due to the rate capacity effect and the amount of charge recovery due to the recovery effect are dependent on the types of batteries.

**(1) Lead-Acid Battery**    As one of the oldest and most developed rechargeable battery technologies, lead-acid batteries have a short cycle life (500–800 cycles) and a low energy density (30–50 Wh/kg) due to the inherent high density of lead as a metal. They also have a poor low-temperature performance, and thus they require a thermal management system. In spite of these disadvantages, their ability to supply high surge currents means that lead-acid cells maintain a relatively high power density (75–300 W/kg). These features, along with their low cost (100–200 $/kWh) and high energy efficiency (70–90%), make lead-acid batteries suitable in motor vehicles to provide the high current demand for automobile starter motors. Lead-acid batteries have also been used in a few large-scale commercial energy management systems. The largest one is a 40 MWh system in Chino, California, built in 1988 [4].

**(2) Li-Ion Battery**    The Li-ion battery, first demonstrated in the 1970s, is a family of rechargeable batteries in which lithium ions move from negative electrode to positive electrode through discharging, and in the opposite direction during charging. It is now the battery of choice in portable electronic devices, and is growing in popularity in military, electric vehicle, and aerospace applications. The growing popularity of the Li-ion battery is mainly due to the following reasons: high energy density (100–250 Wh/kg or 250–360 Wh/L) [3, 5], high cycle efficiency (higher than 90%), long cycle life (as high as 10,000 cycles), no memory effect, and low self-discharge rate (0.1–0.3% per day).

Although this battery has taken over 50% of the small portable devices market [3], there are some challenges in building large-scale Li-ion based energy storage systems. The main hurdle is the high cost (above 600 $/kWh) due to special packaging and internal overcharge protection circuits. Several companies, such as

SAFT and Mitsubishi, are working to reduce the manufacturing cost of Li-ion batteries to capture large new energy markets, especially markets for electrical vehicles.

**(3) NiMH Battery**    The nickel-metal hydride battery, abbreviated NiMH, is a type of rechargeable battery similar to the nickel-cadmium cells (NiCd). The only difference is that the former one uses a hydrogen-absorbing alloy for the negative electrode instead of cadmium. The energy density of NiMH batteries is more than double that of lead-acid batteries and 40% higher than that of NiCd batteries. NiMH batteries are also relatively inexpensive. However, they suffer from the memory effect, although much less pronounced than that in NiCd batteries, and they have a rather high self-discharge rate.

The most significant feature of NiMH batteries is the high power density (250–1000 W/kg), the highest among the existing battery types. Therefore, these batteries are widely used in high current drain consumer electronics, such as digital cameras with LCDs and flashlights. Recently, they have been used in hybrid electric vehicles such as the Toyota Prius, Honda Insight, Ford Escape Hybrid, Chevrolet Malibu Hybrid, and Honda Civic Hybrid [6, 7].

**(4) Metal-Air Batteries**    A metal-air battery has an anode which is made of pure metal and an air cathode that is connected to an inexhaustible source of air. Thus this type of battery can be regarded as "half a fuel cell." Metal-air batteries potentially have the highest energy density (theoretically more than 10 kWh/kg) and are relatively inexpensive to produce. They are also environmentally friendly. However, such batteries suffer from the difficulty and low efficiency of the required electrical recharging. Thus, many manufacturers offer refuel units where the consumed metal is replaced mechanically instead of offering an electrically rechargeable battery.

### 7.1.2 Emerging Technologies

**(1) Supercapacitors**    Electric double layer capacitors, more commonly known as supercapacitors, are widely exploited to mitigate load current fluctuations in batteries. Supercapacitors have a superior cycle efficiency, reaching almost 100%, and a long cycle life [8]. Moreover, supercapacitors exhibit a significant higher volumetric power density but a lower energy density compared with batteries [9]. Thus they are suitable for energy storage in situations with frequent charging/discharging cycles or periodic high current pulses. In a battery-supercapacitor hybrid system, the supercapacitor stores surplus energy from the battery during low demand periods and provides extra energy during peak load current demand periods.

However, a distinct disadvantage of a supercapacitor for electrical energy storage is its large self-discharge rate compared to that of ordinary batteries. A supercapacitor may lose more than 20% of its stored energy per day even if no load is connected to it. Another important concern with supercapacitors is the terminal voltage variation resulting from the nature of a capacitor, whereby the terminal voltage is linearly proportional to its state of charge (SOC). The terminal voltage increases or decreases accordingly as the supercapacitor is charged or discharged. This terminal voltage variation is much higher than that observed in typical batteries. This effect results in a significant conversion efficiency variation in the power converters which are connected to the supercapacitors. Hybrid energy storage systems including a supercapacitor should properly account for these characteristics to be practical.

**(2) Portable Room-Temperature Fuel Cells** An excellent battery substitute is a fuel cell, which can provide the energy densities required by portable computing and communication devices of the future. A fuel cell is an electrochemical device that generates electric energy using hydrogen ($H_2$) and oxygen ($O_2$). Fuel cells are environmentally clean, generating less emission than an internal combustion engine generator. A fuel cell provides instant power (no recharging) and very high energy density compared to batteries. It can be highly modular and in turn configurable for a wide range of outputs.

For instance, the Li-ion battery pack used in cellular phones has a typical volumetric energy density of 250 Wh/l (watt-hours per liter), with a theoretical upper limit of 480 Wh/l. A methanol-based fuel cell and a hydrogen-based fuel cell (using sodium borohydride) have a volumetric energy density of 4780 Wh/l and 7314 Wh/l, respectively. Even assuming 20% efficiency when converting chemical to electrical energy, these fuel cells have energy densities ranging from 1000–1500 Wh/l: more than four times the energy density of modern Li-ion batteries. Small-size room-temperature fuel cells, such as the proton exchange membrane fuel cell (PEMFC) powered by hydrogen, are under development for portable applications where the power range is on the order of 1–100 W. PEMFCs are called room-temperature fuel cells and are suitable for portable applications because they operate at low temperatures ($<70°C$).

PEMFCs transform the chemical energy liberated during the electrochemical reaction of $H_2$ and $O_2$ to electrical energy without combustion [10–12]. Some PEMFCs acquire hydrogen from methanol instead of carrying hydrogen. A reforming process converts methanol into hydrogen, absorbing heat. However, direct methanol fuel cells (DMFCs) directly accept methanol without reforming. In fact, DMFCs can be classified as a kind of PEMFC because the DMFC membrane exchanges protons only. However, a distinction is generally made between the two for convenience, because the membrane accepts methanol rather than hydrogen. So, conventionally PEMFC refers to hydrogen fuel cells only, where hydrogen is fed to the membrane as a fuel.

DMFCs (direct methanol fuel cells) are another type of room-temperature fuel cell. DMFCs are attractive for portable applications, although they have many disadvantages. DMFC systems generally require more elaborate balance-of-plant (BOP) control than PEMFC systems. Furthermore, methanol is flammable and toxic even if it is much easier to store and carry than hydrogen. DMFCs are also subject to problems that cause lower efficiency and reliability, such as fuel (methanol) crossover and catalyst dissolution [13–15].

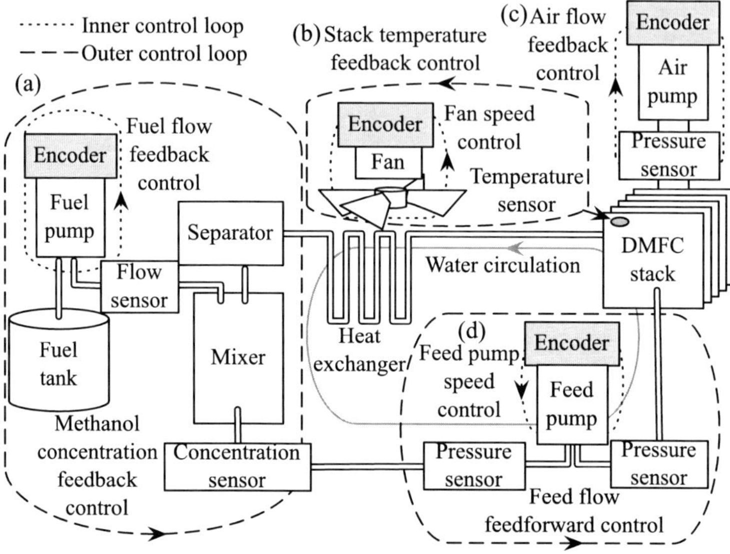



**Fig. 7.2** Generic active DMFC BOP system architecture [13]. © 2010 IEEE

There are two types of DMFCs. Passive DMFCs require virtually no BOP components, but they are limited to applications requiring a few watts only [16–18]. Active DMFCs are superior in power and efficiency, but require stringent BOP control. Practically applicable DMFC systems should provide accurate dilution of methanol, and good maintenance of the target operating temperature, proper fuel and air flow rates, etc. [15, 19, 20].

Active DMFCs provide a much higher power density than passive DMFCs because they operate at near-optimal conditions at all times, in spite of environmental changes. This is achieved by continuous adjustments to their operating parameters, which are made by BOP components, as illustrated in Fig. 7.2. Although the cost of BOP control is significant, the most costly element of a DMFC system is the catalyst in the cell. A reduced requirement for the catalyst makes active DMFCs more cost-effective sources of power.

**(3) Other Emerging Storage Technologies** Cryogenic energy storage (CES) [21, 22], which can be regarded as a special type of low-temperature thermal energy storage (TES), is a new energy storage technology. The principle of CES is to utilize the atmosphere as a heat source to vaporize and superheat a cryogenic fluid (e.g., liquid nitrogen or liquid air) in a thermal power cycle. This is different from the conventional heat engines, which use a high-temperature heat source and the atmosphere as a heat sink. The use of liquid nitrogen or liquid air as an energy storage medium would not pose any environmental burden. In general, CES has a relatively high energy density (150–250 Wh/kg), low capital cost, and a relatively long storage period, but it currently exhibits low efficiency (40–50%). Table 7.1 summarizes the

**Table 7.1** Comparison of energy storage elements [2, 3]

| Storage elements | Capital cost ($/kWh) | Cycle efficiency | Cycle life |
| --- | --- | --- | --- |
| Lead-acid | 100–200 | 70–90% | 500–800 |
| NiCd battery | 800–1,000 | 70–90% | 2,000–2,500 |
| NiMH battery | 450–1,000 | 66% | 500–1,000 |
| Li-ion battery | 600–2,500 | >90% | 1,000–10,000+ |
| NaS battery | 300–500 | 87% | 2,500 |
| Metal-air battery | 10–60 | <50% | 100–300 |
| Supercapacitor | 20,000–50,000 | >90% | 50,000+ |
| Flywheel | 1,000–5,000 | >90% | 20,000+ |
| High-temperature TES | 30–60 | 30–60% | – |
| CES | 3–30 | 40–50% | – |

| Storage elements | Self-discharge per day | Energy density | Power density |
| --- | --- | --- | --- |
| Lead-acid | 0.1–0.3% | 30–50 Wh/kg | 75–300 W/kg |
| NiCd battery | 0.2–0.6% | 50–75 Wh/kg | 150–300 W/kg |
| NiMH battery | 0.5–1% | 60–80 Wh/kg | 250–1,000 W/kg |
| Li-ion battery | 0.1–0.3% | 100–250 Wh/kg | 250–340 W/kg |
| NaS battery | ~20% | 150–240 Wh/kg | 150–230 W/kg |
| Metal-air battery | Very small | 1–10 kWh/kg | – |
| Supercapacitor | 20–40% | 2.5–15 Wh/kg | 100,000+ W/kg |
| Flywheel | 100% | 10–30 Wh/kg | 400–1,500 W/kg |
| High-temperature TES | 0.05–1% | 80–200 Wh/kg | – |
| CES | 0.5–1% | 150–250 Wh/kg | 10–30 W/kg |

characteristics of the diverse energy storage technologies discussed in Sects. 7.1.1 and 7.1.2.

### 7.1.3  Characterization of Batteries

Batteries are nonideal energy sources. In other words, minimizing the energy consumption of a battery-powered system is not equivalent to maximizing its battery life. Battery-aware task scheduling algorithms and power management policies, which try to reduce the unavailable charge at the end of a given workload, have been developed.

High-level battery models [23–27] have been developed that capture battery discharge behavior with reasonable accuracy, but are less computationally intensive and easier to configure than detailed physical models [28]. Such models have been used to develop battery-aware heuristics for task scheduling [26, 29–31], voltage scaling [26, 29], and discharge current shaping [25, 32–34]. These heuristics can be considered as battery-charge optimization techniques and aimed at improving the

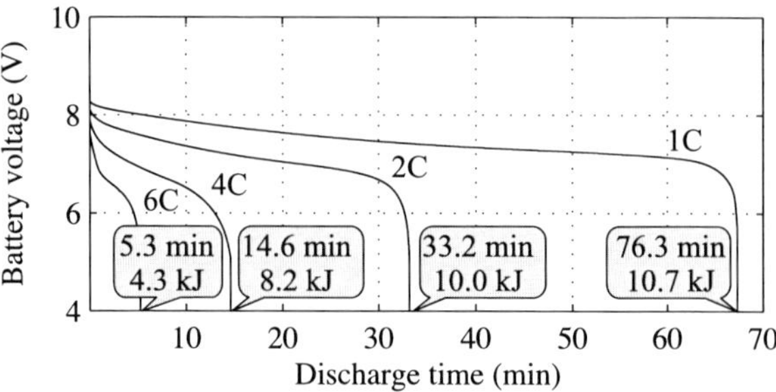

**Fig. 7.3** Discharging a 2-cell Li-ion battery with a constant current of 1C, 2C, 4C, and 6C [35]. © 2011 IEEE



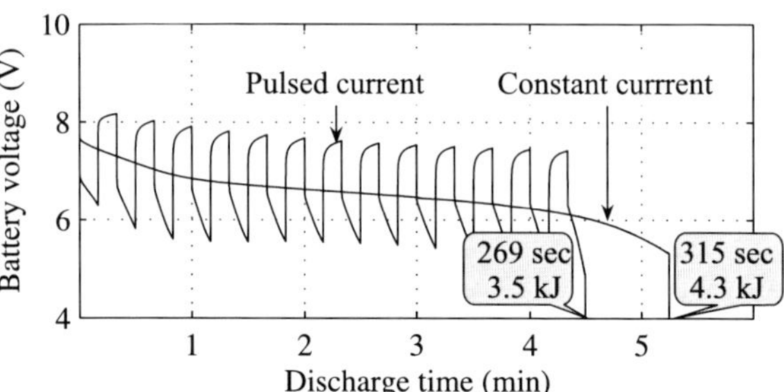

**Fig. 7.4** Discharging a 2-cell Li-ion battery with a 6C constant current and 12C pulsed current with a 20 s period and a 50% duty cycle [35]. © 2011 IEEE



battery lifetime compared to traditional energy optimization techniques. Traditional low-power techniques ignore battery effects like the rate capacity effect. However, none of the workload models used in these works accounts for the rest periods that occur commonly in a wide range of portable battery-operated systems.

### 7.1.3.1 Rate Capacity Effect

The rate capacity effect significantly degrades the battery discharge efficiency under high load currents. Electronic systems commonly exhibit a large amount of load current fluctuation, which challenges the maximum discharge capacity of batteries. Typical electronic systems determine the battery size by average power consumption, and thus a large pulsed discharge current can significantly shorten the battery service life; this can often be underestimated.

Figure 7.3 shows the voltage drop and total amount of delivered energy from the battery with a constant discharge current of 1C, 2C, 4C, and 6C, when using 2-cell series Li-ion GP1051L35 cells. The discharge efficiency at a 6C load current is a mere 40% of the 1C discharge efficiency. As presented in Fig. 7.4, drawing a pulsed current of 12C at a 50% duty, which is 6C on average, results in only 81.3% delivered energy and a shorter lifetime compared with drawing a constant current of 6C. This example clearly demonstrates the need to reduce the peak current draw.

### 7.1.3.2 Recovery Effect

The battery discharge behavior can be interpreted in terms of movement of charge between two different charge components [36]. This is best illustrated with an ex-

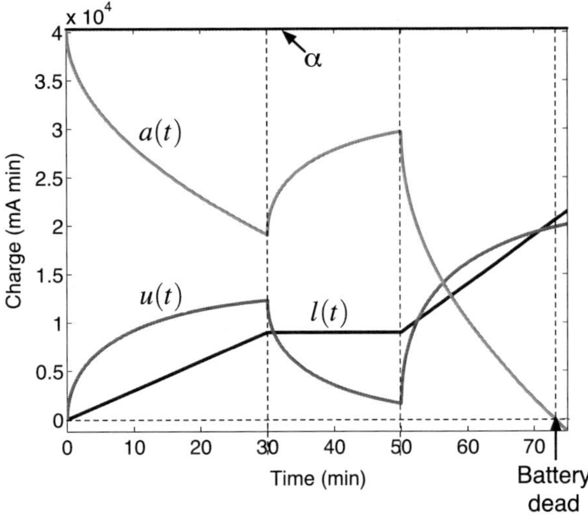

**Fig. 7.5** A plot of $u(t)$, $l(t)$, and $a(t)$ for an example load profile [36]. © 2005 IEEE

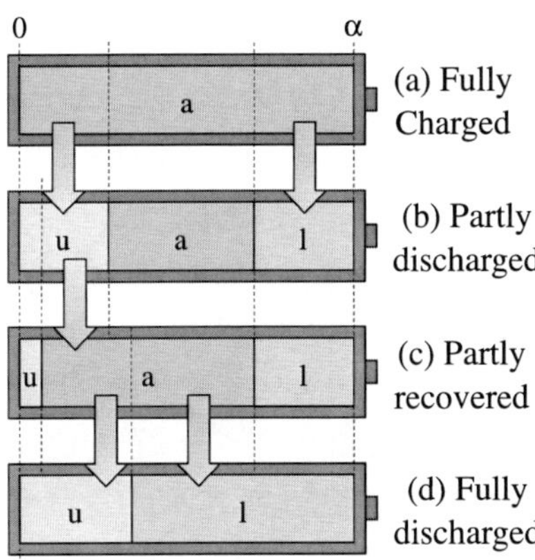



**Fig. 7.6** An illustration of charge movement during discharge and recovery [36]. © 2005 IEEE

ample. Figure 7.5 shows the plots of the $l(t)$, $u(t)$, and $a(t)$ curves for an example load profile such that

$$i(t) = \begin{cases} 300 \text{ mA}, & 0 \text{ min} \leq t \leq 30 \text{ min}, \\ 0 \text{ mA}, & 30 \text{ min} \leq t \leq 50 \text{ min}, \\ 500 \text{ mA}, & 50 \text{ min} \leq t \leq 75 \text{ min}, \end{cases} \tag{7.2}$$

which discharges a battery with an initial capacity $\alpha = 40{,}350$ mA·min. Figure 7.6 shows a graphical representation of the charge movement. At time $t = 0$, the battery is fully charged (Fig. 7.6(a)), or the entire charge is available ($a(0) = \alpha$), and no charge has been lost ($l(0) = 0$) or been made unavailable ($u(0) = 0$). A constant current load of 300 mA is then applied for 30 min. During this period, the available charge is depleted in two ways: it is delivered to the load as $l(t)$ and is permanently lost, it is locked up within the battery as $u(t)$ and can be recovered (Fig. 7.6(b)). At time $t = 30$ min, the load is turned off for a duration of 20 min. During this period, the battery recovers charge; i.e., the unavailable charge gradually becomes available (Fig. 7.6(c)). At time $t = 50$ min, a constant current of 500 mA is applied until the battery is discharged at $t = 73$ min. At this instant, all the available charge at time $t = 50$ min has either been lost to the load or become unavailable (Fig. 7.6(d)).

### *7.1.4 Power Source Modeling*

#### 7.1.4.1 Battery Models

Battery models for electronic systems have been studied intensively during the past decades. We can find analytical models based on electrochemical process modeling and analysis [1], but the electrochemical battery models are too complicated to use in an electronic system-level design. Battery models in the form of an electric circuit are more suitable for electronic system-level design [24, 37–39].

**(1) Electrochemical Models**    The electrochemical models reflect the chemical processes which occur in the battery. The models describe battery discharge processes in great detail, making them the most accurate battery models. However, they are also the most complex and difficult to configure because of their highly detailed descriptions.

A representative electrochemical model for lithium and lithium-ion cells is introduced in [28]. This model consists of six nonlinear differential equations and derives the voltage and current as functions of time based on these equations. The model gives us the potentials in the electrolyte and electrode phases, salt concentration, reaction rate, and current density in the electrolyte as functions of time and position in the cell.

DualFoil is a software program that uses this model to simulate lithium-ion batteries [40]. It computes how all the battery properties change over time for a load profile set by the user. The user sets more than 50 parameters such as the thickness of the electrodes, the initial salt concentration in the electrolyte, and the overall heat capacity. Such parameter setting requires comprehensive information about the battery, and the accuracy of the program is very high if the parameters are properly set. This program is often used as a baseline of comparison against other models rather than using experimental results to check the accuracy.

**(2) Equivalent Circuit Models**    Electric circuits can partly represent battery characteristics [41]. A simple PSpice circuit simulation has been performed to simulate nickel-cadmium, lead-acid, and alkaline batteries. The core of the models for the different types of batteries is the same:

– A capacitor representing the capacity of the battery
– A discharge-rate normalizer determining the lost capacity at high discharge currents

- A circuit discharging the capacity of the battery
- A voltage versus state-of-charge lookup table
- A resistor representing the battery's resistance

Although the models are much simpler than the electrochemical models and therefore computationally less expensive, it still takes some effort to configure the electrical circuit models. Minor changes in the model are usually required to accommodate a specific cell type. The models are also less accurate than electrochemical models, having an error of approximately 10% [1].

**(3) Analytical Models**    Analytical models describe the battery at a higher level of abstraction than the electrochemical and electrical circuit models. The major properties of the battery are modeled using a few equations. This makes this type of model much easier to use than the electrochemical and electrical circuit models.

The simplest model for predicting battery lifetimes that takes into account part of the nonlinear properties of the battery is Peukert's law [42]. It captures the nonlinear relationship between the lifetime of the battery and the rate of discharge, but without modeling the recovery effect. According to Peukert's law, the battery lifetime ($L$) can be approximated by

$$L = \frac{a}{I^b},$$ (7.3)

where $I$ is the discharge current, and $a$ and $b$ are constants which are obtained from experiments. Ideally, $a$ would be equal to the battery capacity and $b$ would be equal to 1.

The model that describes the diffusion process of the active material in the battery is introduced in [26]. The diffusion is considered to be one dimensional in a region of length $w$. $C(x, t)$ is the concentration of the active material at time t and distance $x[0, w]$ from the electrode. We need to compute the time at which the concentration at the electrode surface, $C(0, t)$, drops below the cutoff level $C_{\mathrm{cutoff}}$ to determine the battery lifetime. The one-dimensional diffusion process is described by Fick's laws:

$$-J(x, t) = D\frac{\partial C(x, t)}{\partial x},$$
$$\frac{\partial C(x, t)}{\partial t} = D\frac{\partial^2 C(x, t)}{\partial x^2},$$ (7.4)

where $J(x, t)$ is the flux of the active material at time $t$ and position $x$, and $D$ is the diffusion constant. According to Faraday's law, the flux at the left boundary of the diffusion region ($x = 0$) is proportional to the current $i(t)$. The flux at the right boundary ($x = w$) is zero. This gives the following boundary conditions:

$$D\frac{\partial C(0, t)}{\partial x} = \frac{i(t)}{vFA},$$
$$D\frac{\partial C(w, t)}{\partial x} = 0,$$ (7.5)

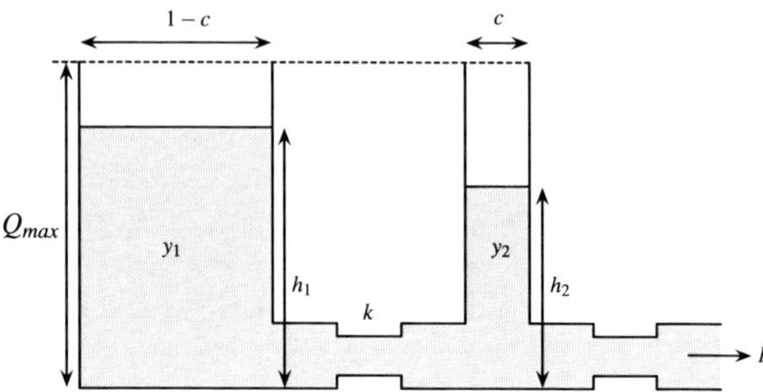

**Fig. 7.7** Two-well model of the Kinetic Battery Model [36]. © 2005 IEEE



where $A$ is the area of the electrode surface, $F$ is Faraday's constant ($96485.31$ C·mol$^{-1}$), and $v$ is the number of electrons involved in the electrochemical reaction at the electrode surface. It is possible to obtain an analytical solution from the differential equations and boundary conditions using Laplace transforms.

The Kinetic Battery Model (KiBaM) [43] is another analytical approach. It uses a chemical kinetics process as its basis. In the model the battery charge is distributed over two wells: the available-charge well and the bound-charge well, as shown in Fig. 7.7. The available-charge well supplies electrons directly to the load, whereas the bound-charge well supplies electrons only to the available-charge well.

The rate at which charge flows between the wells depends on the difference in heights of the two wells, and on a parameter $k$. The parameter $c$ gives the fraction of the total charge in the battery that is part of the analytical models of the available-charge well. The change of the charge in both wells is given by the following system of differential equations:

$$\frac{dy_1}{dt} = -I + k(h_2 - h_1),$$
$$\frac{dy_2}{dt} = -k(h_2 - h_1),$$

(7.6)

with initial conditions $y_1(0) = cC$ and $y_2(0) = (1-c)C$, where $C$ is the total battery capacity. For $h_1$ and $h_2$ we have $h_1 = y_1/c$ and $h_2 = y_2/(1-c)$. When a load $I$ is applied to the battery, the available charge reduces, and the difference in heights between the two wells grows. The differential equations can be solved using Laplace transforms.

Next to the charge in the battery, the KiBaM models the voltage during discharge. The battery is modeled as a voltage source in series with an internal resistance. The level of the voltage varies with the depth of discharge. The voltage is given by

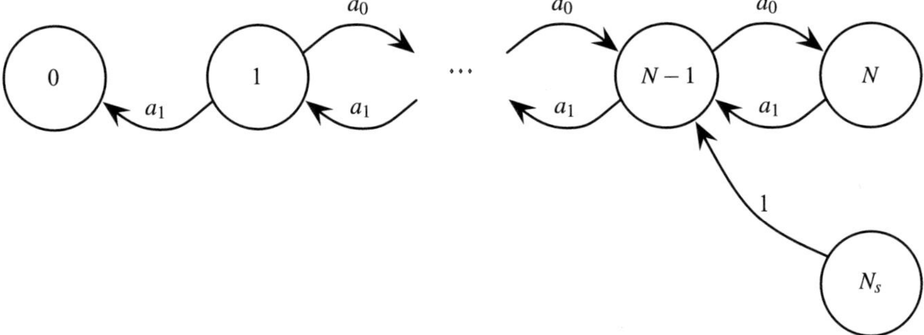

**Fig. 7.8**  The basic Markov chain battery model [44]. © 1999 IEEE

$$V = E - I R_0,  \tag{7.7}$$

where $I$ is the discharge current and $R_0$ is the internal resistance. $E$ is the internal voltage, which is given by

$$E = E_0 - AX + \frac{CX}{D - X},  \tag{7.8}$$

where $E_0$ is the internal battery voltage of the fully charged battery, $A$ is a parameter reflecting the initial linear variation of the internal battery voltage with the state of charge, $C$ and $D$ are parameters reflecting the decrease of the battery voltage when the battery is progressively discharged, and $X$ is the normalized charge removed from the battery. These parameters can be obtained from discharge data. At least three sets of constant discharge data are needed for the nonlinear least square curve fitting, which is described in detail in [43].

**(4) Stochastic Models**    Stochastic models aim to describe the battery in an abstract manner, like the analytical models. However, the discharging and the recovery effect are described as stochastic processes. There are several studies about battery modeling based on discrete-time Markov chains [44].

In the first and simplest model, the battery is described by a discrete-time Markov chain with $N + 1$ states, numbered from 0 to $N$ as illustrated in Fig. 7.8. The state number corresponds to the number of charge units available in the battery. One charge unit corresponds to the amount of energy required to transmit a single packet. $N$ is the number of charge units directly available based on continuous use. In this simple model, for every time step either a charge unit is consumed with probability $a_1 = q$, or recovery of one unit of charge takes place with probability $a_0 = 1 - q$. The battery is considered empty when the absorbing state 0 is reached, or when a maximum of $T$ charge units have been consumed. The number of $T$ charge units is equal to the theoretical capacity of the battery $(T > N)$.

Rao et al. [36] proposed a stochastic battery model based on the analytical Kinetic Battery Model (KiBaM). The stochastic KiBaM is used to model a Ni-MH battery, instead of a lead-acid battery for which the original KiBaM was developed.

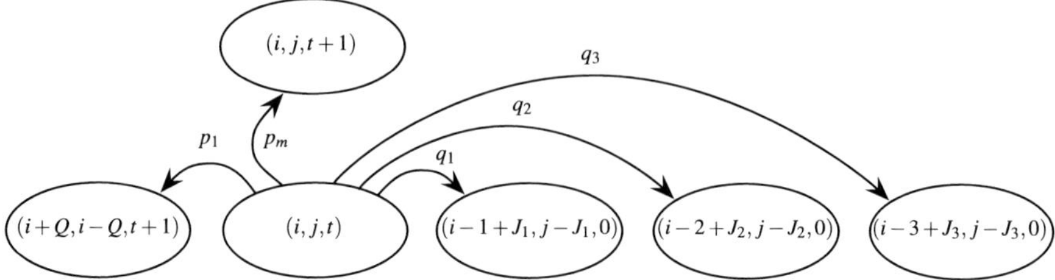



**Fig. 7.9** Part of the state transition diagram of the stochastic KiBaM [36]. © 2005 IEEE

To be able to model this different type of battery, a couple of modifications have been made to the model. First, in the term corresponding to the flow of charge from the bound-charge well to the available-charge well an extra factor $h_2$ is added, changing (7.6) into

$$\frac{dy_1}{dt} = -I + k_s h_2 (h_2 - h_1),$$
$$\frac{dy_2}{dt} = -k_s h_2 (h_2 - h_1). \tag{7.9}$$

This causes the recovery to be slower when less charge is left in the battery. The second modification is that in the stochastic model the possibility of no recovery during idle periods is added.

The battery behavior is represented by a discrete-time transient Markov process. The states of the Markov chain are labeled with three parameters $(i, j, t)$. The parameters $i$ and $j$ are the discretized levels of the available-charge well and bound-charge well, respectively, and $t$ is the length of the current idle slot; this is the number of time steps taken since the last time some current was drawn from the battery.

Figure 7.9 shows a part of the state transition diagram. The transitions are summarized as follows:

$$(i, j, t) \rightarrow \begin{cases} (i + Q, j - Q, t + 1), \\ (i, j, t + 1), \\ (i - I + J, j - J, 0). \end{cases} \tag{7.10}$$

The first two equations correspond to the time steps in which the current is zero. With probability $p_r$, the battery recovers $Q$ charge units, and with probability $p_{nr}$, no recovery occurs. Both $p_r$ and $p_{nr}$ depend on the length of the idle time slot ($t$). The third equation corresponds to the time steps in which a current is drawn from the battery. With probability $q_I$, $I$ charge units are drawn from the available-charge well, and at the same time $J$ charge units are transferred from the bound-charge well to the available-charge well. The probabilities $q_I$ are defined by the load profile. Since the $q_I$ are equal for all states, it is impossible to control in what sequence the currents are drawn from the battery in this model, and thus to fully model a real usage pattern.

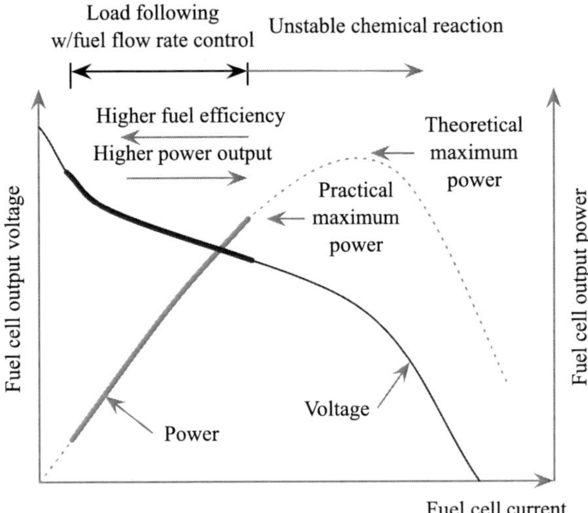


**Fig. 7.10** Polarization curves for a PEMFC [45]. © 2008 IEEE

## 7.1.4.2 Fuel Cell Power Models

Figure 7.10 shows a typical plot of the voltage and power versus current of a fuel cell. As the current density increases, the open-circuit voltage reduces. A PEMFC generates a cell output voltage of about 1.2 V. The maximum fuel cell output current is determined by the current density of the membrane times the area of the membrane. The fuel cell output voltage is determined by the single-cell voltage times the number of stacked cells. The fuel cell output power can be varied with a given range, referred to as the load-following region, by controlling the fuel flow rate. In fact, the fuel cell output power depends on temperature, pressure, and humidity, and these need to be dynamically controlled for maximum efficiency.

**(1) Static Characteristics**   This section presents a few details of how the fuel cell voltage is related to the current. Using thermodynamic values of the standard-state entropy change, the internal voltage or Nernst potential of one fuel cell $E_{\text{fc}}$ is given by [46]

$$E_{\text{fc}} = 1.229 - 0.85 \times 10^{-3}(T_{\text{fc}} - 298.15)$$

$$+ 4.3085 \times 10^{-5} T_{\text{fc}}\left( \ln(P_{\text{H}_2}) + \frac{1}{2}\ln(P_{\text{O}_2}) \right), \qquad (7.11)$$

where $T_{\text{fc}}$ is the temperature of the fuel cell in kelvins (K), and $P_{\text{H}_2}$ and $P_{\text{O}_2}$, respectively, are the partial pressure of $H_2$ and $O_2$, expressed in atmospheres. This is the theoretical maximum voltage which can be harnessed from a cell. However, when some current is drawn from the cell, the voltage drops due to certain losses in it. The characteristic shape of the voltage and current density curve for a fuel cell can be explained using three main losses: activation loss, ohmic loss, and concentration loss. Each of these is summarized here.

The activation loss is caused by the slowness of reactions taking place on the surface of the electrodes. The activation loss or activation overvoltage arises from the need to move electrons, and to break and form the chemical bonds in the anode and cathode. Part of the available energy is lost in driving the chemical reaction that transfers the electrons to and from the electrodes. For low-temperature PEMFCs, the activation overvoltage can be described using the Tafel equation,

$$v_a = \frac{RT}{2\alpha F} \times \ln\left(\frac{i}{i_0}\right), \tag{7.12}$$

where $R$ is the gas constant, $T$ is the stack temperature, $\alpha$ is the charge transfer coefficient, which is different for the anode and the cathode, and $i_0$ is the exchange current density, which is again different for the anode and the cathode. The authors of [47] have used a model of the form

$$v_a = V_0 + V_1\left(1 - e^{-c_1 i}\right), \tag{7.13}$$

where $V_0$ is the open-circuit voltage drop, and $V_1$ and $c_1$ are constants, which can be determined by nonlinear regression with empirical methods. The activation overvoltage depends strongly on the temperature and the $O_2$ partial pressure.

The ohmic loss has two components: resistance to the flow of electrons through the material of the electrode and resistance to the flow of protons ($H^+$) through the membrane. The membrane resistance is a function of the water content in the membrane and the stack current. The voltage drop that corresponds to the ohmic loss, $v_\Omega$, is proportional to the current,

$$v_\Omega = i R_\Omega, \tag{7.14}$$

$$R_\Omega = R_{\text{electron}} + R_{\text{proton}}, \tag{7.15}$$

$$R_{\text{proton}} = \frac{r_M \times l}{A}, \tag{7.16}$$

$$r_M = \frac{181.6(1 + 0.03(\frac{i}{A}) + 0.062(\frac{T}{303})^2(\frac{i}{A})^{2.5})}{(\lambda - 0.634 - 3(\frac{i}{A}))\exp(4.18(\frac{T-303}{T}))}, \tag{7.17}$$

where $\lambda$ is a measure of the water content in the membrane; a value of 14 indicates a well-hydrated membrane. $l$ is the thickness of the membrane, $A$ is the membrane area of the cell, and $i$ is the stack current.

The concentration loss results from the change in concentration of the reactants at the surface of the electrodes as the fuel is used. Because of this, at high current density operation, additional losses occur, so this type of loss is also often called mass transport loss. The voltage drop amount due to the concentration losses is given by [48]

$$v_c = i\left(c_2 \times \frac{i}{i_{\max}}\right)^{c_3}, \tag{7.18}$$

where $c_2$, $c_3$, and $i_{\max}$ are constants that depend on the temperature and the reactant partial pressure. The authors of [49] also suggest another model for the concentration loss:

$$v_{\mathrm{c}} = B \times \ln\left(1 - \frac{i}{i_{\max}}\right), \tag{7.19}$$

where $B$ is a parametric constant in volts, $i$ is the cell current density, and $i_{\max}$ is the limiting current density, another constant for the cell.

Finally, the following equation describes the static operating voltage of a fuel cell at a current density $i$:

$$v_{\mathrm{fc}} = E_{\mathrm{fc}} - v_{\mathrm{a}} - v_{\Omega} - v_{\mathrm{c}}$$

$$= E_{\mathrm{fc}} - \left(V_0 + V_1\left(1 - e^{-c_1 i}\right)\right) - i R_{\Omega} - i\left(c_2 \times \frac{i}{i_{\max}}\right)^{c_3}. \tag{7.20}$$

The static stack operating voltage can be calculated by multiplying the cell voltage $v_{\mathrm{fc}}$ by the number of stacked cells.

**(2) Dynamic Characteristics**    The charge double layer is an important concept that helps in understanding the dynamic electrical behavior of the fuel cell. Whenever two different materials are in contact, there is a charge layer buildup at the interface, or a charge transfer occurs from one layer to the other. In electrochemical systems, the charge double layer forms in part due to diffusion effects and in part due to the reaction between the electrons in the electrodes and the ions in the electrolyte. The layer of charge on or near the electrode–electrolyte interface is a store of electrical charge and energy, and as such behaves like an electrical capacitor. Its capacitance is given by

$$C = \varepsilon \times \frac{A}{d}, \tag{7.21}$$

where $\varepsilon$ is the electrical permittivity, $A$ is the real surface area of the electrode, which is several thousand times greater than its length $\times$ width, and $d$ is the separation of the plates, which is very small, typically only a few nanometers. The result is that the capacitance is of the order of a few farads. Therefore, when the current suddenly changes, it takes some time before the activation overvoltage and the concentration overvoltage follow the change in the current. However, the ohmic voltage drop responds instantaneously to a change in the current. Thus, the equivalent circuit in Fig. 7.11 can be used to model the dynamic behavior of the fuel cell. From (7.13) and (7.18) we can define the activation resistance $R_{\mathrm{a}}$ and the concentration resistance $R_{\mathrm{c}}$:

$$R_{\mathrm{a}} = \frac{1}{i}\left(V_0 + V_1\left(1 - e^{-c_1 i}\right)\right),$$

$$R_{\mathrm{c}} = \left(c_2 \times \frac{i}{i_{\max}}\right)^{c_3}. \tag{7.22}$$

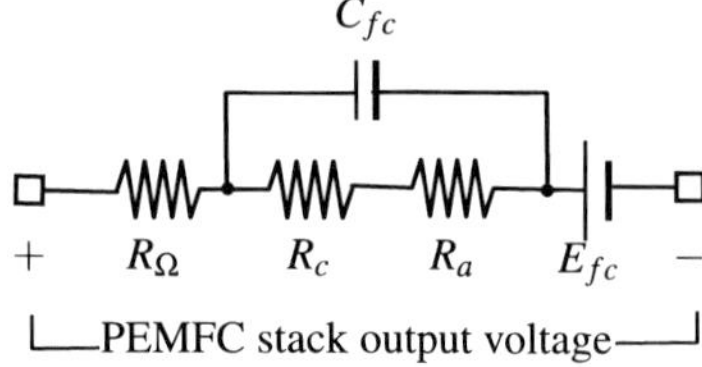



**Fig. 7.11** Equivalent circuit of the PEMFC stack [45]. © 2008 ACM

Therefore, the dynamic FC voltage behavior can be described by

$$i = C_{\text{fc}} \times \frac{dv_{\text{c}}}{dt} + \frac{v_{\text{c}} - V_0}{R_{\text{a}} + R_{\text{c}}},$$
$$v_{\text{fc}} = E_{\text{fc}} - v_{\text{c}} - i R_\Omega, \tag{7.23}$$

where $v_{\text{c}}$ is the voltage of the capacitor.

Figure 7.11 shows an equivalent circuit of the fuel cell that can be used to explain its dynamic characteristics. In the circuit diagram, $R_{\text{a}}$ and $R_{\text{c}}$ are the equivalent resistance of $v_{\text{a}}$ and $v_{\text{c}}$, respectively.

## 7.2 DC–DC Conversion and Efficiency

### 7.2.1 Regulator Basics

Constant semiconductor scaling has resulted in many different technology semiconductor chips on the same board and system. As each technology requires different levels of supply voltage, it is no longer possible to use a single supply voltage such as 5 V or 3.3 V for the whole board or system. Nevertheless, the board or system prefers to be powered by a single voltage power source, typically a battery pack. So, voltage regulators (DC–DC converters) are used to generate various different supply voltage levels for each device, as illustrated in Fig. 7.12. It shows a simplified power supply network for a typical battery-operated embedded system.

The primary role of a DC–DC converter is to provide a regulated power source. Unlike passive components, logic devices do not draw a constant current from their power supply. The power supply current changes rapidly with changes in the internal states of devices. An unregulated power supply is subject to a big IR drop corresponding to the load current, whereas a regulated power supply aims to keep the output voltage constant regardless of variation in the load current. The phenomenon of IR drop is caused by the internal resistance of the power supply.

Almost all modern digital systems are supplied with power through DC–DC converters because high-performance CMOS devices are optimized to specific supply voltage ranges. DC–DC converters are generally classified as two types: (1) linear voltage regulators and (2) switching voltage regulators, according to the circuit implementation.

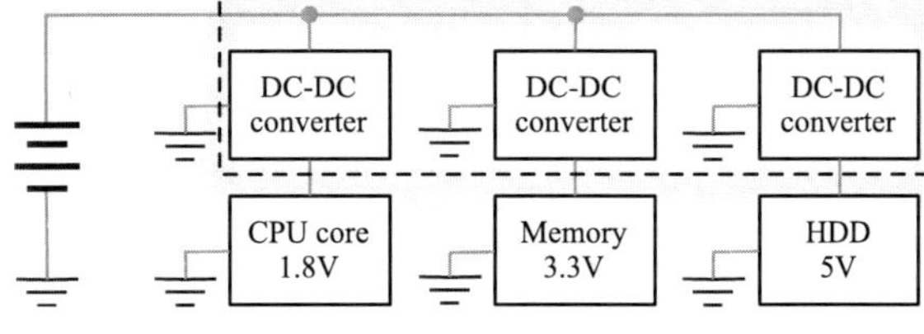

**Fig. 7.12** DC–DC converters generate different supply voltages for the CPU, memory, and hard disk drive from a single battery [50]. © 2007 IEEE



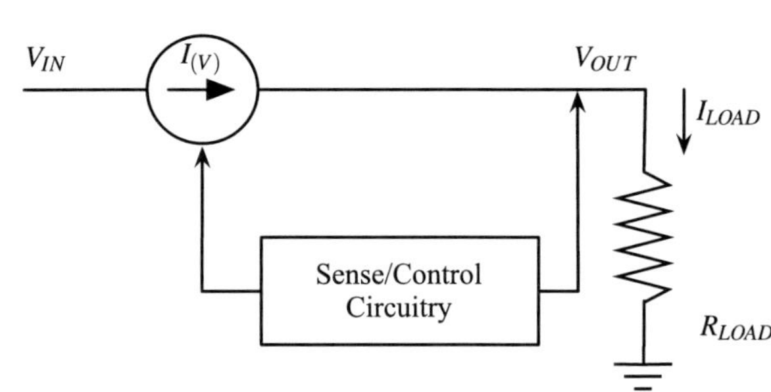

**Fig. 7.13** Linear regulator functional diagram



If the efficiency of a DC–DC converter was constant over its entire operating range, we could ignore the DC–DC converter effect on the total energy consumption of the system. However, in reality, the efficiency of a DC–DC converter is closely correlated with the level of output voltage and load current. Consequently, when a power management scheme such as DVFS, which involves varying the supply voltage, is implemented in an embedded system, it is also essential to properly schedule the output voltage of the DC–DC converter, so that the overall energy consumption of the system is minimized. Note that, in the case of a switching regulator, in addition to the output voltage, its power efficiency is affected by the load current as well.

## 7.2.2 Linear Regulators

A linear regulator operates by using a voltage-controlled current source to force a fixed voltage to appear at the regulator output terminal, as shown in Fig. 7.13. The control circuitry must monitor the output voltage and adjust the current source to hold the output voltage at the desired value. The design limit of the current source defines the maximum load current that the regulator can source and still maintain regulation.

The dropout voltage is defined as the minimum voltage drop required across the regulator to maintain output voltage regulation. A critical point to be considered is that the linear regulator that operates with the smallest voltage across it dissipates the least internal power and has the highest efficiency.

Low dropout (LDO) linear regulators, called LDOs for short, have a series transistor between the power source (input) and the load (output), and the voltage drop across this transistor varies to keep the output voltage constant as the load current changes, so that the regulated voltage must be lower than the input voltage. The power conversion efficiency of an LDO is primarily determined by the voltage drop between the input and the output, as its input and output currents are almost the same. Recent LDOs have a very small voltage drop, around 300 mV, between the input and the output, and the peak efficiency is high. But in general, the output voltage is fixed as part of the design specification, while the input voltage varies within a particular range due to the IR drop across the power source. Therefore, the LDO does not achieve high average efficiency, even though its dropout is small. Linear regulators are more efficient when the power source's voltage is closer to the required output voltage.

There is an upper bound on the efficiency of a linear regulator, which is equal to the output voltage divided by the input voltage. The power lost in a linear regulator, computed as

$$1 - \frac{V_{OUT}}{V_{IN}} \times P_{IN} = (V_{IN} - V_{OUT}) \times I_{OUT}, \tag{7.24}$$

must be dissipated by its package [51].

### 7.2.3 Switching Regulators

A switching regulator uses an inductor, a transformer, or a capacitor as an energy storage element to transfer energy from the power source to the system. The amount of power dissipated by voltage conversion in a switching regulator is relatively low, mainly due to the use of low-resistance MOSFET switches and energy storage elements. Switching regulators can increase (boost), decrease (buck), and invert input voltage with a simple modification to the converter topology. Figure 7.14(a) shows the basic structure of the step-down (buck) switching regulator.

A switching regulator contains a circuit, located on the path between the external power supply and the energy storage element, which controls two MOSFET switches. The switch control techniques most widely used in practical DC–DC converters are pulse width modulation (PWM), which controls the turn-on duty ratio of each MOSFET with a fixed switching frequency, and pulse frequency modulation (PFM), which controls the switching frequency by constraining the peak current flowing through the inductor. Each control technique has its own advantages and shortcomings. DC–DC converters controlled by PWM generate less ripple in the output voltage, their switching noises are easier to filter out, and they are more efficient under heavy loading, whereas DC–DC converters controlled by PFM exhibit higher efficiency with light loads.

Most commercial DC–DC converters use either PWM or hybrid PWM/PFM control. The hybrid technique inherits higher light-load efficiency from the PFM control technique, but PWM is preferred for applications that require low cost, or small size, and for noise-sensitive systems including analog circuits and wireless communication subsystems.

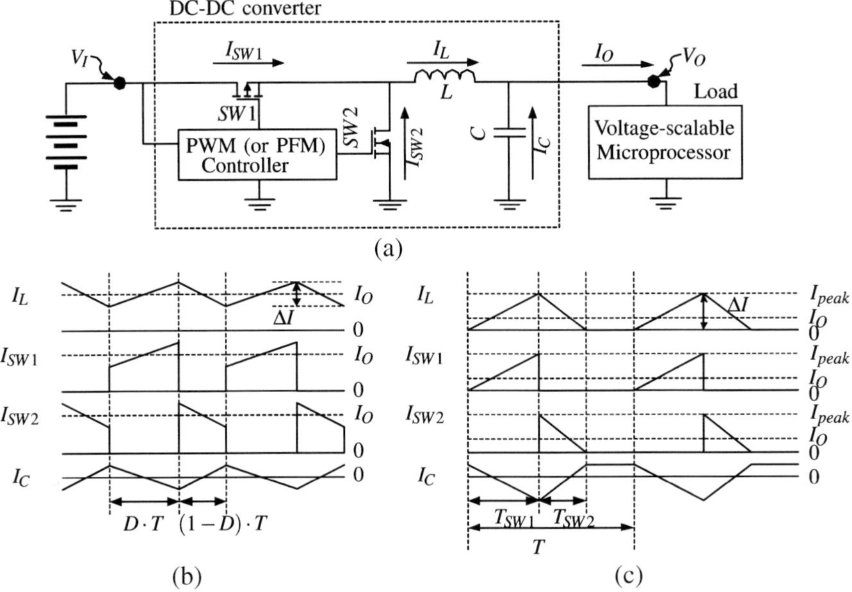


**Fig. 7.14** Simplified block diagram of a buck converter and the current flow thorough each component: (**a**) structure of a buck DC–DC converter, (**b**) current flow in a PWM DC–DC converter, (**c**) current flow in a PFM DC–DC converter [50]. © 2007 IEEE

### 7.2.3.1 Power Dissipation of Switching Regulator

An ideal switching regulator consumes no power, unlike an ideal linear voltage regulator. However, practical DC–DC converters have nonideal characteristics that cause power to be lost. Generally, the major sources of power dissipation in DC–DC conversion are classified into three categories [52], which will be discussed in the remainder of this section. Based on a number of previous studies of power loss in DC–DC converters [52–54], we express the power dissipation due to each source in terms of manufacturing parameters and load conditions such as the output voltage and the output current, which can be controlled by DVS. The power dissipation of the DC–DC converter is expressed as the sum of the following three components:

$$P_{\text{dcdc}} = P_{\text{conduction}} + P_{\text{gate\_drive}} + P_{\text{controller}}. \tag{7.25}$$

**(1) Conduction Power Dissipation**    All the elements of a DC–DC converter, such as switches, inductors, and capacitors, are nonideal and have their own resistive components $R_{\text{ESR}}$. This means that power dissipation $I^2 R_{\text{ESR}}$ due to the current $I$ through these elements is unavoidable.

Although varied amounts of current flow through different components, as shown in Fig. 7.14, these currents are all positively related to the load current $I_O$ of the system. Consequently, the conduction power dissipation of the DC–DC converter can be reduced by reducing the load current, which can be achieved by high-level power management controlling the load power.

Since the two types of DC–DC converters have different ways of switching their two MOSFETs, as shown in Figs. 7.14(b) and (c), their conduction power dissipation has different characteristics. Therefore, different conduction power models are used for the different types of DC–DC converter.

The power consumption of a PWM DC–DC converter can be formulated as

$$
\begin{aligned}
P_{\text{conduction(PWM)}} \\
&= I_O^2 \left( D R_{SW1} + (1 - D) R_{SW2} + R_L \right) \\
&\quad + \frac{1}{3} \left( \frac{\Delta I_{L(\text{PWM})}}{2} \right)^2 \left( D R_{SW1} + (1 - D) R_{SW2} + R_L + R_C \right),
\end{aligned}
\tag{7.26}
$$

where $I_O$ is the output current (i.e., the load) of the DC–DC converter; and $R_{SW1}$, $R_{SW2}$, $R_L$, and $R_C$ are the turn-on resistance of the top MOSFET ($SW1$), the turn-on resistance of the bottom MOSFET ($SW2$), the equivalent series resistance of the inductor $L$, and the equivalent series resistance of the capacitor $C$, respectively. $D$ and $\Delta I_{L(\text{PWM})}$ are the duty ratio (time when the current actually flows through the component/total time) and the ripple of the current flowing through the inductor, respectively, and can be expressed as follows:

$$
D = \frac{V_O}{V_I}, \qquad \Delta I_{L(\text{PWM})} = \frac{V_O(1 - D)}{L_f f_S},
\tag{7.27}
$$

where $V_I$, $V_O$, and $L_f$ are the input voltage, output voltage, and the value of the inductor. $f_S$ is the switching frequency, which is assumed to be constant in a PWM DC–DC converter.

$P_{\text{conduction(PWM)}}$ consists of two terms. The first and second terms represent the conduction power consumptions due to the dc component and the ac component (or current ripple), respectively, of the current flowing through all components (i.e., $SW1$, $SW2$, $L$, and $C$) on the current path. In the first term of $P_{\text{conduction(PWM)}}$ in (7.26), $D R_{SW1} + (1 - D) R_{SW2} + R_L$ is the effective resistance of the current path of the DC–DC converter, considering the duty ratio of each component on that path. The duty ratios for $SW1$, $SW2$, and $L$ are $D$, $(1 - D)$, and 1, respectively. (Since the dc component of the current flowing through $C$ is zero, the term related to $C$ is omitted.) It is well known that the conduction power consumption of some systems can be expressed by $I^2 R$, where $I$ is the current flowing through the system and $R$ is the resistive component of the system. Therefore, the product of this effective resistance and $I_O^2$, where $I_O$ is equivalent to the dc component of the current flowing through each component, can be used to model the dc component of the conduction power consumption of the PWM DC–DC converter. In the second term, $D R_{SW1} + (1 - D) R_{SW2} + R_L + R_C$ is the effective resistance, and $(1/3)(\Delta I_{L(\text{PWM})}/2)^2$ is the square of the ac component (or current ripple) of the current flowing through the components.

A PFM DC–DC converter has a variable switching frequency that depends on the output current, the output voltage, and other factors. Therefore, the switching frequency should be characterized accurately to determine the amount of conduction power dissipation of a PFM DC–DC converter. From [52], the switching frequency can be described as

$$f_{S(PFM)} = \frac{1}{T} = \frac{2I_O}{I_{peak}(T_{SW1} + T_{SW2})}, \tag{7.28}$$

where $I_{peak}$ is the peak inductor current allowed in a given PFM DC–DC converter, and $T_{SW1}$ and $T_{SW2}$ are the turn-on times of the top MOSFET ($SW1$) and the bottom MOSFET ($SW2$), respectively. $T_{SW1}$ and $T_{SW2}$ can be determined as follows:

$$T_{SW1} = \frac{I_{peak}L_f}{V_I - V_O}, \qquad T_{SW2} = \frac{I_{peak}L_f}{V_O}. \tag{7.29}$$

$P_{conduction(PFM)}$ in (7.30) is modeled in the same way as $P_{conduction(PWM)}$. In the first term, $((T_{SW1} + T_{SW2})/T)(I_{peak}/2)^2$ is the square of the dc component of the current flowing through each component, and in the second term, $(1/3)((T_{SW1} + T_{SW2})/T)(\Delta I_{L(PFM)}/2)^2$ is the square of the ac component of that current. The duty ratios for $SW1$ and $SW2$ are $(T_{SW1}/(T_{SW1} + T_{SW2}))$ and $(T_{SW2}/(T_{SW1} + T_{SW2}))$, respectively. These expressions for the current and duty ratios can be found in (or derived from) many references (e.g., [52] and [54]). Replacing $T_{SW1}$ and $T_{SW2}$ with the expressions from (7.29), we can also construct the alternative expression shown in the last two lines of the following equation:

$$
\begin{aligned}
P_{conduction(PFM)} &\\
&= \frac{T_{SW1} + T_{SW2}}{T}\left(\left(\frac{I_{peak}}{2}\right)^2\left(\frac{T_{SW1}R_{SW1}}{T_{SW1} + T_{SW2}} + \frac{T_{SW2}R_{SW2}}{T_{SW1} + T_{SW2}} + R_L\right)\right.\\
&\quad \left. + \frac{1}{3}\left(\frac{\Delta I_{L(PFM)}}{2}\right)^2\left(\frac{T_{SW1}R_{SW1}}{T_{SW1} + T_{SW2}} + \frac{T_{SW2}R_{SW2}}{T_{SW1} + T_{SW2}} + R_L + R_C\right)\right)\\
&= \frac{2I_O}{I_{peak}}\left(\left(\frac{I_{peak}}{2}\right)^2\left(\frac{V_O R_{SW1}}{V_I} + \frac{(V_I - V_O)R_{SW2}}{V_I} + R_L\right)\right.\\
&\quad \left. + \frac{1}{3}\left(\frac{I_{peak}}{2}\right)^2\left(\frac{V_O R_{SW1}}{V_I} + \frac{(V_I - V_O)R_{SW2}}{V_I} + R_L + R_C\right)\right) \tag{7.30}
\end{aligned}
$$

where $\Delta I_{L(PFM)}$ is the ripple of the inductor current, which is almost the same as $I_{peak}$ in the PFM DC–DC converter.

**(2) Gate Drive Power Dissipation**    The gate capacitance of two MOSFET switches is another source of power dissipation in DC–DC converters. A DC–DC converter controls the output voltage and maintains the required load current by opening and closing two switches alternately. This process requires repeated charging of the gate capacitances of the two switches. Thus, the gate drive power dissipation is directly affected by the amount of switching per unit time, which is the switching frequency. Consequently, PWM DC–DC converters with a constant switching frequency consume a fixed gate drive power that is independent of the load condition, whereas PFM DC–DC converters consume less gate drive power as the output current diminishes. The gate drive power dissipation is roughly proportional to the input voltage, the switching frequency, and the gate charge of MOSFETs, as shown in the following equation [54]:

$$P_{\text{gate\_drive}} = V_\text{I} f_\text{S}(Q_{SW1} + Q_{SW2}), \tag{7.31}$$

where $Q_{SW1}$ and $Q_{SW2}$ are the gate charges of the top MOSFET and the bottom MOSFET, respectively.

This gate drive power model can be applied to both PWM and PFM DC–DC converters in the same way, except that $f_\text{S}$ is a constant in the PWM model, but a variable in the PFM model.

**(3) Controller Power Dissipation**  Besides the gate drive power dissipation of the control circuit, the static power dissipation of the PWM or PFM control circuit, and the power lost in miscellaneous circuits in a DC–DC converter should be considered. Generally, controller power dissipation is independent of the load condition, which makes this power dissipation a dominant one under light loads. We characterize the controller power dissipation as

$$P_{\text{controller}} = V_\text{I} I_{\text{controller}}, \tag{7.32}$$

where $I_{\text{controller}}$ is the current flowing into the controller of the DC–DC converter, excluding the current charging the gate capacitance.

Almost all manufacturing parameters ($R_{SW1}$, $R_{SW2}$, $R_L$, $R_C$, $f_\text{S}$, $I_{\text{peak}}$, $I_{\text{controller}}$, etc.) can be obtained from datasheets provided by the manufacturer of each component. A power consumption model for the DC–DC converter can be built with this information. Then, the power model is validated by comparing the power estimated by the model with figures from the datasheets or a circuit simulation.

If there is one parameter whose value is not known, for example, $I_{\text{controller}}$, its value can be estimated from the difference between the energy consumption calculated by the energy model excluding only $I_{\text{controller}}$ terms and the energy consumption obtained from the curve of load current versus efficiency (or power consumption), which is provided by the manufacturers for a specific condition, as follows:

$$P_{\text{dcdc(PWM)}}(v) = i_O(v)^2 \left( \frac{v}{V_\text{I}} R_{SW1} + \left(1 - \frac{v}{V_\text{I}}\right) R_{SW2} + R_L \right)$$

$$+ \frac{1}{3}\left( \frac{1}{2} \frac{v}{L_f f_\text{S}} \left(1 - \frac{v}{V_\text{I}}\right) \right)^2$$

$$\times \left( \frac{v}{V_\text{I}} R_{SW1} + \left(1 - \frac{v}{V_\text{I}}\right) R_{SW2} + R_L + R_C \right)$$

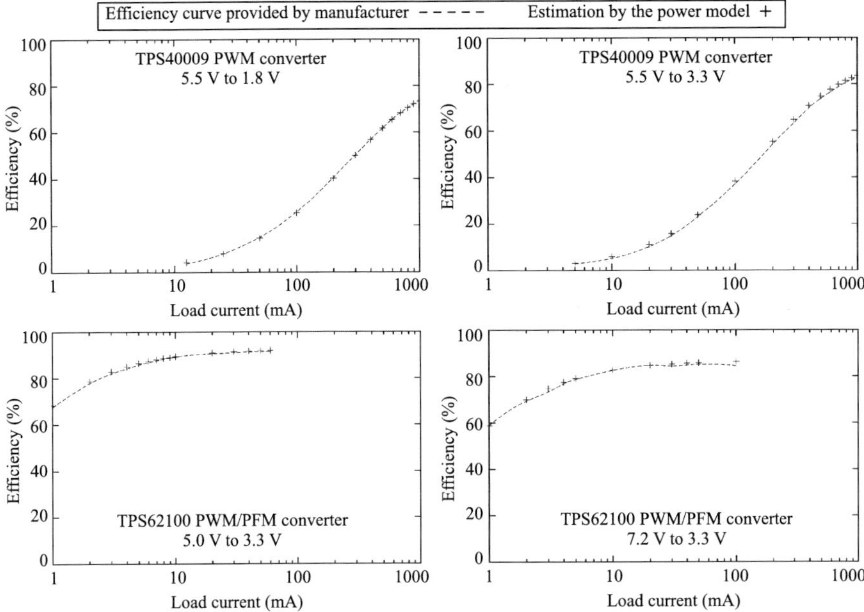


**Fig. 7.15** Comparison of DC–DC converter efficiency curves provided by a manufacturer with the estimates from our DC–DC converter power model [50]. © 2007 IEEE

$$+ V_{\mathrm{I}} f_{\mathrm{S}} (Q_{SW1} + Q_{SW2}) + V_{\mathrm{I}} I_{\mathrm{controller}}$$

$$= P_{\mathrm{dcdc\_except\_I_{controller}(PWM)}}(v) + V_{\mathrm{I}} I_{\mathrm{controller}}, \tag{7.33}$$

$$I_{\mathrm{controller}} = \frac{(P_{\mathrm{dcdc(PWM)}}(v) - P_{\mathrm{dcdc\_except\_I_{controller}(PWM)}}(v))}{V_{\mathrm{I}}}. \tag{7.34}$$

Since all circuit parameters except $I_{\mathrm{controller}}$ are given, we can obtain the value of $P_{\mathrm{dcdc\_except\_I_{controller}(PWM)}}(v)$ from the power model and the value of $P_{\mathrm{dcdc(PWM)}}(v)$ from the datasheet for a specific output voltage ($v$ in the previous equations) and, thus, estimate the value of $I_{\mathrm{controller}}$. Finally, to verify the validity of the power model and the estimated parameter, we incorporate this parameter value into the power model and then see whether it estimates the power consumption of the DC–DC converter accurately for different output voltages and changed values of other parameters.

To validate these power models, we compared the efficiency curves provided by manufacturers with the curves estimated by our model for two commercial DC–DC converters: (1) the TPS40009 [55] and (2) the TPS62100 [56], which use PWM and PWM/PFM hybrid control, respectively. All manufacturing parameters have been extracted from datasheets. As shown in Fig. 7.15, the power models for both the PWM and PFM converters are accurate enough to allow us to estimate the power dissipation of real DC–DC converters.

## 7.3 Applications

### *7.3.1 Passive Voltage Scaling*

#### 7.3.1.1 Power Conversion Efficiency for Ultra Low-Power Microprocessors

Although modern semiconductor devices are fabricated to consume low power due to various device- and circuit-level low-power techniques, high-level power managements significantly affect the lifetime of a sensor node [57]. Among them, dynamic voltage scaling (DVS) is one of the most effective high-level power management schemes for microprocessors [58]. However, modern medium- to high-performance microprocessors allow very limited ranges of supply voltage scaling, and their leakage power as well as other power-hungry off-chip memory and peripheral devices decrease the energy gain, thus often discouraging employment of DVS [59, 60].

However, ultra low-power (ULP) microcontrollers support a wide range of supply voltage scaling (3.6 V to 1.8 V for MSP430), and their energy-voltage-frequency characteristics are very close to ideal for DVS with negligible amounts of leakage power. Although ULP microcontrollers have a great potential for efficient DVS, practically it is not easy to apply DVS to ULP microcontrollers used in wireless sensor nodes. A generic DVS scheme has been observed for a wireless sensor network, but the target processor is a 32-bit RISC, StrongARM, whose power density is more than 1 W at 206 MHz, which is out of the scope of this paper [61]. In fact, although many new low-power sensor nodes have been developed, most of them do not support DVS because of the area and cost overhead of an output-adjustable DC–DC converter [62]. Even if a sensor node supports DVS, low efficiency of the DC–DC converter may offset the DVS gain. The load current of the ULP microcontroller decreases according to the square of the supply voltage, which makes the light current consumption of the ULP microcontroller even lighter, and thus results in unusually low DC–DC converter efficiency [50].

The TI MSP430F1611 is an ULP RISC microcontroller that consumes only 300 µA at 1 MHz clock frequency at 2.2 V supply voltage. It operates up to 8 MHz when 3.6 V supply voltage is applied. It is a 16-bit RISC microcontroller with 48 KB embedded flash memory and 10 KB SRAM. It is equipped with rich peripherals for sensing and actuating analog devices such as a 12-bit ADC, three operational amplifiers, dual 12-bit DACs, comparators, a hardware multiplier, serial communication interfaces, a delay-locked loop (DLL), and a supply voltage supervisor.

We measured the power consumption of MSP430F1611 by changing the supply voltage and the clock frequency. As illustrated in Fig. 7.16, we can scale the supply voltage down to 1.8 V, and the MSP430F1611 can safely run at 4 MHz. Since MSP430F1611 does not allow a supply voltage below 1.8 V, only clock scaling can further reduce its power consumption, as shown in Fig. 7.16(a). We obey the manufacturer's recommendation, though actual experiments have shown that MSP430F1611 tolerates quite higher clock frequencies for a given supply voltage. Figure 7.16(b) shows a design space considering both the supply voltage and

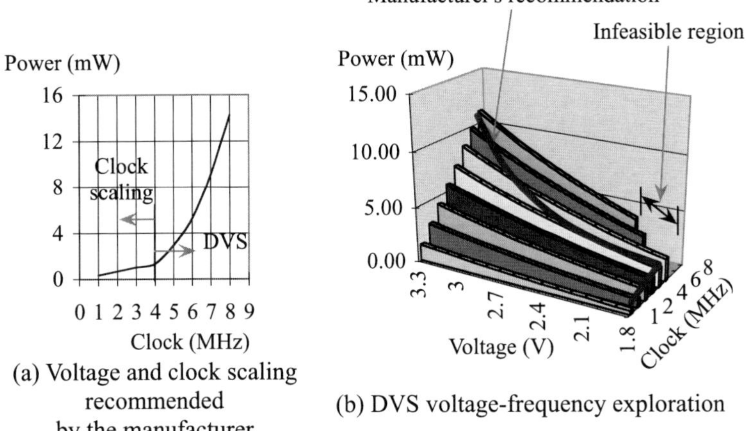


(a) Voltage and clock scaling
recommended
by the manufacturer

(b) DVS voltage-frequency exploration

**Fig. 7.16** TI MSP430 ULP microcontroller shows an almost ideal power–voltage–frequency relation for DVS [63]. © 2007 IEEE

clock frequency. It turns out that MSP430F1611 has virtually ideal power-voltage-frequency characteristics for DVS; i.e., the power consumption is almost exactly proportional to the square of the supply voltage.

### 7.3.1.2 PVS: Passive Voltage Scaling

**Principle of Operation**  PVS is a new supply voltage scaling that eliminates a DC–DC converter by hooking up a battery directly to a microcontroller. The battery connection is the same as the one to the recent wireless sensor nodes [64], which operate the microprocessor at a fixed low clock frequency at all times. But the substantial difference is that we scale the clock frequency according to the battery voltage. Figure 7.17 illustrates the concept of PVS. When the battery has a high state of charge level, $V_B$ is high enough to run the ULP core at the maximum speed. Generally, $V_B$ is decreasing by the state of charge loss (Fig. 7.17(a)), and it is monitored by an embedded voltage supervisor (Fig. 7.17(b)). The voltage supervisor generates interrupts and helps the ULP core slow down the clock frequency according to $V_B$. The clock scaling can be done continuously or discretely. In this paper, we introduce a discrete PVS. As a result, the throughput of the sensor nodes is a function of the battery voltage, i.e., the state of charge of the battery (Fig. 7.17(c)).

A discrete PVS can be easily implemented with the embedded peripheral devices of MSP430F1611 as shown in Fig. 7.17, which does not incur additional hardware overhead. In a discrete PVS, the supply voltage monitor checks to see if $V_B$ crosses the boundary of the predefined subranges. Whenever $V_B$ crosses the boundaries of the subranges, it generates an interrupt. The ULP core then reprograms the DLL to scale down the clock frequency. In fact, a discrete PVS adjusts the clock frequency

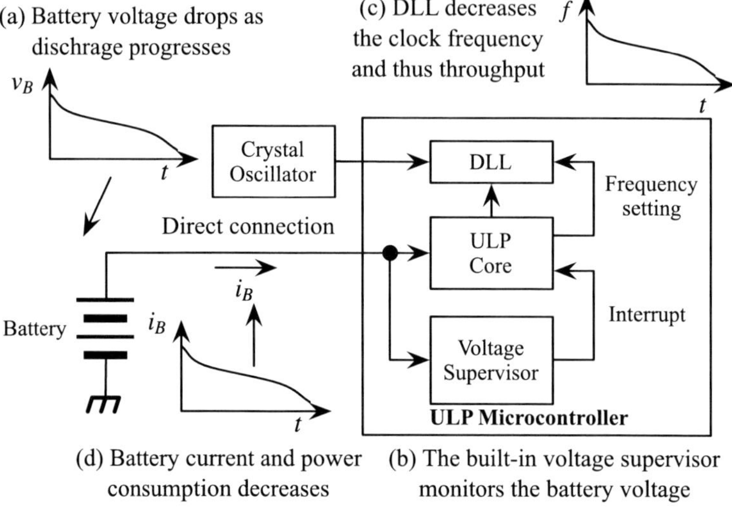



**Fig. 7.17** PVS without a DC–DC converter: the clock frequency changes by the battery voltage, in a passive manner [63]. © 2007 IEEE

only a few times over the whole lifetime of a sensor node, which incurs only negligible overhead.

The clock frequency is adjusted by the following delay equation:

$$\frac{1}{f} \propto \frac{V_{DD}}{(V_{DD} - V_t)^\alpha},\tag{7.35}$$

where $V_{DD}$ is the supply voltage such that $V_{DD} = V_B$, $V_t$ is the threshold voltage of the CMOS logic that composes the ULP microcontroller, and $\alpha$ is a velocity saturation coefficient.

**Throughput Characteristics**   Sensor nodes with DC–DC converters regulate the battery voltage, and thus the microcontrollers can run with any desired clock frequencies over the entire lifetime. In other words, the performance of the microcontrollers has nothing to do with the battery state of charge (SOC). Recent wireless sensor nodes have removed the DC–DC converter and simply fixed the clock frequency low enough so that the ULP microcontrollers tolerate the lowest battery voltage. Thus, such a sensor node also exhibits throughput which is independent of the battery SOC [64]. However, as the sensor node experiences the lowest battery voltage only when the battery is about to die out, this is a very conservative method that wastes the performance of the ULP microcontrollers during most of its lifetime.

In contrast, as PVS continuously decreases the clock frequency according to (7.35), the throughput of the microcontroller also decreases. This is a new characteristic of a digital system, which, however, was found in the old analog systems. For example, old portable radios could make a loud sound output when they had new batteries, but they made a softer sound output as the battery discharge progressed.

**Energy Gain of PVS**   DVS achieves energy gain from a microprocessor by reducing the dynamic energy for a clock cycle. At the same time, the leakage energy and other peripheral energy increase as the execution time of a task is lengthened [59, 60]. The baseline energy consumption of a system, with a DC–DC converter and a fixed $f_c$, is given by

$$E(\text{Baseline}) = E(\text{CPU}) + E(\text{Peripherals}) + E(\text{DC–DC}), \qquad (7.36)$$

so the energy gain from DVS can be written as follows:

$$E_{\text{gain}}(\text{DVS}) \approx E_{\text{gain}}^{\text{DVS}}(\text{CPU}) - E_{\text{loss}}(\text{Peripherals}). \qquad (7.37)$$

The primary advantage of PVS is longer battery life, but the sources of energy gain are completely different from those of DVS; PVS does not aim at only CPU energy reduction. First, elimination of the DC–DC converter can restore the energy loss, which ranges from typically 20% to 40% in conventional DVS for ULP microcontrollers. Second, decreasing the battery current such that $i_B \propto V_B$ increases the total energy that can be drawn from the battery. Third, a DC–DC converter is a noisy analog component that requires passive components such as an inductor and a bulk capacitor. The layout and circuit board pattern design is not cheap in terms of minimizing and isolating the switching noise. PVS helps reduce the complexity, area, and cost of a system. Fourth, PVS is ideal for cheap alkaline batteries whose voltage drop is significant by the state of charge loss. Most previous battery-aware power management assumed the use of Li-ion batteries, but alkaline batteries are more appropriate for a disposable and low-cost system. Fifth, PVS recovers potential performance of the CPU and reduces the execution time (or increase throughput). This consequently does not incur appreciable peripheral energy overhead, as opposed to DVS. The energy gain from PVS can be written as follows:

$$E_{\text{gain}}(\text{PVS}) \approx E_{\text{gain}}^{\text{PVS}}(\text{CPU}) + E_{\text{gain}}(\text{Battery}) + E(\text{DC–DC}). \qquad (7.38)$$

Thus, the energy gain of PVS is not limited to the CPU, and has great potential to reduce the entire system energy when the CPU power portion is not dominant. We demonstrate the amount of energy savings in the following example.

### 7.3.1.3 Example

In this section, we evaluate PVS considering the whole sensor network. We demonstrate how PVS achieves longer lifetime with traditional performance-driven routing for a distributed data processing. A necessary condition to justify the effectiveness of PVS is that network latency (not power consumption) should be bounded by computational delay, not by communication delay, which is common in wireless sensor networks [65].

We borrowed TinyOS [66] and real applications and compared the actual battery SOC and lifetime of PVS with DC–DC and Telos. We used TOSSIM and a

**Fig. 7.18** Residual energy snapshots of the sensor network with DC–DC, Telos, and PVS [63]. © 2007 IEEE

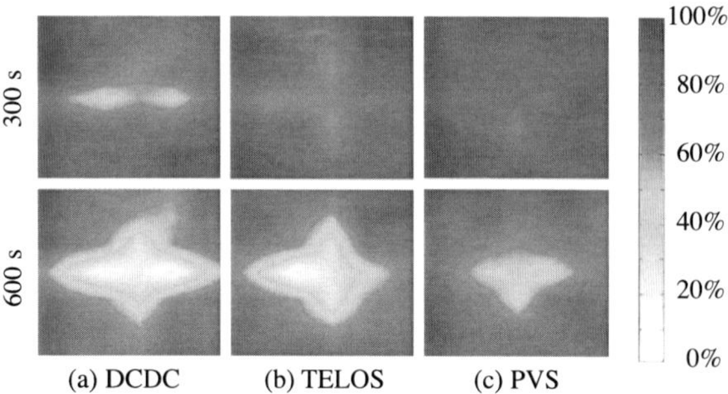



customized MultiHopRouter component to include the battery, DC–DC converters, and detailed MSP430F161 energy and performance models. We have employed a discrete battery model considering the rate capacity and recovery effects [67]. We assume a $7 \times 7$ sensor network where each node is spaced at an equal distance, forming a grid. The center node is a sink node having a storage element; it only receives packets. The other nodes sample ambient noise with a 7.5 kHz sampling rate with 8-bit resolution, and forward it to the sink node. Once a node is triggered, it samples the noise for 4 s. We assume that a noise level that exceeds the threshold occurs every 8 s with uniform spatial probability.

Once a node is triggered, it captures and forwards the noise data to an adjacent node without compression. The adjacent node receives the data and compresses it with a ratio of 2, and forwards it to its adjacent node again, until the data reaches the sink node. At this point, we do not attempt any further compression. If a node receives another data packet during data compression, it forwards the newly ar-rived data packet to the next adjacent node without compression. The compression throughput of a node is 80 Kbps at 8 MHz. The radio bandwidth is 250 Kbps, and the actual data transmission rate is 60 Kbps. We scale down the capacity of the Toshiba LR-44 button alkaline battery to 0.2 mAh (500:1) for faster simulation. The TX power and RX power are 52.2 mW and 56.4 mW, respectively [66]. We assume ideal scheduled wake-up for the RX power management. The underlying routing scheme is a performance-driven routing: a tree-based routing [68].

Figure 7.18 shows the snapshots of the residual energy of the sensor network for the three different schemes: DC–DC, Telos, and PVS. DC–DC shows rapid decrease of the residual energy compared to Telos and PVS. PVS shows the best residual en-ergy savings and even wear of the sensor nodes. Figure 7.19 shows the network performance comparison among DC–DC, Telos, and PVS. Among them, DC–DC shows 0.56 s average latency, but the lifetime is much shorter, around 689 s. Telos lasts more than 817 s, but its average latency is 0.73 s. In contrast, PVS lasts up to 1,095 s and the average latency is 0.54 s. In conclusion, PVS combines the ad-vantages of both Telos and DC–DC, in terms of residual energy savings and high performance. Most of all, PVS offers even wear of the network with a performance-driven routing.

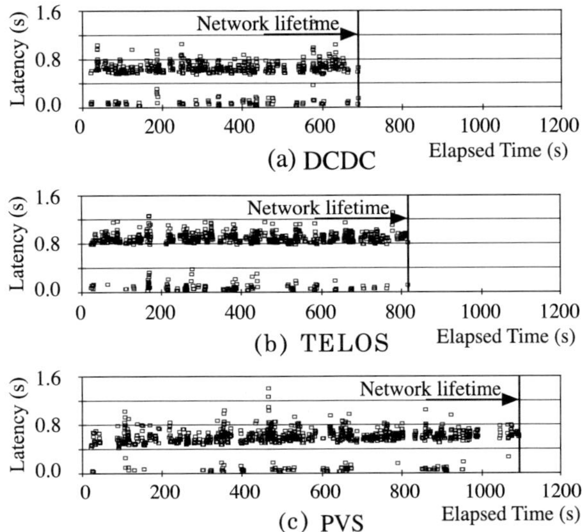



**Fig. 7.19** Latency distribution of the sensor nodes [63]. © 2007 IEEE

## 7.3.2 Dynamic Regulator Scheduling

### 7.3.2.1 Efficiencies of Voltage Regulators

DC–DC converters control the output voltage level by chopping the DC input and applying the pulses to a transformer or a charge storage element. This allows a DC–DC converter to reduce or increase the input voltage, and its efficiency is not strongly related to the input and output voltage difference, as it is in an LDO. Nevertheless, the efficiency of a DC–DC converter is compromised by IR, switching, and control losses. A well-designed switching regulator has an efficiency of 80–90% at a typical load current, and this is generally a much better performance than that of a linear regulator. However, if the load current is small, the efficiency of a DC–DC converter becomes significantly worse because the switching and control loss dominate the total power consumption [50].

Passive voltage scaling (PVS) is a new method of supply voltage and frequency scaling, in which the supply voltage of the CPU is determined by the battery voltage in a passive manner. There is no direct power loss because there is no regulator, but the CPU voltage must always be equal to the battery voltage. This can increase the power consumption of the CPU to a level that is significantly larger than that of the same CPU with an LDO or a DC–DC converter. When the battery has a large amount of residual energy, the PVS may supply voltage to the CPU that is higher than the energy-optimal CPU voltage. In such a situation, the only possible way to reduce the power requirement is dynamic clock frequency scaling.

Figure 7.20 shows how the efficiency of a DC–DC converter, an LDO, and PVS vary with output voltage, where the maximum possible CPU frequency is selected for each voltage. A DC–DC converter is generally more efficient than an LDO within a typical operating range with a medium to large load current. But when

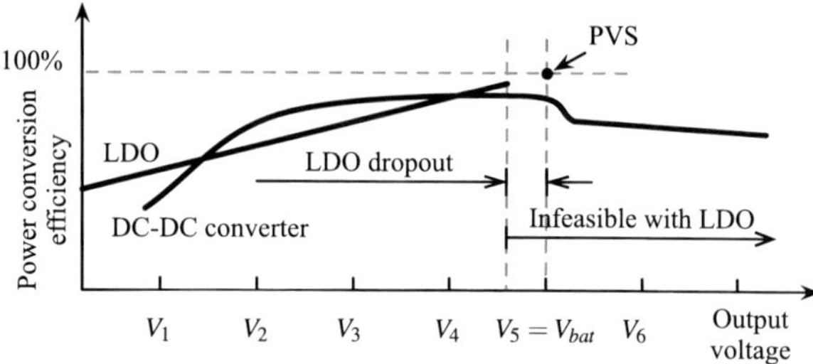

**Fig. 7.20** Efficiencies of various voltage regulation methods ($V_{\mathrm{bat}}$ is the battery voltage, or the voltage input to the regulator) [69]. © 2008 IEEE



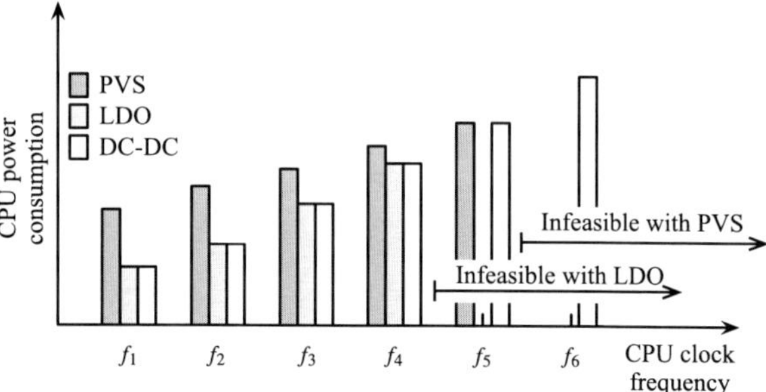

**Fig. 7.21** Power consumption of the CPU with different voltage regulators (note that $V_{\mathrm{bat}}$ is the minimum CPU voltage required for operation at $f_5$) [69]. © 2008 IEEE



the load current is very low, the LDO becomes relatively more efficient. PVS has no converter loss, but it sometimes results in unnecessary high power consumption. Figure 7.21 shows how the power consumption of a CPU varies with its clock frequency, if it runs at the lowest allowable voltage. Note that a CPU supplied by PVS uses more power than a CPU supplied by an LDO or a DC–DC converter for frequencies less than $f_5$. This is because the PVS supply voltage cannot be reduced to match lower frequencies.

### 7.3.2.2 Dynamic Regulator Scheduling (DRS) Problem

DRS can be defined as a three-tuple $(\Gamma, \mathbf{r}, \mathbf{s})$, where $\Gamma = (T_1, \ldots, T_N)$ is a set of $N$ tasks, $\mathbf{r} = (r_1, \ldots, r_M)$ is a set of $M$ voltage regulators, and $\mathbf{s} = (s_1, \ldots, s_L)$ is a set of scaling factors $T_i \in \Gamma$. We wish to consider the practical aspects of DRS as well as create a solvable formulation, and we therefore consider discrete DVFS, in which each $r_k$, where $k = 1, \ldots, M$, can generate $L$ voltage levels. A task $T_i$ is also defined by a three-tuple $T_i = (a_i, d_i, \tau_i)$, where $a_i$ is the arrival time, $d_i$ is the deadline, and $\tau_i$ is the worst-case execution time with a scaling factor of 1. Figure 7.22 illustrates the generalized DRS problem.

We denote the battery voltage by $V_{\mathrm{bat}}$. The power consumption of the system, including the voltage regulator and CPU, is a function of $P(V_{\mathrm{bat}}, s_j)$, where $j = 1, \ldots, L$, which will be described in detail in the next section. Finally, $x_{ij}$, where $i = 1, \ldots, N$ and $j = 1, \ldots, L$, is the binary variable of a 0–1 integer linear programming (ILP) formulation, which is to be determined. The value of $x_{ij}$ indicates whether the $i$th task will execute at the $j$th voltage level: one value of $x_{ij}$ must be 1, and the others must be zero.

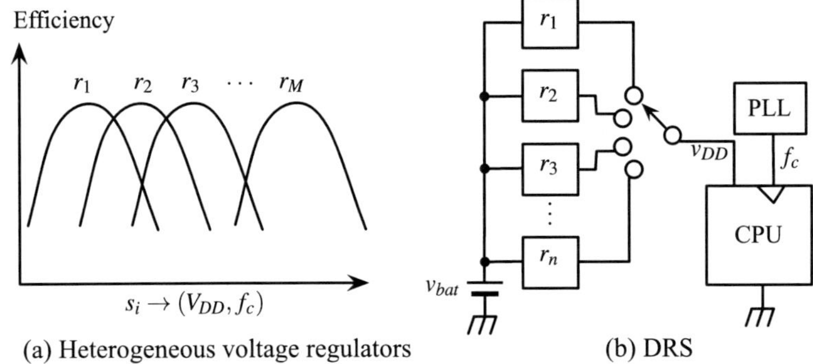



(a) Heterogeneous voltage regulators                (b) DRS

**Fig. 7.22**  Generalized DRS problem [69]. © 2008 IEEE

**Fig. 7.23**  Arrangement of heterogeneous voltage regulators [69]. © 2008 IEEE

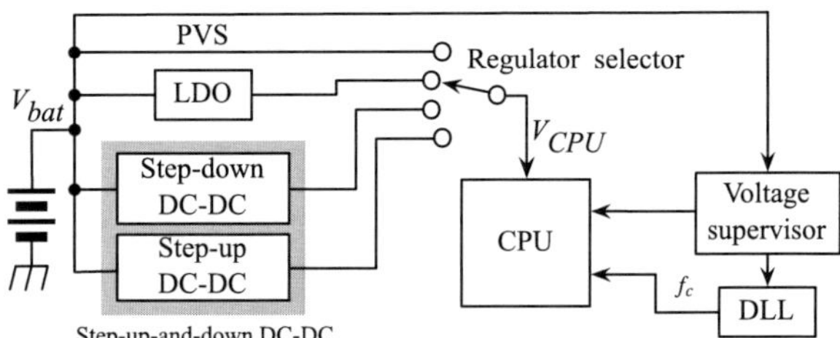



### 7.3.2.3 Example

An example arrangement of heterogeneous voltage regulators is shown in Fig. 7.23. Figure 7.24 shows how the joint optimization considering the regulator efficiency may result in a different scheduling result. We obtained an energy cost function from a system with an MSP430 microprocessor, a TI TPS62100 DC–DC converter, and a TI TPS78601 LDO. Figure 7.25 shows the power consumption of the CPU and the power loss in the regulators versus $V_{bat}$ and $s_j$, for each selected regulator. The clock frequency of the CPU, $f_c$, is determined by $\frac{f_{max}}{s_j}$, where $f_{max}$ is the clock frequency for $s_j = 1$. For the DC–DC and LDO regulators, $V_{CPU}$ is set to the lowest value allowable for each scaling factor; otherwise $V_{CPU}$ is the same as $V_{bat}$. A white bar means that a configuration is not feasible. For instance, when $V_{bat} = 2.7$ V and $s_j = 1$, neither PVS nor LDO can operate the CPU. We can use Fig. 7.25 to find the most effective voltage regulator for a given $V_{bat}$ and $s_j$. For example, when $V_{bat} = 3$ V and $s_j = 1.14$, we choose LDO since it achieves the least overall power consumption.

Table 7.2 shows the best regulator and the power consumption, including the regulator loss, for each combination of battery voltage and scaling factor. It turns out that each regulator has its distinct energy-efficient region and constraints. The PVS configuration is only useful when the input voltage is similar to the optimal CPU voltage. LDO performs the best when the input voltage is slightly higher than the optimal CPU supply voltage. The DC–DC converter is the only feasible solution

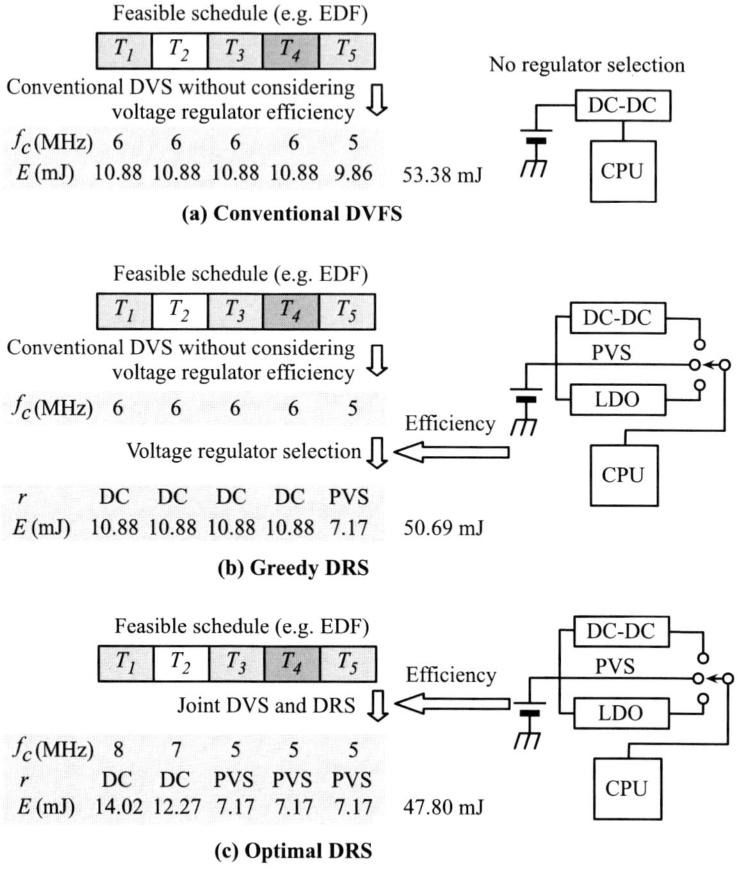



**(a) Conventional DVFS**

**(b) Greedy DRS**

**(c) Optimal DRS**

**Fig. 7.24** Example results from DVFS, greedy DRS, and optimal DRS scheduling. The variables $r$ and $E$ are the type of regulator and the energy consumption, respectively [69]. © 2008 IEEE

when we need to step up the supply voltage, or scale it down by a large factor. The energy cost function for optimal DRS varies over time; as the battery discharges, the best choice of regulator is likely to change. Furthermore, the energy cost function is nonconvex. Although most CPUs have convex energy cost functions, the addition of heterogeneous regulators can result in a nonconvex system energy cost function, as shown in Fig. 7.25.

### 7.3.3 Battery and Supercapacitor Hybrid

#### 7.3.3.1 Parallel Connection

A battery-supercapacitor hybrid, as shown in Fig. 7.26, is a simple way of reducing the effect of load fluctuation on the supplied voltage level. The supercapacitor

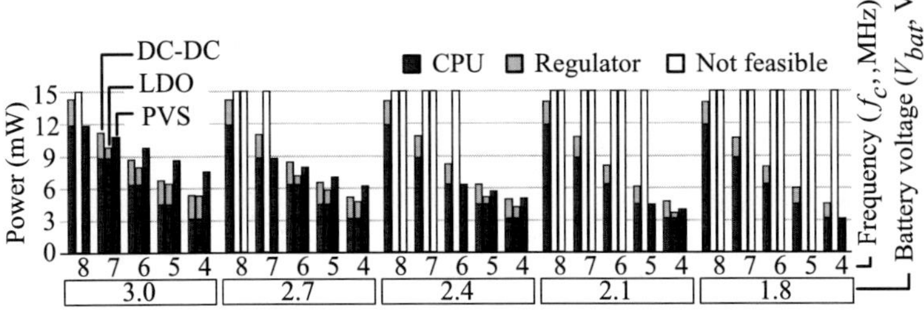

**Fig. 7.25** Total power consumption (CPU power plus voltage regulator loss) versus $V_{bat}$ and $f_c$ [69]. © 2008 IEEE

**Table 7.2** Power consumption (mW) of a multiple heterogeneous voltage regulator system with 1 mA leakage. The cell color denotes the chosen regulator [69]. © 2008 IEEE

| $V_{bat}$ | Scaling factor, $s_j$ | | | | | | | |
|---|---|---|---|---|---|---|---|---|
| | 1.00 | 1.14 | 1.34 | 1.60 | 2.00 | 2.67 | 4.00 | 8.00 |
| 3.3 | 13.11 | 10.81 | 8.83 | 7.00 | 5.61 | 5.25 | 4.68 | 4.02 |
| 3.2 | 13.55 | 10.48 | 8.56 | 6.92 | 5.54 | 5.09 | 4.54 | 3.90 |
| 3.1 | 12.69 | 10.15 | 8.30 | 6.71 | 5.47 | 4.93 | 4.40 | 3.78 |
| 3.0 | 11.83 | 9.82 | 8.03 | 6.49 | 5.32 | 4.77 | 4.26 | 3.66 |
| 2.9 | 14.30 | 10.10 | 7.76 | 6.27 | 5.15 | 4.61 | 4.11 | 3.53 |
| 2.8 | 14.26 | 9.41 | 7.49 | 6.06 | 4.97 | 4.46 | 3.97 | 3.41 |
| 2.7 | 14.22 | 8.76 | 7.22 | 5.84 | 4.79 | 4.30 | 3.83 | 3.29 |
| 2.6 | 14.18 | 10.94 | 7.39 | 5.62 | 4.61 | 4.14 | 3.69 | 3.17 |
| 2.5 | 14.14 | 10.90 | 6.86 | 5.41 | 4.44 | 3.98 | 3.55 | 3.05 |
| 2.4 | 14.11 | 10.85 | 6.35 | 5.19 | 4.26 | 3.82 | 3.41 | 2.93 |
| 2.3 | 14.08 | 10.81 | 8.25 | 5.27 | 4.08 | 3.66 | 3.26 | 2.80 |
| 2.2 | 14.05 | 10.77 | 8.21 | 4.87 | 3.90 | 3.50 | 3.12 | 2.68 |
| 2.1 | 14.02 | 10.74 | 8.16 | 4.48 | 3.73 | 3.34 | 2.98 | 2.56 |
| 2.0 | 14.00 | 10.70 | 8.12 | 6.11 | 3.73 | 3.30 | 2.89 | 2.44 |
| 1.9 | 13.99 | 10.68 | 8.08 | 6.07 | 3.43 | 3.05 | 2.69 | 2.29 |
| 1.8 | 13.98 | 10.65 | 8.04 | 6.02 | 3.14 | 2.81 | 2.50 | 2.14 |

☐ LDO   ▨ DC–DC   ☐ PVS

connected in parallel acts as a lowpass filter that prunes out rapid voltage changes. The battery-supercapacitor hybrid is thus effective in mitigating the rate capacity effect for intermittent (rather than continuous) high load current. The supercapacitor shaves the short-duration, high-amplitude load current spikes and makes a wider duration but lower amplitude current, which would result in better energy efficiency due to lower rate capacity effect in Li-ion batteries.

In the parallel connection configuration, the filtering effect of the supercapacitor is largely dependent on its capacitance: a larger capacitance results in a better filter-

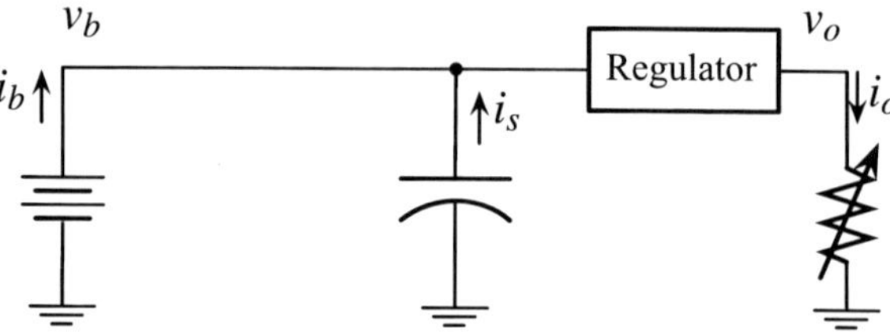

**Fig. 7.26** Parallel connection battery-supercapacitor hybrid system [35]. © 2011 IEEE

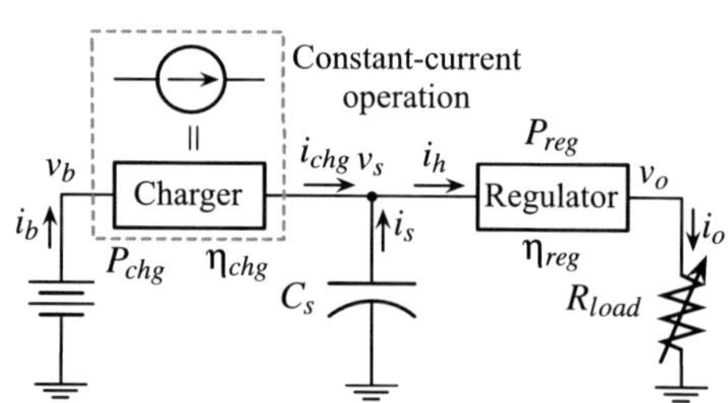

**Fig. 7.27** Battery-supercapacitor hybrid system using constant-current charger [35]. © 2011 IEEE

ing effect. As a result, the parallel connection has a limited ability to reduce the rate capacity effect in the Li-ion battery when the capacitance value of the supercapacitor is not sufficiently large. Unfortunately, because of the volumetric energy density and cost constraints in its practical deployment, the supercapacitor capacitance is generally rather small.

### 7.3.3.2 Constant-Current Charger-Based Architecture

A new hybrid architecture using a constant-current charger was introduced to overcome the disadvantage of the conventional parallel connection hybrid architecture [35]. The constant-current charger-based system is illustrated in Fig. 7.27. The constant-current charger separates the battery and the supercapacitor. It maintains a desired amount of the charging current regardless of the SOC of the supercapacitor; whereas, in the conventional parallel connection configuration, the charging current is not controllable and varies greatly as a function of the SOC of the supercapacitor. Consequently, the proposed hybrid architecture can reduce variation in the battery discharging current even with a small supercapacitor.

### 7.3.3.3 Constant-Current Charger-Based Architecture Design Considerations

Several problems must be addressed in order to develop a constant-current charger-based system. With a fixed supercapacitor charging current, the amount of delivered power from the battery to the load will depend on the terminal voltage of the supercapacitor. Therefore, we need to charge the supercapacitor up to a certain voltage at the initial state to give it enough power before supplying power to the load, as illustrated in interval ⓐ of Fig. 7.28. We will use part of the pre-charged energy in the

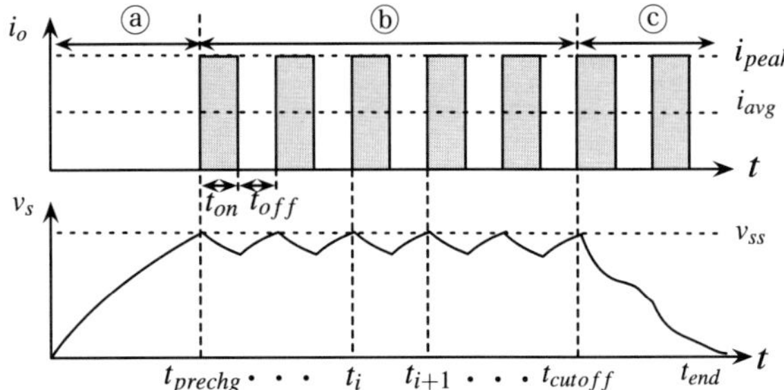



**Fig. 7.28** Pre-charged and steady-state operation of supercapacitor [35]. © 2011 IEEE

supercapacitor after the battery cutoff time (interval ⓒ in Fig. 7.28). We consider this amount of energy to calculate the end-to-end energy efficiency.

Next, we carefully control the charging current and the terminal voltage of the supercapacitor to achieve efficient and stable operation. The value of the supercapacitor charging current must be carefully set because it largely affects the efficiency of the system. The terminal voltage of the supercapacitor determines the amount of power which is transferred from the battery to the supercapacitor because of the fixed charging current. This terminal voltage must thus be maintained within a proper range in order to meet the load power demand. It is to supply sufficient power transfer to the load, which is determined by the capacitor voltage and the charging current. If the supercapacitor's terminal voltage is too high, excessive power will be transferred from the battery to the supercapacitor; in this case, the terminal voltage continuously rises until some other circuit element pinches off the voltage rise at that terminal. On the other hand, if the supercapacitor's terminal voltage is too low, the load demand may not be met. During interval ⓑ in Fig. 7.28 we have steady-state operation with periodic pulsed load and constant charging current. We need to make the supercapacitor voltage at the start time of the high current time period equal to the voltage at the end time of the low current time period.

Finally, we need to consider the system-level power and energy density. In general, the larger the supercapacitor, the higher the energy efficiency. However, supercapacitors have a significant disadvantage in terms of their volumetric energy density and dollar cost per unit of stored energy compared to batteries. For portable applications where the size is a constraint and cost is a factor, the size of the supercapacitor should be minimized while still achieving a reasonable energy efficiency.

#### 7.3.3.4 Design Example

The overall energy efficiency from the battery to the load in the hybrid system is given by

$$\eta_{system} = \frac{E_{deliver}}{E_{stored}}, \tag{7.39}$$

where $E_{stored}$ and $E_{deliver}$ denote the stored energy in the battery and delivered energy to the load, respectively. The energy density of the system may be calculated

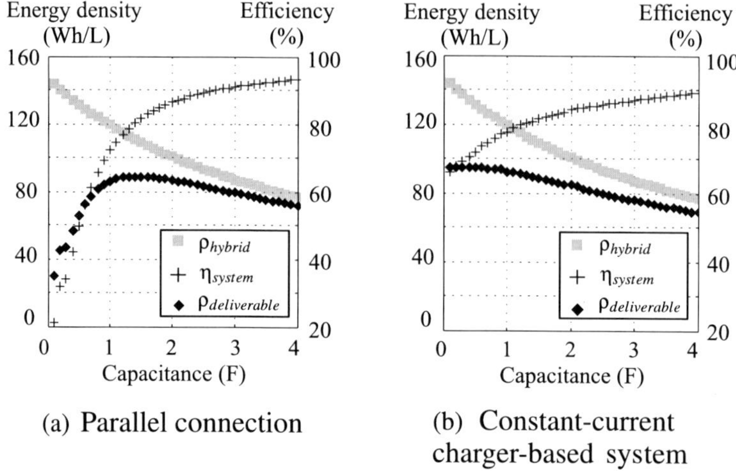



(a) Parallel connection

(b) Constant-current charger-based system

**Fig. 7.29** System efficiency, energy density, and deliverable energy density of the parallel connection and the constant-current charger-based system with pulsed load [35]. © 2011 IEEE

as

$$\rho_{\text{hybrid}} = \frac{E_{\text{b}} + E_{\text{s}}}{H_{\text{b}} + H_{\text{s}}}, \tag{7.40}$$

where $E_{\text{b}}$, $E_{\text{s}}$, $H_{\text{b}}$, and $H_{\text{s}}$ are the amount of stored energy in the battery, amount of stored energy in the supercapacitor, volume of the battery, and volume of the supercapacitor, respectively. Finally, we get the deliverable energy density, which is given by

$$\rho_{\text{deliver}} = \eta_{\text{system}}\rho_{\text{hybrid}}. \tag{7.41}$$

Based on definitions of the system efficiency, delivered energy, volumetric energy density, and deliverable energy density, we can address the optimization problem as determining $(C_{\text{s}}, i_{\text{chg}})$ to achieve the maximum deliverable energy density $\rho_{\text{deliver}}$ for a given battery and load profile, considering the maximum voltage rating of the supercapacitor as well as the current and voltage rating of the converters as constraints. The key parameters of the supercapacitors in the proposed system are the voltage rating and capacitance; they are directly related to the energy transfer efficiency and energy density of the proposed system. The capacitance value of the supercapacitor affects the efficiency due to its filtering effect on the pulsed load. Moreover, the volume of the supercapacitor is determined by its capacitance value and voltage rating. Because different amounts of charging current result in different steady states, the value of the charging current results in different requirements for the voltage rating for the supercapacitor. These two design parameters, the supercapacitor capacitance and the charging current, also strongly influence the efficiency of the charger and regulator. The maximum ratings of the switching converters, battery, and supercapacitor should be considered as constraints.

We have obtained the energy efficiency and deliverable energy density of the system by simulation using the parameters of the Li-ion battery and supercapacitor. We use a 10 s period, 10C discharge rate, and a pulse current with 10% duty cycle as a pulsed load. From Fig. 7.29(a), the energy efficiency, $\eta_{system}$, increases as the capacitance of the supercapacitor increases. However, the overall volumetric energy density of the system decreases as the capacitance increases because the energy density of the supercapacitor is much lower than that of the battery. The constant-current charger-based hybrid architecture achieves higher deliverable energy density with small capacitance because the conversion efficiency is less affected by the capacitance, as shown in Fig. 7.29(b).

# References


1. Jongerden, M., Haverkort, B.: Battery modeling. University of Twente (2008)
2. Henson, W.: Optimal battery/ultracapacitor storage combination. J. Power Sources (2008)
3. Chen, H., Cong, T.N., Yang, W., Tan, C., Li, Y., Ding, Y.: Progress in electrical energy storage system: a critical review. Prog. Nat. Sci. (2009)
4. Moore, T., Douglas, J.: Energy storage, big opportunities on a smaller scale. EPRI J. (2006)
5. OEM Li-ion batteries. www.panasonic.com
6. Kopera, J.J.: Inside the nickel metal hydride battery (2004)
7. Fetcenko, M., Ovshinsky, S., Reichman, B., Young, K., Fierro, C., Koch, J., Zallen, A., Mays, W., Ouchi, T.: Recent advances in NiMH battery technology. J. Power Sources **165**(2), 544–551 (2007)
8. Simjee, F., Chou, P.H.: Everlast: long-life, supercapacitor-operated wireless sensor node. In: Proceedings of the International Symposium on Low Power Electronics and Design, pp. 197–202 (2006)
9. Atwater, T., Cygan, P., Leung, F.: Man portable power needs of the 21st century: I. applications for the dismounted soldier. ii. enhanced capabilities through the use of hybrid power sources. J. Power Sources **91**(1), 27–36 (2000)
10. Gamburzev, S., Appleby, A.J.: Recent progress in performance improvement of the proton exchange membrane fuel cell (PEMFC). J. Power Sources **107**(1), 5–12 (2002)
11. Costamagna, P., Srinivasan, S.: Quantum jumps in the PEMFC science and technology from the 1960s to the year 2000: Part I. Fundamental scientific aspects. J. Power Sources **102**(1–2), 242–252 (2001)
12. Costamagna, P., Srinivasan, S.: Quantum jumps in the PEMFC science and technology from the 1960s to the year 2000: Part II. Engineering, technology development and application aspects. J. Power Sources **102**(1–2), 253–269 (2001)
13. Chang, N., Seo, J., Shin, D., Kim, Y.: Room-temperature fuel cells and their integration into portable and embedded systems. In: Proceedings of IEEE Asia South Pacific Design Automation Conference, ASP-DAC, January 2010, pp. 69–74 (2010)
14. Aricó, A.S., Srinivasan, S., Antonucci, V.: DMFCs: From fundamental aspects to technology development. Fuel Cells **1**(2), 133–161 (2001)
15. Kim, Y., Shin, D., Seo, J., Chang, N., Cho, H., Kim, Y., Yoon, S.: System integration of a portable direct methanol fuel cell and a battery hybrid. Int. J. Hydrog. Energy **35**(11), 5621–5637 (2010)
16. Kim, D., Cho, E.A., Hong, S.-A., Oh, I.-H., Ha, H.Y.: Recent progress in passive direct methanol fuel cells at KIST. J. Power Sources **130**(1–2), 172–177 (2004)
17. Chen, R., Zhao, T.: Performance characterization of passive direct methanol fuel cells. J. Power Sources **167**(2), 455–460 (2007)

18. Guo, Z., Faghri, A.: Miniature DMFCs with passive thermal-fluids management system. J. Power Sources **160**(2), 1142–1155 (2006)

19. Steele, B.C.H.: Material science and engineering: the enabling technology for the commercialisation of fuel cell systems. J. Mater. Sci. **36**, 1053–1068 (2001)

20. Qian, W., Wilkinson, D.P., Shen, J., Wang, H., Zhang, J.: Architecture for portable direct liquid fuel cells. J. Power Sources **154**(1), 202–213 (2006)

21. Knowlen, C., Matick, A., Bruckner, A.: High efficiency energy conversion systems for liquid nitrogen automobiles. In: SAE Future Transportation Technology Conference (1998)

22. Wen, D., Chen, H., Ding, Y., Dearman, P.: Liquid nitrogen injection into water: pressure build-up and heat transfer. Cryogenics **46**(10), 740–748 (2006)

23. Manwell, J.F., Rogers, A., Hayman, G., Avelar, C.T., McGowan, J.G.: Hybrid 2: A Hybrid System Simulation Model: Theory Manual. University of Massachusetts Press, Amherst (1998)

24. Benini, L., Castelli, G., Macii, A., Macii, E., Poncino, M., Scarsi, R.: Discrete-time battery models for system-level low-power design. IEEE Trans. Very Large Scale Integr. **9**(5), 630–640 (2002)

25. Chiasserini, C., Rao, R.: Energy efficient battery management. IEEE J. Sel. Areas Commun. **19**(7), 1235–1245 (2002)

26. Rakhmatov, D., Vrudhula, S.: Energy management for battery powered embedded systems. ACM Trans. Embed. Comput. Syst. **2**, 277–324 (2003)

27. Rong, P., Pedram, M.: An analytical model for predicting the remaining battery capacity of lithium-ion batteries. In: Proceedings of Conference on Design, Automation and Test in Europe, pp. 11–48 (2003)

28. Doyle, M., Fuller, T.F., Newman, J.: Modeling of galvanostatic charge and discharge of the lithium/polymer/insertion cell. J. Electrochem. Soc. **140**, 1526–1533 (1994)

29. Luo, J., Jha, N.K.: Battery-aware static scheduling for distributed real-time embedded systems. In: Proceedings of Design Automation Conference, pp. 444–449 (2001)

30. Khan, J., Vemuri, R.: An iterative algorithm for battery-aware task scheduling on portable computing platforms. In: Proceedings of Conference on Design, Automation and Test in Europe, pp. 622–627 (2005)

31. Chowdhury, P., Chakrabarti, C.: Battery aware task scheduling for a system-on-a-chip using voltage/clock scaling. In: Proceedings of IEEE Workshop on Signal Processing Systems, pp. 201–206 (2002)

32. Pedram, M., Wu, Q.: Battery-powered digital CMOS design. IEEE Trans. Very Large Scale Integr. **10**, 601–607 (2002)

33. Rong, P., Pedram, M.: Battery-aware power management based on Markovian decision processes. In: Proceedings of International Conference on Computer-Aided Design, pp. 707–713 (2002)

34. Benini, L., Castelli, G., Macii, A., Scarsi, R.: Battery-driven dynamic power management. IEEE Des. Test Comput. **18**, 53–60 (2001)

35. Shin, D., Wang, Y., Kim, Y., Seo, J., Pedram, M., Chang, N.: Battery-supercapacitor hybrid system for high-rate pulsed load applications. In: Proceedings of IEEE Design Automation and Test in Europe (2011)

36. Rao, V., Singhal, A.K.G., Navet, N.: Battery model for embedded systems. In: Proceedings of the 18th IEEE International Conference on VLSI Design, pp. 105–110 (2005)

37. Gao, L., Liu, S., Dougal, R.: Dynamic lithium-ion battery model for system simulation. IEEE Trans. Compon. Packag. Technol. **25**(3), 495–505 (2002)

38. Chen, M., Rincon-Mora, G.: Accurate electrical battery model capable of predicting runtime and I–V performance. IEEE Trans. Energy Convers. **21**(2), 504–511 (2006)

39. Erdinc, O., Vural, B., Uzunoglu, M.: A dynamic lithium-ion battery model considering the effects of temperature and capacity fading. In: International Conference on Clean Electrical Power, pp. 383–386 (2009)

40. DualFoil. http://www.cchem.berkeley.edu/jsngrp/fortran.html

41. Hageman, S.C.: Simple PSpice models let you simulate common battery types. Electron. Des. News **38**, 117–129 (1993)

42. Rakhmatov, D., Vrudhula, S.: An analytical high-level battery model for use in energy management of portable electronic systems. In: Proceedings of the IEEE/ACM International Conference on Computer-Aided Design, pp. 488–493 (2001)
43. Manwell, J.F., McGowan, J.G.: Lead acid battery storage model for hybrid energy systems. Sol. Energy **50**, 399–405 (1993)
44. Chiasserini, C., Rao, R.: Pulsed battery discharge in communication devices. In: Proceedings of the 5th IEEE International Conference on Mobile Computing and Networking, pp. 88–95 (1999)
45. Lee, K., Chang, N., Zhuo, J., Chakrabarti, C., Kadri, S., Vrudhula, S.: A fuel-cell-battery hybrid for portable embedded systems. ACM Trans. Des. Autom. Electron. Syst. **13**(1) (2008)
46. Amphlett, J., Baumert, R., Mann, R., Peppley, B., Roberge, P.: Performance modeling of the Ballard Mark IV solid polymer electrolyte fuel cell. J. Electrochem. Soc. **142**, 1–8 (1995)
47. Pukrushpan, J.T., Stefanopoulou, A.G., Peng, H.: Control of Fuel Cell Power Systems: Principles, Modeling Analysis, and Feedback Design. Springer, Berlin (2004)
48. Guzzella, L.: Control oriented modeling of fuel-cell based vehicles. In: Presentation in NSF Workshop on the Integration of Modeling and Control for Automotive Systems (1999)
49. Larminie, J., Dicks, A.: Fuel Cell Systems Explained. Wiley, New York (2000)
50. Choi, Y., Chang, N., Kim, T.: Dc-dc converter-aware power management for low-power embedded systems. IEEE Trans. Comput.-Aided Des. Integr. Circuits Syst. **26**(8), 1367–1381 (2007)
51. Falin, J.: A 3-A, 1.2-Vout linear regulator with 80% efficiency and Plost 1 W. Analog Appl. J. 10–12 (2006)
52. Stratakos, A.: High-efficiency low-voltage dc-dc conversion for portable applications. Ph.D. dissertation, University of California, Berkeley, CA, 1999
53. Kursun, V., Narendra, S., De, V., Friedman, E.: Monolithic DC-DC converter analysis and MOSFET gate voltage optimization. In: Proceedings of IEEE International Symposium on Quality Electronic Design, pp. 279–284 (2003)
54. Linear Technology: LTC3445—I2C controllable buck regulator with two LDOs in a 4 mm × 4 mm QFN
55. Texas Instruments: TPS400009—low-input high-efficiency synchronous buck controller
56. Texas Instruments: TPS62100—multimode low-power buck converter
57. Min, R., Bhardwaj, M., Cho, S.-H., Shih, E., Sinha, A., Wang, A., Chandrakasan, A.: Low-power wireless sensor networks. In: Proceedings of the 14th International Conference on VLSI Design, p. 205 (2001)
58. Benini, L., de Micheli, G.: Dynamic Power Management: Design Techniques and CAD Tools. Kluwer Academic, Dordrecht (1998)
59. Jejurikar, R., Pereira, C., Gupta, R.: Leakage aware dynamic voltage scaling for real-time embedded systems. In: Proceedings of the 41st Annual Design Automation Conference, pp. 275–280 (2004)
60. Cho, Y., Chang, N.: Memory-aware energy-optimal frequency assignment for dynamic supply voltage scaling. In: Proceedings of the 2004 International Symposium on Low Power Electronics and Design, pp. 387–392 (2004)
61. Sinha, A., Chandrakasan, A.: Dynamic power management in wireless sensor networks. IEEE Des. Test Comput. **18**, 62–74 (2001)
62. Raghunathan, V., Schurgers, C., Park, S., Srivastava, M.: Energy-aware wireless microsensor networks. IEEE Signal Process. Mag. **19**(2), 40–50 (2002)
63. Cho, Y., Kim, Y., Chang, N.: PVS: passive voltage scaling for wireless sensor networks. In: Proceedings of IEEE International Symposium on Low Power Electronics and Design, ISLPED, August 2007, pp. 135–140 (2007)
64. Polastre, J., Szewczyk, R., Culler, D.: Telos: enabling ultra-low power wireless research. In: Proceedings of the 4th International Symposium on Information Processing in Sensor Networks, pp. 364–369 (2005)
65. Abdelzaher, T., Blum, B., Cao, Q., Chen, Y., Evans, D., George, J., George, S., Gu, L., He, T., Krishnamurthy, S., Luo, L., Son, S., Stankovic, J., Stoleru, R., Wood, A.: Envirotrack: towards

an environmental computing paradigm for distributed sensor networks. In: Proceedings of 24th International Conference on Distributed Computing Systems, pp. 582–589 (2004)

66. Levis, P., Madden, S., Polastre, J., Szewczyk, R., Whitehouse, K., Woo, A., Gay, D., Hill, J., Welsh, M., Brewer, E., Culler, D.: TinyOS: an operating system for sensor networks. In: Ambient Intelligence, pp. 115–148 (2005)

67. Benini, L., Castelli, G., Macii, A., Macii, E., Poncino, M., Scarsi, R.: A discrete-time battery model for high-level power estimation. In: Proceedings of the Conference on Design, Automation and Test in Europe, pp. 35–41 (2000)

68. Levis, P., Madden, S., Gay, D., Polastre, J., Szewczyk, R., Woo, A., Brewer, E., Culler, D.: The emergence of networking abstractions and techniques in TinyOS. In: Proceedings of the 1st Conference on Symposium on Networked Systems Design and Implementation—Volume 1, pp. 1–1 (2004)

69. Cho, Y., Kim, Y., Joo, Y., Lee, K., Chang, N.: Simultaneous optimization of battery-aware voltage regulator scheduling with dynamic voltage and frequency scaling. In: Proceedings of the IEEE International Symposium on Low Power Electronics and Design, ISLPED, Bangalore, India, August 2008, pp. 309–314 (2008)

# Chapter 8
# 3-D ICs for Low Power/Energy

Kyungsu Kang, Chong-Min Kyung, and Sungjoo Yoo

**Abstract** System-level integration is expected to gradually become a reality because of continuing aggressive device scaling (for 2-D dies) and launch of 3-D integration technology. Thus, chip power density, which is already a serious issue due to its exponential increase every year, and the related thermal issues in 3-D IC chips, especially microprocessors, may pose serious design problems unless properly addressed in advance. Such temperature-related problems include material as well as electrical reliability, leakage power consumption and possible regenerative phenomena such as avalanche breakdown, and the resultant manufacturing/packaging costs. In this chapter, various temperature-aware power management methods, e.g., voltage and frequency scaling, power gating, and thread scheduling, are introduced, with a special emphasis on performance improvement and energy-saving features.

## 8.1 Introduction

During the past four decades, semiconductor technology scaling has resulted in a sharp growth in transistor density. Figure 8.1 shows the number of transistor counts of Intel processors since 1971 [1]. Recently, the transistor density has reached 0.4 billion transistors/cm$^2$. As the number of transistor counts in a chip increases, the interconnect wires have become major sources of latency, area, and power consumption in the chip. Various studies have been explored to address the interconnect problem such as the use of on-chip network communication [2]. The use of three-dimensional integrated circuits (3-D ICs) is now gaining attention as one of the most promising solutions of the interconnect problem [3].

K. Kang (✉) · C.-M. Kyung
KAIST, Daejeon, Republic of Korea
e-mail: kyungsu.kang@gmail.com

C.-M. Kyung
e-mail: kyung@ee.kaist.ac.kr

S. Yoo
POSTECH, Pohang, Republic of Korea
e-mail: sungjoo.yoo@postech.ac.kr

C.-M. Kyung, S. Yoo (eds.), *Energy-Aware System Design*,
DOI 10.1007/978-94-007-1679-7_8, © Springer Science+Business Media B.V. 2011

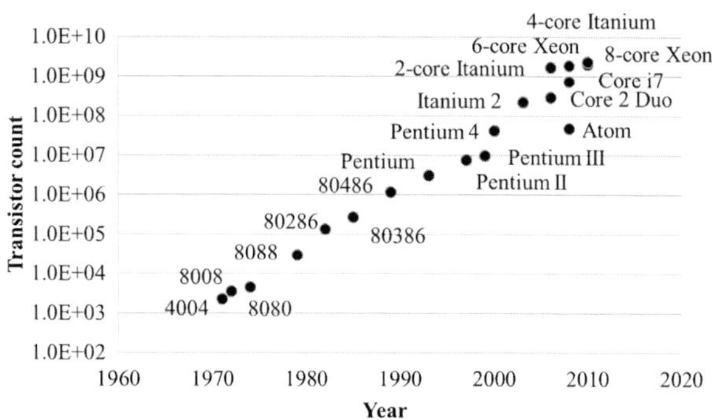

**Fig. 8.1** The number of transistor counts of Intel processors since 1971

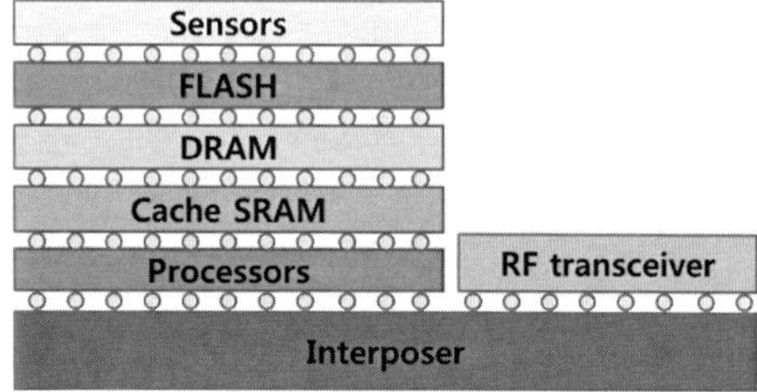

**Fig. 8.2** An example of 3-D integration circuits for a smart image sensor system



3-D integration is a new technology that stacks multiple dies on a single chip with direct vertical interconnects such as through-silicon via (TSV). Compared to the traditional two-dimensional (2-D) chip designs, 3-D integration has the following advantages:

- *Reduction in interconnect wire length*: Joyner et al. [4] have shown that 3-D interconnection reduces the interconnect wire length by a factor of the square root of the number of layers stacked. This reduction in wire length results in improved performance and reduced power consumption [5].
- *Improvement in channel bandwidth*: Very dense vertical interconnects in 3-D ICs allow more data channels among stacked dies. In particular, 3-D stacked memory is one of the most promising applications of 3-D integration technology to resolve the well-known "memory wall" problem [6].
- *Smaller form factor*: The addition of a third dimension allows higher integration density and smaller footprint than those in 2-D integration. For example, the integration density for a four-layer 3-D integration would increase, on average, four times compared with 2-D integration.
- *Heterogeneous integration*: 3-D integration can stack various materials, technologies, and functional components together, as shown in Fig. 8.2.

Despite the advantages of 3-D integration, thermal problems are significant challenges for developing 3-D ICs. As more dies are stacked vertically, the power density (i.e., power dissipation per unit volume) increases drastically, which incurs many temperature-related problems such as reliability (negative bias temperature instability, electro-migration, time-dependent dielectric breakdown, thermal cycling,

**Table 8.1** DVFS and clock throttling comparisons

| DTM solution | Area overhead | Response time | Performance degradation |
|---|---|---|---|
| DVFS | High | Slow | Low |
| Clock throttling | Low | Fast | High |

etc.), power consumption (temperature-induced leakage power), performance (increased circuit delay as the temperature increases), and system cost (e.g., cooling and packaging costs).

There are several dynamic thermal management (DTM) solutions such as stopping with power/clock gating, running at low frequencies and/or low voltages through dynamic voltage and frequency scaling (DVFS), and issuing fewer instructions/functions to control chip temperature [7–9]. When the operating temperature approaches the thermal limit, these DTM solutions can effectively reduce the temperature by reducing chip power consumption, but they inevitably lead to performance degradation. Table 8.1 compares two widely used DTM techniques, DVFS and clock throttling [10]. From a hardware implementation point of view, DVFS has higher area overhead than clock throttling because it involves more voltage regulators and power planes. DVFS also has a slower response time than clock throttling due to the phase-locking time. Isci et al. [11] showed that modern high-performance microprocessors, which support DVFS, perform voltage transitions in the range of 10 mV/μs. However, DVFS has less performance degradation than clock throttling, as more power reduction can be achieved due to the cubic relation of dynamic power on voltage.

The performance effects of DTM are different depending on the memory access behavior of the software program. Generally, the execution time of an application can be divided into two parts. One is the time spent for running CPU instructions. The other is the time spent waiting for data to be fetched from external memory, which is determined by the number of cache misses; this is closely related to the memory access behavior of the application. Given a hardware architecture (CPU frequency, $f_{cpu}$, and memory subsystem frequency, $f_{mem}$) and an application (consisting of $w_{on\text{-}chip}$, the number of CPU clock cycles for the ideal CPU operation of the application, and $w_{off\text{-}chip}$, the number of memory clock cycles spent by the CPU waiting for the completion of the external memory accesses), the total execution time, $t_{ex}$, of the application is calculated as follows:

$$t_{ex}(f_{cpu}) = \frac{w_{on\text{-}chip}}{f_{cpu}} + \frac{w_{off\text{-}chip}}{f_{mem}} \tag{8.1}$$

When the CPU frequency is changed, the change in the execution time depends only on the first part of the execution time ($w_{on\text{-}chip}/f_{cpu}$) because $f_{mem}$ is independent of $f_{cpu}$ and is not scaled. Performance improvement with increase of the clock frequency represents speedup (SU), which is defined as the ratio of the original ex-

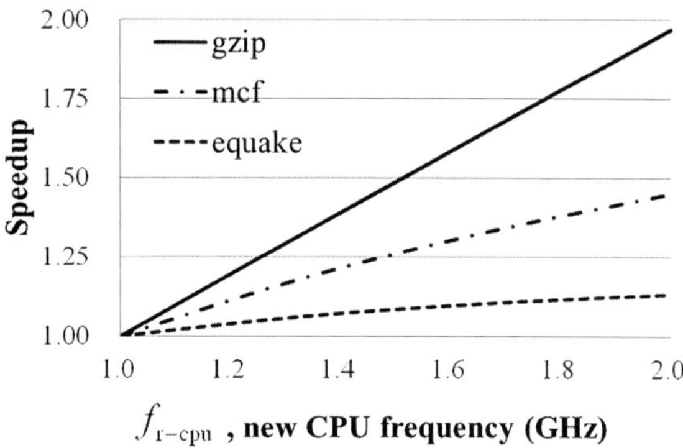

**Fig. 8.3** Speedup vs. new CPU frequency when $f_{\text{r-cpu}}$, reference CPU frequency, is 1 GHz [24]. © 2011 IEEE



$f_{\text{r-cpu}}$ , **new CPU frequency (GHz)**

ecution time (obtained from running the application at $f_{\text{r-cpu}}$) to the new (reduced) execution time (obtained from running the application at $f_{\text{n-cpu}}$) as follows:

$$\text{SU} = \frac{t_{\text{ex}}(f_{\text{r-cpu}})}{t_{\text{ex}}(f_{\text{n-cpu}})} \tag{8.2}$$

Figure 8.3 illustrates SU for three programs in the SPEC2000 benchmark. For example, "gzip" shows more significant performance improvement according to the CPU frequency than other programs. This implies that gzip is a CPU-bound program. By contrast, "equake" has little performance improvement as the frequency increases, because equake is a memory-bound program. Based on these observations, we can conclude that memory-bound programs have much smaller performance degradation than CPU-bound programs when DTM solutions are invoked to reduce chip power density and, then, chip temperature. Thus, in 3-D multi-core systems, we should exploit the memory access behavior of the running programs as well as the operating temperature of each core.

This chapter presents an overview of the challenges related to thermal-aware design for 3-D ICs. We first focus on these challenges for high-performance systems (e.g., desktops and servers), discussing methods for thermal analysis, and then presenting a runtime solution to maximize performance (in terms of instructions per second) of 3-D multi-core systems while satisfying both peak constraints of power consumption and temperature. Next, we describe the thermal-related challenges for low power systems (notebooks, cell phones, etc.) and outline solutions to overcome them.

## 8.2 Thermal Characteristics of 3-D Multi-core Systems

Figure 8.4 illustrates a basic *face-to-face* stacking structure of 3-D chip multi-processors. Face-to-face stacking, which is one of the combinations of strata orientations within the chip stack, focuses on joining the front sides of two wafers. A key potential advantage of face-to-face stacking is a very dense interface between adjacent dies, enabling higher communication bandwidth. The other stacking combination of strata orientations is *face-to-back* stacking. Face-to-back is based on

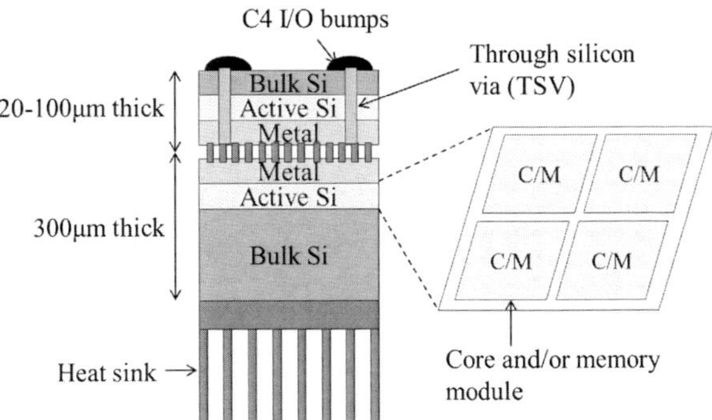

**Fig. 8.4** A face-to-face 3-D die-stacking structure for multi-core systems



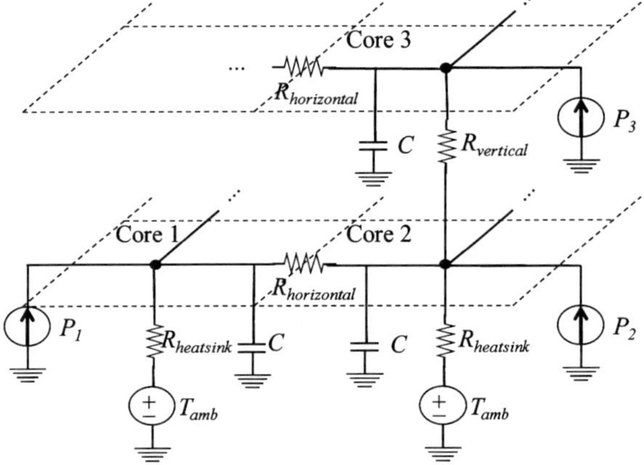

**Fig. 8.5** Simplified thermal model of a general 3-D multi-core system in which each core has uniform temperature distribution [24]. © 2011 IEEE



bonding the front side of the bottom die (the closet die from the heat sink) with the back side of the top die (the furthest die from the heat sink), which gives better scalability in die (layer) count. In both stacking methodologies, the top layer is thinned for better electrical characteristics and improved physical construction of TSVs for power delivery and I/O, as shown in Fig. 8.4 [5]. In Fig. 8.4, each active silicon layer contains processor cores and/or memory modules. One side of the multi-core chip is connected to a substrate through the C4 I/O bumps on the printed circuit board (PCB). The other side of the chip is attached to the heat sink, where most of the heat is dissipated. Heat flow within a package can be modeled by the well-known analogy between heat transfer and electric circuit phenomena in an RC network [12].

Figure 8.5 illustrates a coarse-grained thermal model of a 3-D multi-core system in which each core is represented with a thermal model element (i.e., a thermal resistance, a thermal capacitance, and a current source). In Fig. 8.5, the heat sink is located at the bottom of the chip stack. For a thermal model to be accurate, each block represented as a thermal model element must be small enough for the temperature to be assumed uniform within each block. In this section, we explain two thermal characteristics that are essential for understanding 3-D multi-core power management: (1) *heterogeneous thermal coupling* and (2) *heterogeneous cooling efficiency*.

### 8.2.1 Heterogeneous Thermal Coupling

In the thermal resistance-capacitance (RC) network shown in Fig. 8.5, the power consumption of a core influences the temperatures of other cores as well as its own temperature. Particularly in 3-D multi-core systems, vertically adjacent cores have a much larger thermal influence on each other than horizontally adjacent cores because heat dissipates mostly in the vertical direction. As shown in Fig. 8.4, the thickness of a silicon layer is very small ($<100$ μm). This thinning makes vertical neighbors have a stronger mutual thermal coupling than horizontal neighbors.

Thermal coupling can be explained in terms of thermal resistance. In [12], thermal resistance is estimated based on the relationship between a material's thermal conductivity $k$, area $A$, and thickness $H$, i.e., $H/(kA)$, which is analogous to electrical resistance. Thus, the thermal resistance among horizontally adjacent cores (cores 1 and 2 in Fig. 8.5) is much larger than that among vertically adjacent cores (cores 2 and 3 in Fig. 8.5) since $A$ is smaller among horizontally adjacent cores (and the effective $H$ is larger). When applying the models in [10, 13], in Fig. 8.5, the thermal resistance between cores 1 and 2 ($R_{\text{horizontal}}$) is approximately 2.44 K/W and the thermal resistance between cores 2 and 3 ($R_{\text{vertical}}$) is approximately 0.15 K/W. $R_{\text{horizontal}} \approx 16 R_{\text{vertical}}$, which means that most heat is propagated vertically.

### 8.2.2 Heterogeneous Cooling Efficiency

In 3-D multi-core systems, the cores in a stack are not only strongly correlated in their temperatures, but also the cores near the heat sink have higher cooling efficiency than those far from the heat sink. This has also been noted in the literature for steady-state temperature. Ignoring the heat flow in the horizontal direction, the steady-state temperatures of cores 2 and 3 shown in Fig. 8.5 is expressed as follows:

$$T_3^{\text{ss}} = P_3 R_{\text{vertical}} + T_2^{\text{ss}} \tag{8.3}$$

$$T_2^{\text{ss}} = (P_2 + P_3) R_{\text{heatsink}} + T_{\text{amb}} \tag{8.4}$$

where $T_2^{\text{ss}}$ and $T_3^{\text{ss}}$ are the steady-state temperatures of cores 2 and 3, $P_2$ and $P_3$ are the power consumptions of cores 2 and 3, respectively, $T_{\text{amb}}$ is the ambient temperature, and $R_{\text{heatsink}}$ is the thermal resistance from core 2 to the ambient through the cooling solution (i.e., heat sink). As shown in (8.3) and (8.4), core 3 is hotter than core 2 by $P_3 R_{\text{vertical}}$, showing that a core closer to the heat sink has a larger cooling efficiency. Zhou et al. [14] give a more detailed analytical thermal analysis for a two-layer 3-D multi-core system, taking into account the thermal capacitance as well. For a thin die within 100 μm, the thermal capacitance between the top die and the ambient, which represents how quickly temperature changes from the top die, is reported as 23.6–34.7 mW·s/K. In [14], Zhou et al. demonstrated that the top core always has a higher temperature than that of the bottom core.

## 8.3 Temperature-Aware Power Management Techniques for 3-D Multi-core Systems

### 8.3.1 Definition of Power Management Problem

In this section, runtime per-core frequency/voltage assignment and thread migration for 3-D multi-core systems are described. In the system, the operating system (OS) is assumed to allocate threads dynamically across cores and to change their frequency/voltage levels on a discrete time interval. The discrete time intervals for thread migration and DVFS are mostly determined by the scheduler in the OS. In this section, we denote the time interval of thread migration as $\Delta l$, and that of DVFS as $\Delta t$. Figure 8.6 shows a general floorplan of 3-D multi-core systems. The system consists of $N\ (=XYZ)$ cores where $X$ and $Y$ are, respectively, the number of cores in the $x$-axis and $y$-axis, and $Z$ is the number of core layers. Each core is assumed to have the same size and is denoted as $(i, j, k)$ for its $x$, $y$, and $z$ positions.

The problem is to find (1) the frequency/voltage schedule of each core and (2) a thread schedule such that the instruction throughput of the system is maximized while the peak power ($P_{\max}$) and temperature ($T_{\max}$) constraints are met. The instruction throughput of a multi-core system is defined as the total number of instructions executed by all the cores per second as follows:

$$\text{IPS}_{\text{total}} = \sum_{i=1}^{X}\sum_{j=1}^{Y}\sum_{k=1}^{Z}\text{IPS}_{i,j,k} \tag{8.5}$$

where $\text{IPS}_{i,j,k}$ is the instructions per second (IPS) of core $(i, j, k)$.

As described in Sect. 8.2, cores in the same horizontal position (i.e., the same $x$ and $y$ positions with different $z$ positions as shown in Fig. 8.6) have strong thermal correlations. Thus, in this chapter, we call cores in the same horizontal position a "core-set." Also, we refer to core $(i, j, Z)$ as a "top core $(i, j)$," which is the farthest core from the heat sink in core-set $(i, j)$. Note that the core that is farthest from the heat sink is the hottest core in the core-set, as explained in Sect. 8.2.

### 8.3.2 Runtime Temperature-Aware Thread Migration Techniques

Thread migration is the act of reassigning currently running jobs (i.e., threads) on the multi-core system. For temperature reduction, it is important to distribute the

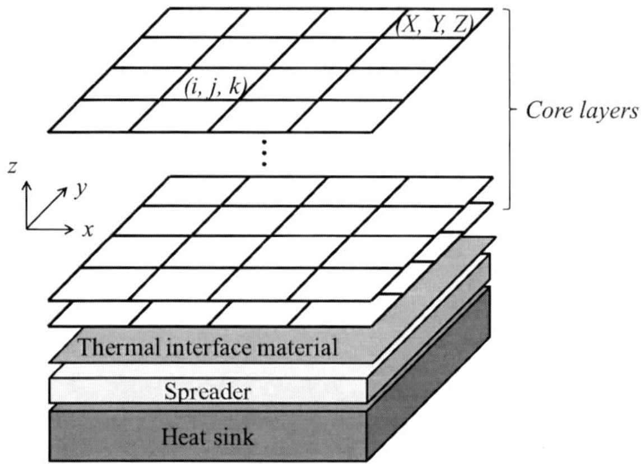


**Fig. 8.6** 3-D stacked-die floorplan and indexes of cores [24]. © 2011 IEEE

heat generated by running threads more uniformly across the multi-core chip. In this section, we introduce several possible approaches of thread migration for this purpose.

#### 8.3.2.1 IPC-Aware Thread Migration Without Considering the Thermal Characteristics in 3-D Multi-core Systems (I-Migr2D)

In a complex high-performance superscalar processor, a dominant portion of the switching power is consumed by circuits employed to exploit the instruction-level parallelism [15, 16]. Thus, switching power consumption is presented as a function of instructions per cycle (IPC) because effective switching capacitance of a core is proportional to the switching activity, i.e., IPC of threads running on the core.

At every thread migration interval, I-Migr2D allocates (reassigns) threads according to their power consumption (IPCs) and the temperatures of cores running their own threads to balance the temperature among the cores in the 3-D multi-core system. Essentially, a thread with high power consumption should be assigned to a low-temperature core. By defaults, cores switch to threads every 100 ms. At each thread migration point, the scheduler sorts the IPCs of all threads and the current temperature of each core. Then, it assigns the thread with the highest IPC to the coolest core, as shown Fig. 8.7. This technique can be considered as an extension of the core hopping algorithm applied to 2-D multi-core systems [17].

This technique requires hardware performance monitors, which are provided by most modern processors, to estimate the IPC of each running thread at runtime. PAPI [18] is a portable interface in the form of a library to invoke hardware performance monitors. It is being widely used to collect low-level performance metrics such as clock cycles, instruction counts, cache misses, pipeline stalls, and cache translation lookaside buffer (TLB) misses of CPUs running UNIX/Linux operating systems. Thus, at runtime, the average IPC of each thread can be estimated by gathering the number of execution clock cycles and instructions from PAPI. Although the absolute values of these metrics depend on microarchitecture characteristics, their

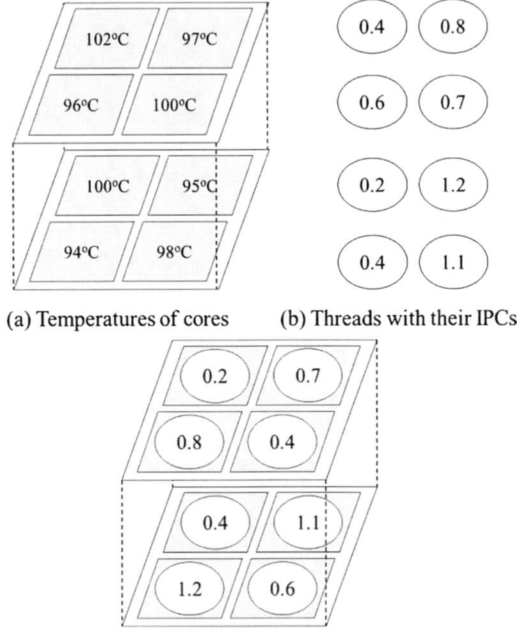

**Fig. 8.7** An example of I-Migr2D approach



(a) Temperatures of cores        (b) Threads with their IPCs

(c) Thread re-assignment result of *I-Migr2D*

relative differences in the set of applications running on the multi-core system are mostly independent of the microarchitecture.

### 8.3.2.2 IPC-Aware Thread Migration Considering the Thermal Characteristics in 3-D Multi-core Systems (I-Migr3D)

In a 3-D multi-core system, there is a strong thermal correlation among vertically adjacent cores, as explained in Sect. 8.2. This implies that if a core-set contains the hottest core, it might also contain the second hottest core, as shown in Fig. 8.8(a). When I-Migr2D is applied to 3-D multi-core systems, it is possible that the threads with the lowest and the second lowest IPCs are mapped on the hottest core-set, which has the maximum sum of temperatures of cores within the core-set. Similarly, the threads with the highest and the second highest IPCs might be assigned to the coolest core-set. This can lead to temperature oscillations between the hottest core-set and the coolest core-set at every thread migration. Thus, for thread migration in 3-D multi-core systems, cores in the same core-set should not be treated independently, but be considered together.

I-Migr3D reallocates threads according to the 3-D floorplan as well as their power consumption (i.e., IPCs) and the cores' temperatures in order to balance the temperatures among core-sets. At each thread migration point, the scheduler sorts the IPCs of all threads and forms thread-sets, which are sets of threads to be mapped on each core-set. When forming thread-sets, the power consumed by each thread-set

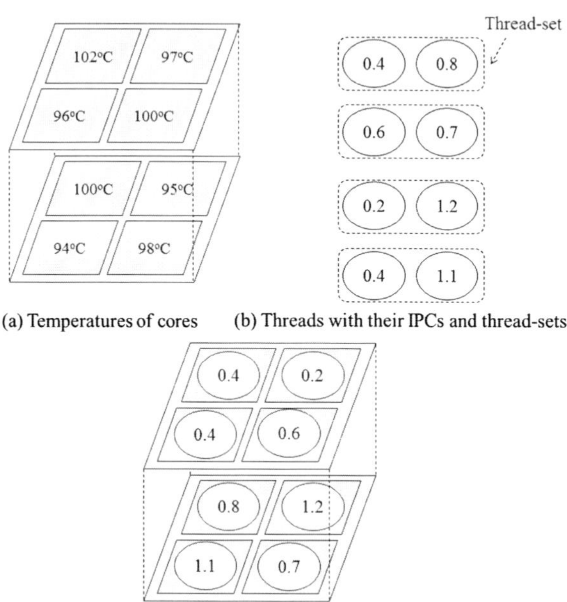



(a) Temperatures of cores    (b) Threads with their IPCs and thread-sets

(c) Thread re-assignment result of *I-Migr3D*

**Fig. 8.8** An example of I-Migr3D approach

is made to be balanced so that the temperatures among core-sets are balanced. Then, the scheduler assigns the thread-set with the highest IPC sum to the coolest core-set. Next, within a thread-set, it assigns threads with higher IPCs to cores closer to the heat sink among vertically stacked cores. Figure 8.8 shows an example result of I-Migr3D. This technique can be considered as an extension of the work in [10] and [30]. (In Fig. 8.8, the heat sink is located at the bottom of the chip stack.)

### 8.3.2.3 IPC- and SU-Aware Thread Migration Considering the Thermal Characteristics in 3-D Multi-core Systems (IS-Migr3D)

Without DTM solutions such as DVFS, I-Migr3D is good enough to exploit the thermal characteristics in 3-D multi-core systems. However, in modern high-performance systems, DTM must be supported to prevent thermal emergencies. Conventionally, when a core temperature increases above the given threshold temperature, the core will be put to a low power state through DVFS. In 3-D multi-core systems, since the cores farther from the heat sink are usually hotter, thermal emergencies usually occur in these cores, and they then can run at a lower clock frequency. To consider the memory stall time in clock frequency scaling, memory-bound programs have much smaller performance degradation than CPU-bound programs as the clock frequency decreases, as explained in Sect. 8.1.

The basic idea of IS-Migr3D is to map threads with higher memory boundness (i.e., low SU) onto cores farther from the heat sink. this is done because the cores far from the heat sink tend to run at lower clock frequencies due to their lower cooling

---

Procedure of thread migration

---

(*Thread-set formation*)

1. Assume $XY$ empty sets which can store maximally $Z$ threads.
2. Sort all threads according to the descending order of IPC.
3. Put the thread with the highest IPC into a set with the lowest sum of IPCs.
4. Repeat 3 until all threads are assigned to one of $XY$ sets.

(*Thread-set allocation*)

5. Sort all thread-sets according to the ascending order of IPC sum.
6. Assign the thread-set with the lowest IPC sum to the core-set with the highest temperature sum.
7. Sort threads in each core-set according to the ascending order of memory-boundness.
8. Assign the thread with the lowest memory-boundness to the core closest to the heatsink.
9. Repeat 8 until all threads in the core-set are assigned.
10. Repeat 6–9 until all thread-sets arc assigned.

---

**Fig. 8.9**  Procedure of IS-Migr3D solution [24]. © 2011 IEEE

efficiency; thus, we can reduce performance degradation, i.e., reduce the number of stall cycles, on these cores (due to low frequency operation) by allocating threads with high memory boundness to them. To do that, IS-Migr3D assigns a thread-set with the highest IPC sum to the coolest core-set, and assigns the thread with the highest SU to the core closest to the heat sink in the core-set.

Figure 8.9 shows the overall flow of IS-Migr3D. The method is largely divided into thread-set formation (lines 1–4) and thread allocation (lines 5–10) steps, which are invoked periodically with the period of thread migration step, $\Delta l$ (100 ms in our experiment). In the step of thread-set formation, thread-sets are formed according to the IPC of each thread to minimize the difference among IPC sums of the thread-sets. To do that, $XY$ empty sets, each of which can maximally contain $Z$ threads, are prepared (line 1) and threads are inserted into thread-sets in a descending order of IPC (lines 2–4). In the thread allocation step, thread-sets are allocated to core-sets to balance the temperature among core-sets (lines 5 and 6). Each thread in a core-set, which is ordered in an ascending order of memory boundedness, is allocated to an available (i.e., not yet allocated) core nearest to the heat sink (lines 7–9).

SU can be estimated at runtime by using a runtime performance monitor such as PAPI [18]. The execution time, which is used to calculate SU, is presented as (8.2). In (8.1), $w_{\text{on-chip}}$ is an application-specific constant which can be obtained in design time and $f_{\text{mem}}$ is also a constant. $w_{\text{off-chip}}$ is a time-varying parameter that depends on program phases and can be estimated by $w_{\text{off-chip}} = (aN_{\text{miss}} + b)f_{\text{cpu}}$ where $N_{\text{miss}}$ is the number of L2 cache misses [21]. $a$ and $b$ are constants that depend on the system memory hierarchy. We can obtain the number of L2 cache misses with PAPI and we can estimate $w_{\text{off-chip}}$ and, thus, SU at runtime.

### *8.3.3 Runtime Temperature-Aware Frequency/Voltage Scaling Techniques with Peak Power Constraint*

Using runtime thermal management techniques for both 2-D and 3-D multi-core systems such as dynamic voltage and frequency scaling (DVFS), chip packages with cooling solutions do not need to be designed for the worst-case power consumption scenarios. The package cost for chip cooling can thereby be significantly reduced. However, most previous studies on runtime thermal management have two limitations. First, they control only the *steady-state* chip temperature, not the *instantaneous* one, by controlling the chip power so that the expected steady-state temperature does not exceed the thermal limit. This happens because the computational complexity of the runtime algorithm that manages instantaneous temperature can be much larger than the one that only controls steady-state temperature. Second, previous papers have focused on either peak power or peak temperature, even though both constraints are key factors to be considered. Considering that the total chip power consumption continues to increase as the number of cores increases and the supply voltage reduction stagnates, peak power constraint is another important limiter in improving system performance. In this section, we introduce several possible approaches of power budgeting (i.e., the DVFS method) to maximize the performance (in terms of IPS) of 3-D multi-core systems while satisfying both peak constraints of power consumption and temperature.

#### 8.3.3.1 DVFS with Steady-State Temperature Analysis (S-DVFS)

To maximize 3-D chip multi-processor throughput, cores should operate at different voltage and frequency levels due to heterogeneous thermal characteristics as well as heterogeneous runtime workloads. Dynamic power consumed by a core always varies over time according to the phases of programs running the core. For example, a thread which is in a memory-intensive phase might dissipate less power than another thread in a compute-intensive phase. IPC is a good metric to find whether the thread is in a memory-intensive phase or compute-intensive phase at runtime. In addition, among cores within a core-set, the core farthest from the heat sink has the highest thermal resistance between the core and the ambient and, thus, the lowest cooling efficiency. Therefore, given a thermal constraint, both cooling efficiency and IPC need to be considered when assigning voltage and frequency level to each core in the 3-D chip multi-processor.

As explained in Sect. 8.2, vertically aligned cores have strongly correlated temperatures as well as heterogeneous cooling efficiency in a core-set. The strong thermal correlation can transform the problem of DVFS for three-dimensionally distributed cores into the DVFS problem for two-dimensionally distributed core-sets such that the temperature of the farthest core from the heat sink in each core-set does not exceed the maximum temperature limit, $T_{\mathrm{max}}$. Thus, the problem of S-DVFS can be presented as follows. (For simplicity, the symbols $i$ and $j$ are omitted because the cores within single core-set are considered below in this section.)

$$\text{Find } f_k \quad \text{for all } k = 1, 2, \ldots, Z$$

where $f_k$ is the frequency of core located on layer $k$

$$\text{such that} \quad \text{IPS}^{\text{cs}}_{\text{total}} = \sum_{k=1}^{Z} f_k \text{IPC}_k \quad \text{is maximized} \tag{8.6}$$

$$\text{subject to} \quad \sum_{k=1}^{Z} R_k P(f_k) + T_{\text{amb}} = T_{\text{max}} \tag{8.7}$$

where $P(f_k)$ is power consumption of a core running at $f_k$ and $R_k$ is thermal resistance between the core on layer $k$ and the ambient.

The thermal resistance, $R_k$, is calculated by the sum of thermal resistances of vertically stacked cores,

$$R_k = \sum_{l=1}^{k} r_l \tag{8.8}$$

where $r_l$ is the thermal resistance from the core on layer $l - 1$ to the core on layer $l$. Note that layer 0 indicates the ambient in (8.8).

In order to maximize $\text{IPS}^{\text{cs}}_{\text{total}}$ in (8.6) while satisfying (8.7), a Lagrange function can be defined as follows by using the Lagrange multipliers [22];

$$L(f_1, f_2, \ldots, f_Z, \lambda) = \sum_{k=1}^{Z} f_k \text{IPC}_k + \lambda \left( \sum_{k=1}^{Z} R_k P(f_k) + T_{\text{amb}} - T_{\text{max}} \right) \tag{8.9}$$

where $\lambda$ is a Lagrange multiplier. In [22], $\text{IPS}^{\text{cs}}_{\text{total}}$ in (8.6) is maximized when $(f_1, f_2, \ldots, f_Z, \lambda)$ is a stationary point of (8.9) where the partial derivatives of (8.9) are zero. The partial derivative of the Lagrange function with respect to frequency, $f_k$, is equaled to zero as follows

$$\frac{\mathrm{d}L(f_1, f_2, \ldots, f_Z, \lambda)}{\mathrm{d}f_k} = \text{IPC}_k + \lambda \left( R_k \frac{\mathrm{d}P(f_k)}{\mathrm{d}f_k} \right) = 0 \tag{8.10}$$

Thus, from (8.10), $\text{IPS}^{\text{cs}}_{\text{total}}$ is maximized when $f_k$ satisfies the following equation because $\lambda$ is a constant

$$\forall k; \quad \frac{R_k}{\text{IPC}_k} \frac{\mathrm{d}P(f_k)}{\mathrm{d}f_k} = M \tag{8.11}$$

where $M$ is a constant.

To satisfy both peak power and temperature constraints, S-DVFS allocates anequal power budget to cores and sets the frequency/voltage levels of each core within its own power budget based on (8.11) while keeping the steady-state temperatures of all the top cores (the left-hand side of (8.7)) under $T_{\text{max}}$.


### 8.3.3.2 DVFS with Instantaneous Temperature Analysis (I-DVFS)

In a single core system, the instantaneous temperature of the core is affected by the initial temperature and steady-state temperature of the core as follows [23]:

$$T(t) = \left( T^{\text{init}} - T^{\text{ss}} \right) e^{-t/RC} + T^{\text{ss}} \tag{8.12}$$

where $T^{\text{init}}$, $T^{\text{ss}}$, and RC are the initial temperature, steady-state temperature, and thermal RC time constant of the core, respectively. The thermal RC time constant is usually on the order of milliseconds, and millions of times larger than the order of clock cycle. Therefore, the voltage and frequency schedule based on the steady-state temperature analysis fails to exploit the temperature slack (defined as peak temperature constraint minus current temperature) because it is too conservative [24].

The basic idea of I-DVFS is that the amount of power budget of a core is allocated in proportion to its dynamic temperature slack, i.e., $T_{\text{max}} - T_{\text{current}}$. The higher the temperature slack is, the more power budget is assigned to the core, while the total power consumption of all cores does not exceed $P_{\text{max}}$ and the instantaneous temperature of each core does not exceed $T_{\text{max}}$.

In (8.12), the steady-state temperature of the core, $T^{\text{ss}}$, is calculated as

$$T^{\text{ss}} = RP + T_{\text{amb}} \tag{8.13}$$

$T^{\text{ss}}$ increases with the power consumption of the core, as shown in (8.13). Thus, from (8.12) and (8.13), the temperature of a core, $T(t)$, increases with the steady-state temperature of the core, $T^{\text{ss}}$, and the power consumption of the core. According to the relationship between the power consumption and steady-state temperature in (8.13), power budgeting among core-sets in the 3-D multi-core system is equivalent to assigning steady-state temperatures to each core-set. Maximizing performance (i.e., instruction throughput) for the 3-D multi-core system requires that the more temperature slack a core-set has, the higher the steady-state temperature, i.e., the larger the power budget that is assigned to the core-set, i.e.,

$$\frac{T_{\text{max}} - T_{i,j,Z}(t)}{T^{\text{ss}}_{i,j,Z} - T_{i,j,Z}(t)} = K \quad \text{for all } i \text{ and } j \tag{8.14}$$

where $K$ is a constant [24]. In (8.14), the numerator represents the temperature slack of core-set $(i, j)$ since the core $(i, j, Z)$, i.e., top core $(i, j)$, is the hottest in the corresponding core-set $(i, j)$. Thus, the core-set with the largest temperature slack is assigned the highest steady-state temperature and the largest power budget.

In (8.14), the constant $K$ must be determined so that the performance is maximized while the assigned total power budget does not violate both the peak power consumption constraint, $P_{\text{max}}$, and the peak temperature constraint, $T_{\text{max}}$:

$$\forall i, \ \forall j; \quad T_{i,j,Z}(t + \Delta t) \leq T_{\text{max}} \quad \text{and} \tag{8.15}$$

$$\sum_{i=1}^{X} \sum_{j=1}^{Y} P_{i,j}^{\text{cs}} \leq P_{\text{max}} \tag{8.16}$$

where $P_{i,j}^{\text{cs}}$ is the power consumption of core-set $(i, j)$. In inequality (8.15), the temperature of top core $(i, j)$ in the next time step, i.e., $T_{i,j,Z}(t + \Delta t)$, is estimated from (8.17) as follows:

$$T_{i,j,Z}(t + \Delta t) = \left( T_{i,j,Z}(t) - T^{\text{ss}}_{i,j,Z} \right) \rho + T^{\text{ss}}_{i,j,Z} \tag{8.17}$$

where $\rho$ $(= e^{-\Delta t/\mathrm{RC}})$ is a constant since $\Delta t$ is a constant. Then, the relation among power consumption, thermal resistance, and steady-state temperature of a core-set can be modeled as

$$P_{i,j}^{\mathrm{cs}} = \frac{T_{i,j,Z}^{\mathrm{ss}} - T_{\mathrm{amb}}}{R_{i,j}^{\mathrm{cs}}} \qquad (8.18)$$

where $P_{i,j}^{\mathrm{cs}}$ is sum of the power consumptions of all cores within core-set $(i, j)$ and $R_{i,j}^{\mathrm{cs}}$ is the thermal resistance between core-set $(i, j)$ and the ambient. The number of running threads in a core-set can be smaller than the number of cores in the core-set. In that case, all threads are assigned to cores near to the heat sink, and the cores where threads are not allocated are power-gated. Thus, the thermal resistance of a core-set is modeled as follows;

$$R_{i,j}^{\mathrm{cs}} = R_{\mathrm{hs}} + (N_{i,j}^{\mathrm{cs}} - 1)\,R_{\mathrm{inter}} \qquad (8.19)$$

where $N_{i,j}^{\mathrm{cs}}$ is the number of running threads in core-set $(i, j)$. Given a set of core temperatures, as $K$ decreases in (8.14), more power budget, i.e., higher steady-state temperatures, are assigned to cores. However, the allocated power budgets (steady-state temperatures) must satisfy inequalities (8.15) and (8.16); thus, the optimal $K$ is the smallest $K$ that satisfies inequalities (8.15) and (8.16). In order to find the optimal $K$, a binary search algorithm is a good candidate; the computational complexity is $O(\log_2 N_K)$, where $N_K$ is the number of possible $K$ values. After finding the optimal $K$ and, thus, optimal steady-state temperature of each core for power budgeting, the voltage/frequency levels are assigned to each core based on the same approach of S-DVFS, except that the left-hand side of (8.7) is not equal to $T_{\mathrm{max}}$, but equal to the optimal steady-state temperature of each top core determined by (8.14).

### 8.3.4 Experimental Results

In this section, the temperature-aware power management techniques for 3-D multi-core systems, which are combinations of the thread migration and DVFS techniques, are presented. Then, the experimental results with detailed quantitative comparisons among the power management techniques are provided. Table 8.2 shows four techniques, each of which is a combination of one of the three thread migration methods (I-Migr2D, I-Migr3D, IS-Migr3D) and one of the two power budgeting methods (S-DVFS, I-DVFS).

The results of IPS for each algorithm (described in Table 8.2) are shown in Fig. 8.10. The results of IPS are normalized with respect to [I-Migr2D + S-DVFS]. In Fig. 8.10, hipc, lipc, and mipc represent benchmarks that have high IPC, low IPC, and mixed IPC, respectively. Similarly, hm, lm, and mm represent benchmarks that have high memory boundedness (i.e., low SU), low memory boundedness, and mixed memory boundedness, respectively. Details of the experimental setups and results are described in [24].

**Table 8.2** Comparison of temperature-aware optimization techniques

| Algorithm | Consideration for thread migration | Consideration for DVFS |
|---|---|---|
| I-Migr2D + S-DVFS | IPC and 2-D floorplan | Steady-state temperature analysis |
| I-Migr3D + S-DVFS | IPC and 3-D floorplan | Steady-state temperature analysis |
| IS-Migr3D + S-DVFS | IPC, SU, and 3-D floorplan | Steady-state temperature analysis |
| IS-Migr3D + I-DVFS | IPC, SU, and 3-D floorplan | Instantaneous temperature analysis |

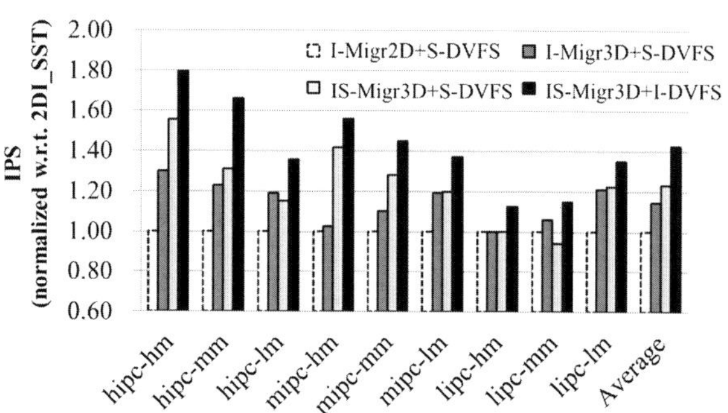


**Fig. 8.10** IPS result for each algorithm described in Table 8.2 [24]. © 2011 IEEE

[IS-Migr3D + I-DVFS] achieves 17.0% and 5.6% IPS improvement, on the average, compared to [I-Migr2D + S-DVFS] and [I-Migr3D + S-DVFS], respectively. This is because [I-Migr3D + S-DVFS] utilizes memory boundedness and per-core heterogeneous cooling efficiency for thread migration. Figure 8.10 shows that [IS-Migr3D + S-DVFS] yields more improvement of IPS with higher memory boundedness in most cases. [IS-Migr3D + I-DVFS] improves IPS by 28.3% on average (ranging from 11.2% to 44.4%) compared with [I-Migr2D + S-DVFS], and by 18.5% on average (ranging from 7.7% to 34.2%) compared with [I-Migr3D + S-DVFS]. The difference between [IS-Migr3D + I-DVFS] and [I-Migr3D + S-DVFS], i.e., 18.5% performance improvement, shows that exploiting the instantaneous temperature is effective in achieving further performance improvement under the given peak power and temperature constraints. The reason is that global power budgeting based on instantaneous temperature allows a higher amount of power consumption and, thus, allows a higher clock frequency to each core than a budgeting based on steady-state temperature. Table 8.3 shows the average power dissipation and average temperature slack of [IS-Migr3D + I-DVFS] and [I-Migr3D + S-DVFS]. [IS-Migr3D + I-DVFS] consumes more power and has less temperature slack for all test cases than [I-Migr3D + S-DVFS]. Table 8.3 and Fig. 8.10 show that, if the difference of power consumption between [I-Migr3D + S-DVFS] and [IS-Migr3D + I-DVFS] becomes larger, improvement in terms of IPS becomes larger.

The experimental results show that [IS-Migr3D + I-DVFS] offers up to 34.2% (average 18.5%) performance improvement compared with existing thermal-aware power management solutions for 3-D multi-core systems.

**Table 8.3** Average power dissipation ($P_{\text{AVG}}$) and average temperature slack ($T_{\text{SLACK}}$) of [IS-Migr3D + I-DVFS] and [I-Migr3D + S-DVFS] [24]. © 2011 IEEE

| SET | I-Migr3D + S-DVFS | | IS-Migr3D + I-DVFS | |
|---|---|---|---|---|
| | $P_{\text{AVG}}$ (W) | $T_{\text{SLACK}}$ (°C) | $P_{\text{AVG}}$ (W) | $T_{\text{SLACK}}$ (°C) |
| hipc-hm | 91.63 | 7.29 | 123.40 | 2.46 |
| hipc-mm | 113.86 | 5.08 | 118.97 | 4.29 |
| hipc-lm | 99.00 | 6.95 | 104.63 | 6.02 |
| mipc-hm | 104.30 | 7.32 | 150.53 | 2.30 |
| mipc-mm | 105.94 | 5.17 | 115.32 | 4.25 |
| mipc-lm | 84.56 | 7.53 | 87.77 | 6.99 |
| lipc-hm | 69.98 | 10.44 | 70.14 | 10.40 |
| lipc-mm | 85.03 | 7.37 | 85.70 | 6.94 |
| lipc-lm | 119.34 | 5.74 | 124.64 | 4.85 |

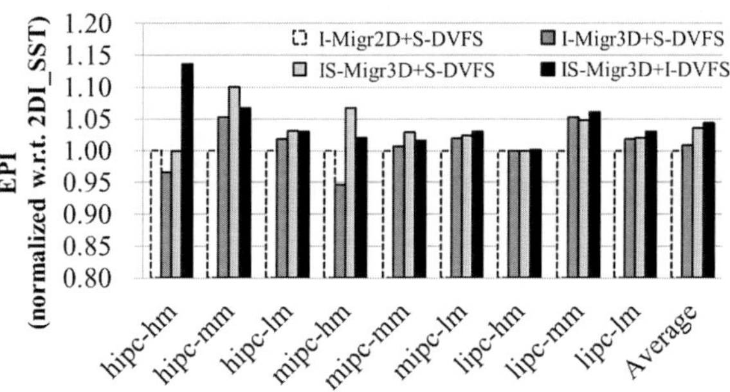


**Fig. 8.11** EPI result for each algorithm described in Table 8.2 [24]. © 2011 IEEE

[IS-Migr3D + I-DVFS] gives higher performance while consuming more energy than the other methods because [IS-Migr3D + I-DVFS] assigns higher frequencies to cores through aggressive power budgeting by exploiting temperature slack. Thus, the overhead in energy consumption must be minimized. Figure 8.11 shows the comparison of energy per instruction (EPI). The figure shows that [IS-Migr3D + I-DVFS] gives slight degradation in EPI by about 4.1% and 3.2% on average compared with [I-Migr2D + S-DVFS] and [I-Migr3D + S-DVFS], respectively. However, comparing both IPS and EPI in Figs. 8.10 and 8.11, [IS-Migr3D + I-DVFS] gives a significant performance improvement (by 28.3% and 18.5%) with a minimal overhead (by 4.1% and 3.2%) in energy efficiency.

## 8.4 Temperature-Induced Energy Minimization Techniques

At high temperature, chip cooling is crucial due to the exponential temperature dependence of leakage current. However, traditional cooling methods, e.g., power/clock gating applied when a temperature threshold is reached, often cause excessive performance degradation. Therefore, runtime power management needs

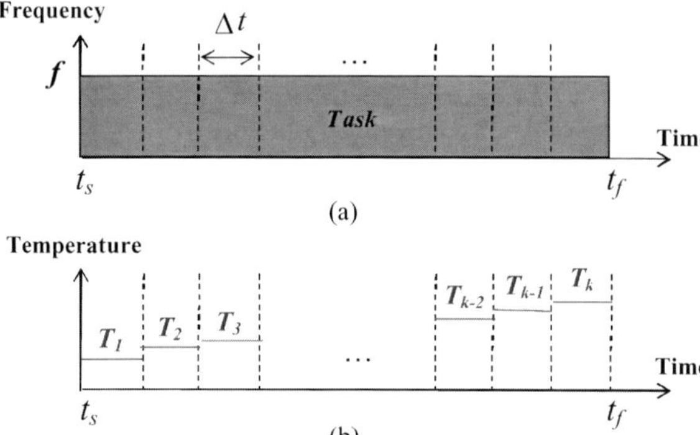

**Fig. 8.12** Temperature-aware energy calculation of a task at a fixed clock frequency ($f$) given a time period $[t_s, t_f]$. (**a**) Task execution time is divided into a time interval $\Delta t$ that is small enough to consider the temperature during $\Delta t$ to be constant. (**b**) The temperature-aware energy consumption is calculated by (8.20) [27]. © 2010 IEEE



to undergo integrated cooling (i.e., power gating) and running (i.e., DVFS) in a temperature-aware manner without incurring a performance penalty. Although temperature-aware DVFS and DTM are self-sufficient techniques, applying them together will help fully exploit the potential of both techniques. In this section, several runtime power management methods, which exploit a program's runtime distribution and a core's instantaneous temperature to minimize the total energy consumption of a periodic hard real-time system, are explained.

### 8.4.1 Definition of Energy Minimization Problem

System power consumption can vary during task execution while the operating frequency and the switching power consumption are constant, because the temperature can vary with time (due to the switching power consumption), thereby changing the leakage power consumption. However, because the thermal time constant of a system is usually on the order of milliseconds, it is not necessary to update the temperature for every clock cycle for temperature-aware power estimation. During power estimation, the temperature can be obtained from a thermal model such as HotSpot [25] at each time step, $\Delta t$. The time interval, $\Delta t$, must be sufficiently small so that the temperature during $\Delta t$ can be considered constant.

Given a time period $[t_s, t_f]$, where $t_f - t_s = K \Delta t$, temperature-aware energy consumption is calculated by summing the energy consumptions of all $K$ intervals as follows. (Figure 8.12 illustrates how to calculate energy consumption in a temperature-aware manner.)

$$E_{\text{active}} = \sum_{k=1}^{K} \left( P_s(f) + P_l(f, T_k) \right) \Delta t \tag{8.20}$$

where $P_s$ and $P_l$ denote switching power and leakage power consumption, respectively. Note that we assume that clock frequency is proportional to $V_{dd}$, i.e., $f \propto V_{dd}$ in (8.20). In summary, switching power consumption is a function of clock frequency ($f$), while leakage power consumption is a function of clock frequency and temperature ($T_k$).

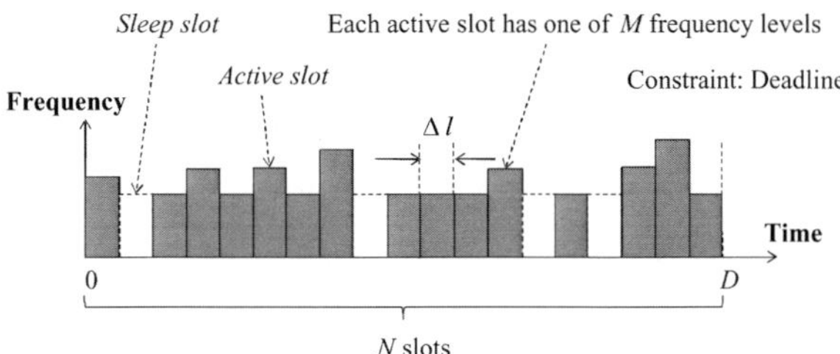



**Fig. 8.13** Problem definition to minimize total energy consumption [27]. © 2010 IEEE

In real environments, the processor changes operation states and/or frequency levels on a discrete time basis. The length of the discrete time period, $\Delta l$, is mostly determined by the scheduler in the OS. Within $\Delta l$, the operation state and frequency level are fixed. The time interval $[0, D]$, where $D$ denotes the task deadline, is divided into $N$ slots of $\Delta l$ duration each, as shown in Fig. 8.13. A slot where the processor is in the active (sleep) state is called the *active (sleep) slot*. In the active slot, the power consumption is the sum of $P_s$ and $P_l$. In the sleep slot, only a small amount of leakage power, $P_{\text{sleep}}$ $(\ll P_l)$, is consumed via power gating. The processor energy consumption, $E_{\text{total}}$, is calculated by the following equation:

$$E_{\text{total}} = \sum_{i=1}^{N} \left( E_{\text{active}}[i]x[i] + E_{\text{sleep}}(1 - x[i]) \right) \tag{8.21}$$

where

$$E_{\text{active}}[i] = P_s\big(f[i]\big)\Delta l + \sum_{k=1}^{L} P_l\big(f[i], T_k\big)\Delta t \tag{8.22}$$

and

$$E_{\text{sleep}} = P_{\text{sleep}}\Delta l \tag{8.23}$$

where $\Delta l = L \cdot \Delta t$ $(L = 20 \gg 1)$, and $f[i]$ is the clock frequency of the $i$th slot. $x[i]$ is 1 if the $i$th slot is an active slot, and 0 if it is a sleep slot.

**Problem definition** The problem of energy minimization can be defined as follows. Given an initial temperature $(T_{\text{init}})$, a task with a deadline $D$, and its runtime distribution (including the information of worst-case execution cycle, WCEC), and assuming that the interval, $[0, D]$ is divided into $N$ slots and the WCEC is divided into $B$ bins,[1] the problem is to find $f[i]$ and $x[i]$ such that the total energy consumption $E_{\text{total}}$ in (8.21) is minimized. The problem complexity is $O((M + 1)^N)$ because each slot can be at one of the $M$ frequency levels or sleep state. The complexity can be extremely high because $M \sim 20$ and $N \sim 500$ for even a deadline of a second $(\Delta l = 2$ ms$)$.

---

[1] Each bin consists of $C_{\text{bin}} = \lceil \text{WCEC}/B \rceil$ cycles.

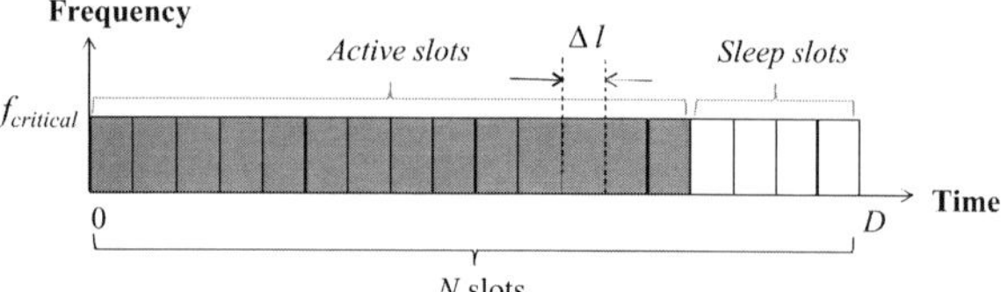


**Fig. 8.14** An example result of T-DVFS when the task completes earlier than the deadline [27]. © 2010 IEEE

## 8.4.2 Runtime Power Management Techniques to Minimize Total Energy Consumption

### 8.4.2.1 Temperature-Aware DVFS only (T-DVFS)

One of the preferred approaches for reducing the overall energy consumption of an embedded system is dynamic voltage frequency scaling (DVFS) because reduction in the supply voltage quadratically reduces switching energy consumption. In this method, given an available time slack (defined as time-to-deadline minus remaining worst execution time), the performance level (i.e., clock frequency and supply voltage) is set at the minimum value not to violate the deadline constraint in the real-time system. However, this method causes increased leakage energy by lengthening the execution time of the task. When a core runs at high temperature, the leakage power consumed by the core is comparable to the core's switching power due to the exponential dependency of leakage current on temperature. Therefore, the problem of DVFS has to be addressed based on performance requirement as well as temperature-induced leakage power.

T-DVFS, which can be considered as an extension of the work in [26], defines a critical supply voltage/speed such that if the speed (clock frequency) is scaled down below the critical speed, $f_{critical}$, the reduction in switching energy will be surpassed by the increase in leakage energy, resulting in more total energy consumption. Thus, T-DVFS operates the core at a higher speed than the critical one and puts the core into sleep (power-gated) state when the task completes earlier than the deadline. This can trade the switching energy savings for more leakage energy savings in order to obtain the maximal total energy reduction. When finding the critical speed, T-DVFS takes the steady-state temperature into account in order to estimate the temperature-induced leakage energy. Figure 8.14 shows an example result of T-DVFS when the task completes earlier than the deadline.

### 8.4.2.2 Temperature-Aware Power Gating at Critical Speed (T-PG)

Without consideration of cooling policy, the most simple and energy-efficient method is to completely execute the task and then switch the core to the sleep mode in order to save energy, as shown in Fig. 8.14 when the task's worst execution time at critical speed is smaller than the time-to-deadline. However power gating is one of the most effective methods to lower the chip temperature. Thus, inserting sleep slots

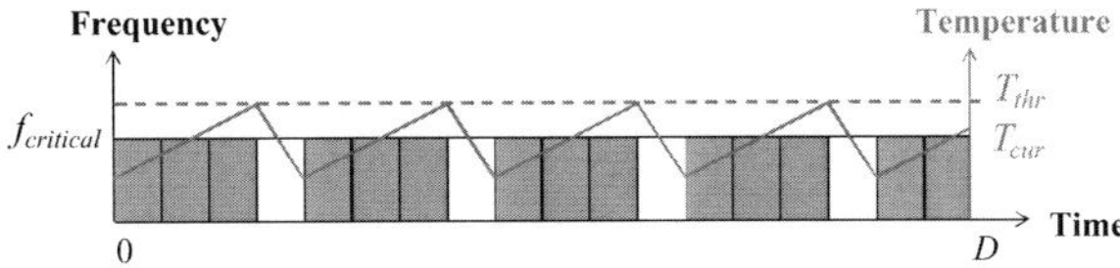

**Fig. 8.15** Determining the sleep slot location using the threshold temperature, $T_{thr}$ [27]. © 2010 IEEE

between active slots reduces the operating temperature properly, thereby reducing the leakage power consumption.

Determining the locations of active and sleep slots is a permutation problem. The number of permutations can be $N!/(N_a!N_s!)$, where $N_a$ and $N_s$ are the number of active slots and sleep slots, respectively. When the critical speed, $f_{critical}$, is assigned to active slots and the task finishes before deadline, $N_a$ and $N_s$ are determined by the critical speed, $f_{critical}$, as follows:

$$N_a(f = f_{critical}) = \left\lceil \frac{w/f_{critical}}{\Delta l} \right\rceil \tag{8.24}$$

$$N_s(f = f_{critical}) = N - N_a = N - \left\lceil \frac{w/f_{critical}}{\Delta l} \right\rceil \tag{8.25}$$

where $w$ is the number of execution clock cycles of the task.

The permutation problem can be easily solved by using the following lemma, which proves that a uniform distribution of sleep slots gives the optimal leakage energy consumption (proof given in [27]).

**Lemma** *Given $N_a$ active slots and $N_s$ sleep slots for a task that is running at a fixed clock frequency during active slots, the energy consumption is minimal when idle slots are evenly distributed across the task execution.*

This lemma shows that a uniform location of idle slots gives optimal leakage energy consumption. However, there can be different possibilities of uniform location. For instance, if there are 8 active slots and 3 idle slots, there can be three different cases of uniform location.[2] In order to simplify the solution while maintaining the same uniform idle slot insertions, T-PG performs a heuristic approach, as Fig. 8.15 illustrates. If the current temperature exceeds a threshold temperature $T_{thr}$, T-PG inserts a sleep slot, hopefully between active slots, to drop the temperature, as illustrated in Fig. 8.15. $T_{thr}$ must be determined to minimize the maximal operating temperature. For instance, if $T_{thr}$ is set too high, too few sleep slots are inserted between active slots. In this case, most sleep slots may be used only after all the active slots, thereby failing to contribute to lowering the temperature during task execution. If $T_{thr}$ is set too low, most sleep slots are inserted in the beginning of the task, causing the temperature to rise later in the remaining active slots after all available sleep slots are exhausted. To find the optimal $T_{thr}$ that distributes sleep slots evenly between active slots, a binary search can be used to sweep $T_{thr}$ between $T_{amb}$ and

---

[2]The three cases are (1) A, A, S, A, A, A, S, A, A, A, S; (2) A, A, A, S, A, A, S, A, A, A, S; and (3) A, A, A, S, A, A, A, S, A, A, S, where A and S represent active and sleep slots.

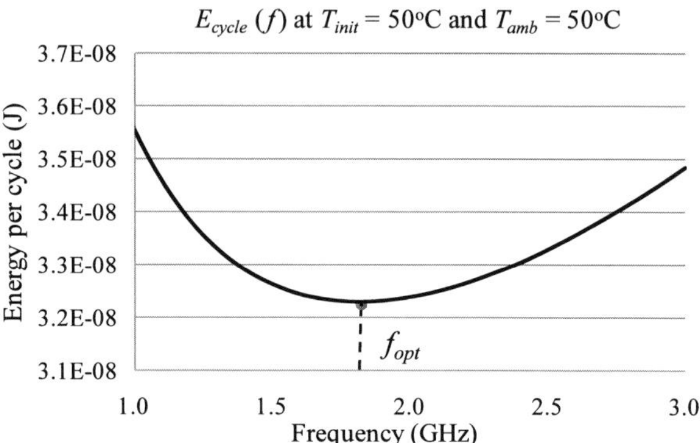

**Fig. 8.16** Total energy consumption (=switching + leakage) per cycle as a function of clock frequency [27]. © 2010 IEEE

$T_{\mathrm{max}}$, which is the allowed maximum junction temperature of a general processor. T-PG can be considered as an extension of the work of [28].

### 8.4.2.3 Temperature-Aware Integrated DVFS and Power Gating (T-INT)

Increasing active slots by lowering supply voltage/frequency obviously yields less heat generation and less temperature increase due to less switching power during active state. However, increasing sleep slots also helps lower the temperature by inserting sleep slots between active slots. Thus, at high temperature, it is sometimes beneficial to borrow the time budget from active states (by running at higher frequency than the critical speed) to allow more sleep states for temperature drop. This works because reducing the temperature and, thus, leakage power consumption by cooling can be more urgent than running the task when the temperature is very high.

Figure 8.16 shows the total energy consumption per cycle according to the frequency when the initial and ambient temperatures are the same, 50°C. In order to obtain $E_{\mathrm{cycle}}(f)$, at each frequency $f$, the total energy is estimated by simulating the task on the power/thermal model until the temperature reaches steady state, assuming that the deadline is sufficiently larger than the thermal time constant of the processor. Note that, in Fig. 8.16, the steady-state temperature is estimated from the average power consumption, as shown in Fig. 8.17(b). In Fig. 8.17(b), the average power and steady-state temperature are expressed as follows:

$$P(f, T) = \frac{P_{\mathrm{active}}(f, T)N_{\mathrm{a}}(f) + P_{\mathrm{sleep}}N_{\mathrm{s}}(f)}{N_{\mathrm{a}}(f) + N_{\mathrm{s}}(f)} \tag{8.26}$$

$$T^{\mathrm{new}}(P) = RP + T_{\mathrm{amb}} \tag{8.27}$$

where $R$ is the thermal resistance between the core and the ambient. As Fig. 8.17(b) shows, the estimation of $E_{\mathrm{cycle}}$ needs iterations of power and temperature estimation because of the dependency of leakage power on temperature. In our experiments, five iterations have usually been sufficient to yield a convergence in both the estimated temperature and power consumption with a variation of less than 0.1%.

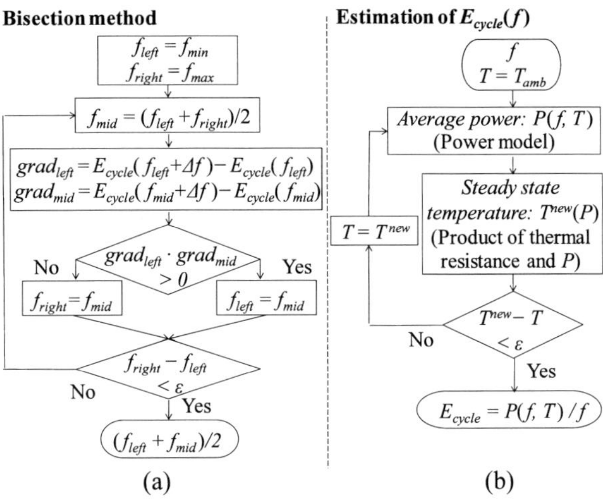



**Fig. 8.17** Algorithm flow of energy-optimal frequency decision: (**a**) flow of bisection method, (**b**) flow of $E_{cycle}$ estimation [27]. © 2010 IEEE

Figure 8.16 shows that the curve of $E_{cycle}$ is convex. Thus, the frequency for minimal energy consumption, called the optimal frequency, $f_{opt}$, is obtained when the slope of the curve is zero. A root-finding algorithm such as the *bisection method* (bisection) is a good solution to find the optimal frequency; the algorithm flow is shown in Fig. 8.17(a). In Fig. 8.17(a), the bisection method performs, at each intermediate frequency, four times of estimation of $E_{cycle}$ to obtain the gradients of $E_{cycle}$ at $f_{left}$ and $f_{mid}$. When the optimal frequency, $f_{opt}$, is determined, the increased sleep slots are evenly distributed during active slots based on the T-PG method.

### 8.4.2.4 Runtime Distribution-Aware DVFS only (R-DVFS)

In most instances, task execution time is not a fixed value and has runtime distribution even with a constant operating frequency. There are two sources of runtime distribution: data-dependent behavior and conflicts in accessing architectural resources, e.g., cache and memory. Programs with loops and if/else statements tend to have different loop counts and take different execution paths depending on (input) data. For instance, in the case of a video codec, object movements, i.e., input picture data, can determine the execution time of time-consuming functions, e.g., motion estimation. Cache misses and DRAM access scheduling are typical sources of runtime variation in the architectural area.

A hard real-time system does not allow any deadline to be missed. Consequently, for a task with runtime distribution, runtime power management solutions must consider the worst-case execution time (WCET) to guarantee meeting the deadline. However, this can be too conservative if the number of average execution clock cycles is much smaller than the worst case. Alternatively, the probability information of the number of clock cycles can be used for better runtime voltage/frequency scaling. If a periodic task consumes different numbers of clock cycles in different execution instances, the later cycles have lower probabilities to be consumed than the earlier cycles.

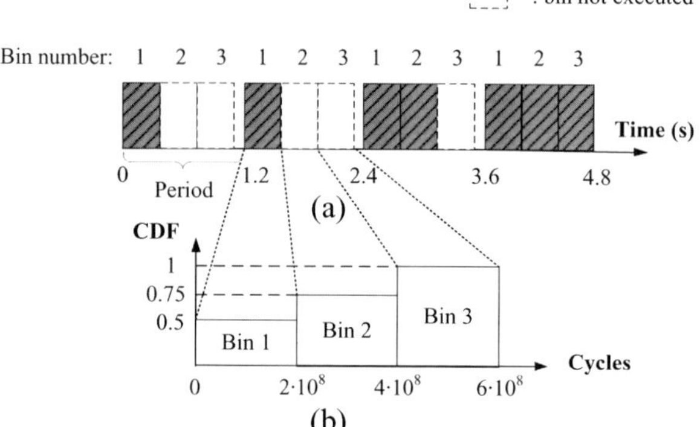


**Fig. 8.18** (a) Worst-case execution cycles of each task instance in a period are divided into three bins. (b) Cumulative distribution function for each bin $j$, where $\mathrm{CDF}(j+1) - \mathrm{CDF}(j)$ denotes the probability of the task being finished with bin $j$ [27]. © 2010 IEEE

Figure 8.18 shows an example of executing a periodic task with runtime distribution where the worst-case execution cycle (WCEC) for a task period is divided into three bins as shown in [20]. The task has one execution instance (shown as "bin executed" at the top of Fig. 8.18) during each period of 1.2 second. In Fig. 8.18, the four instances require two, two, four, and six hundred million cycles, respectively. We can compute the probability of each bin with respect to the total number of instances. The first bin has a probability of 100% (=4/4) because the first bin is executed in every instance (there are four instances in Fig. 8.18). The second and third bins have probabilities of 50% (=2/4) and 25% (=1/4), respectively. Therefore, the probabilities of each task instance being finished with the execution of the first, second, and third bin are 50% (=100 − 50), 25% (=50 − 25), and remaining 25%, respectively. The cumulative distribution function (CDF) of each bin is shown in Fig. 8.18(b).

The key idea of R-DVFS is to assign lower clock frequencies to the lower index bins in proportion to their probabilities of occurrence. Thus, the higher their probabilities, the lower the clock frequency that is assigned. The lower clock frequency assigned to the lower index bins must be compensated by assigning higher clock frequencies to the higher index bins. However, the total energy consumption is reduced because the eventual energy consumption is the sum of the products, over all bins, of consumed energy per bin and the associated bin probability.

The frequency assigned to the $j$th bin is $f_j$ and the execution time for this bin is $C_{\mathrm{bin}}/f_j$. The switching energy for this bin is proportional to $f_j^3 C_{\mathrm{bin}}/f_j$ when assuming that the supply voltage is proportional to frequency. The optimization goal is to find a frequency schedule, i.e., $f_1, f_2, \ldots, f_B$, to minimize the total expected energy. This is formulated as follows.

$$\text{Find} \quad f_j \quad \text{for all } j = 1, 2, \ldots, B$$

$$\text{such that} \quad \mathrm{Cost} = \sum_{j=1}^{B} p_j f_j^2 C_{\mathrm{bin}} \quad \text{is minimized} \tag{8.28}$$

**Fig. 8.19** Total energy consumption per cycle vs. time-to-deadline [27]. © 2010 IEEE

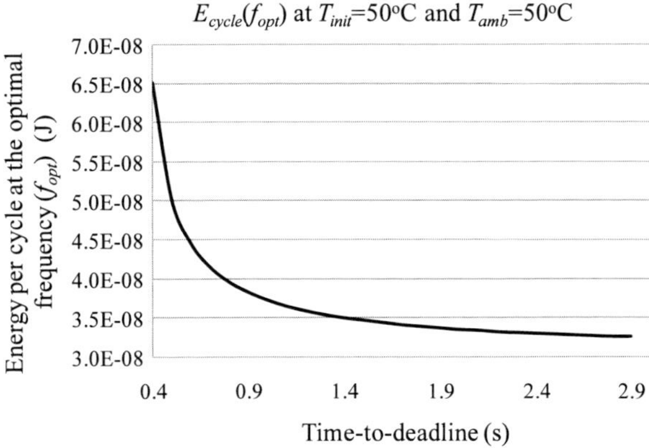



where $p_j$ is the bin probability to be executed,

$$\text{subject to} \quad \sum_{j=1}^{B} \frac{b}{f_j} \leq D \tag{8.29}$$

Based on the earlier study [20], the optimal solutions can be obtained by assigning $f_j$ as follows:

$$f_j = \frac{\sum_{h=1}^{B} b \sqrt[3]{p_h}}{D \sqrt[3]{p_j}} \tag{8.30}$$

The frequency schedules are nondecreasing because $p_j$ is always equal to or larger than $p_{j+1}$.

### 8.4.2.5  Temperature and Runtime Distribution-Aware Integrated DVFS and Power Gating (TR-INT)

As mentioned before, at high temperature, it is important to reduce temperature-induced leakage power for total energy minimization. T-INT, which exploits the integration of power gating and DVFS, is a good method to minimize the total energy consumption. However, in T-INT, worst-case execution time is assumed in applying power gating and DVFS. On the other hand, R-DVFS only focuses on reducing the switching energy through runtime voltage/frequency scaling, although it considers the task's runtime distribution. The basic idea of incorporating the statistical information, i.e., runtime distribution into the integrated DVFS and power gating is to allocate a time budget (as given by the time-to-deadline) to the bins in proportion to their bin probabilities. To do that, each bin is considered as a virtual sub-task, and the time budget of the bin is regarded as the time-to-deadline of the virtual sub-task. Thus, allocating more time budget to a low-index bin corresponds to assigning a longer time-to-deadline to the bin.

Figure 8.19 illustrates, assuming a fixed execution cycle of a (virtual) task, the energy consumption per cycle vs. time-to-deadline (i.e., the given time budget) at the minimal-energy frequency, $f_{\text{opt}}$ (explained in T-INT). Increasing time-to-deadline (more time budget) allows for operation at lower optimal frequency ($f_{\text{opt}}$) operation, thereby decreasing energy consumption by lowering the supply voltage. Note that the increased time-to-deadline reduces leakage power consumption as well as switching power consumption by lowering the supply voltage. This is so because the reduced switching power consumption gives less temperature increase and thereby less leakage power consumption than the case when the time-to-deadline is not extended. For TR-INT problem formulation, $E_{\text{cycle}}(f_j)$ in Fig. 8.19 is approximated with a function, $aS^{-b}+c$, where $S$ is the given time-to-deadline (i.e., time budget), where $a,b,$ and $c$ are fitting parameters depending on the target system conditions such as processor, heat sink, and ambient temperature. The TR-INT problem is formulated as follows.

$$\text{Find } S_j \quad \text{for all } j = 1, 2, \ldots, B$$

where $S_j$ is the time budget allocated to bin $j$ and $B$ is the number of bins

$$\text{such that} \quad \text{Cost} = \sum_{j=1}^{B} p_j E_{\text{cycle}}(f_j)$$

$$= \sum_{j=1}^{B} p_j \left(aS_j^{-b} + c\right) \quad \text{is minimized} \tag{8.31}$$

$$\text{subject to} \quad \sum_{j=1}^{B} S_j \leq D \tag{8.32}$$

where $D$ is the time-to-deadline.

In (8.31), $p_j E_{\text{cycle}}(f_j)$ represents the expected energy consumption per cycle contributed by bin $j$. Note that each bin has the same number of execution cycles. Thus, if the execution cycle is taken into account, then the function Cost in (8.31) becomes the expectation of the total energy consumption of the task. Because $a$ and $c$ are constants, the function Cost in (8.31) can be replaced by Cost* as follows:

$$\text{Cost}^* = \sum_{j=1}^{B} p_j S_j^{-b} \tag{8.33}$$

Note that, as shown in Fig. 8.19, $E_{\text{cycle}}(f_j)$ is monotonically decreasing as the time budget is increasing. $b$ must be positive to let $S_j^{-b}$ convex in (8.33). Jensen's inequality [29] can be applied to find the condition minimizing (8.33). To do that, (8.33) can be rewritten as follows:

$$\text{Cost}^* = \sum_{j=1}^{B} \left(p_j^{-1/b} S_j\right)^{-b} \tag{8.34}$$

**Table 8.4** Comparison of energy consumption results for H.264 decoder, ray tracing, and equake [27]. © 2010 IEEE

| Benchmark | T-DVFS (J) | R-DVFS | T-PG | T-INT | TR-INT |
|---|---|---|---|---|---|
| WETC $\geq D$ (14 fps for H.264 decoder) | | | | | |
| H.264 | 15.34 | 1.06 | N/A | 0.85 | 0.79 |
| Ray tracing | 4.31 | 1.08 | N/A | 0.91 | 0.81 |
| Equake | 122.47 | 1.07 | N/A | 0.89 | 0.78 |
| WETC $< D$ (7 fps for H.264 decoder) | | | | | |
| H.264 | 18.48 | N/A | 0.95 | 0.84 | 0.74 |
| Ray tracing | 5.10 | N/A | 0.98 | 0.86 | 0.76 |
| Equake | 143.19 | N/A | 0.96 | 0.83 | 0.73 |

According to Jensen's inequality, Cost* in (8.34) can be minimized when $p_j^{-1/b} S_j$ has the same value for all $j$'s, i.e., $p_j^{-1/b} S_j = p_{j+1}^{-1/b} S_{j+1}$ for all $j$'s. $S_j$ needs to satisfy the following relation:

$$S_{j+1} = \sqrt[b]{\frac{p_{j+1}}{p_j}} S_j \tag{8.35}$$

Cost* has a lower bound given as

$$B \left( \frac{\sum_{j=1}^{B} p_j^{-1/b} S_j}{B} \right)^{-b} \tag{8.36}$$

In order to calculate $S_j$ using (8.35), $S_1$ must be determined, and is used to calculate the remaining $S_j$'s by iterative applications of (8.35). $S_1$ is obtained by iterative binary subdivision, i.e., by sweeping $S_1$ values over the interval $[0, D]$ such that Cost in (8.31) is minimized while satisfying the constraint in (8.32).

### 8.4.3 Experimental Results

Experiments were performed with two real examples, i.e., *H.264 decoder* with different frame rates (14 f/s and 7 f/s in our experiments) and *ray tracing*, and a benchmark program, *equake* from SPEC2000. For *ray tracing* and *equake*, let us assume the time-to-deadline according to a utilization ratio (i.e., WETC $/D = 1.13^3$ and WETC$/D = 0.70$ in our experiment). The input picture for H.264 decoder was extracted from the movie *Harry Potter*. Details of the experimental setups such as power and temperature models are described in [27].

Table 8.4 shows the comparison of energy consumption for *H.264 video decoding*, *ray tracing*, and *equake*. All the energy consumption data are normalized with

---

[3] In our paper, we define WETC (worst-case execution time at critical speed) as WCEC divided by the critical speed.

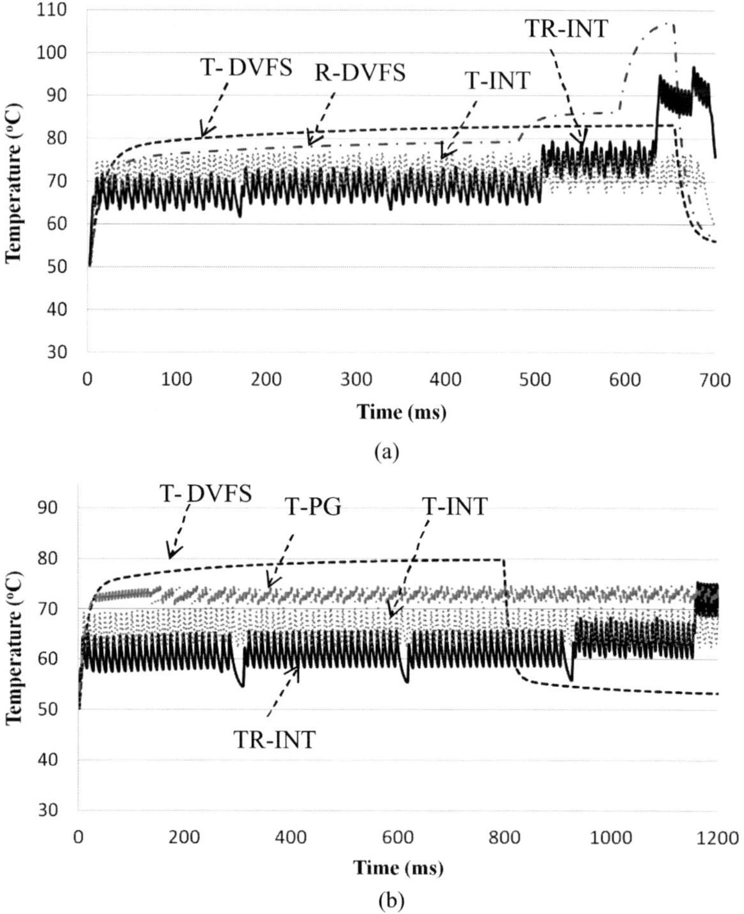





**Fig. 8.20** Thermal behavior of the hottest block in the processor for executing H.264 decoder running (**a**) T-DVFS, R-DVFS, T-INT, and TR-INT when WETC $\geq D$, and (**b**) T-DVFS, T-PG, T-INT, and TR-INT when WETC $< D$. (WCET: worst-case execution time at critical speed) [27]. © 2010 IEEE

respect to the energy consumption of the T-DVFS method. Note that T-PG is not applicable when the frequency needs to be set at a higher level than the critical speed (WETC $\geq D$). On the contrary, R-DVFS is not applicable (a plain DVFS is sufficient) when WETC $< D$. As shown in Table 8.4, TR-INT gives 19.4%–27.2% further reductions in energy consumption compared to the three methods (R-DVFS, T-DVFS, and T-PG). Such a significant improvement is obtained because TR-INT performs the temperature-aware trade-off between power gating and DVFS exploiting the runtime distribution.

Figure 8.20 shows the thermal behavior of the hottest block in the processor for the power management methods. As shown in Fig. 8.20, T-DVFS enters the sleep state only after the completion of task execution. In addition, T-DVFS does not consider the runtime distribution, but is based only on the worst case execution time. Thus, it runs at a high frequency, thereby yielding higher temperature distribution than TR-INT in the lower index bins (i.e., in the beginning of task execution), which incurs more leakage energy consumption for T-DVFS. In the case of R-DVFS, it assigns lower frequencies to the earlier bins and higher frequencies to the later bins in order to exploit runtime distribution. Thus, as shown in Fig. 8.20(a), R-DVFS gives a similar temperature trend as TR-INT. The temperature increases as the bin index increases. However, TR-INT achieves, for most of the runtime, lower temperature levels than R-DVFS because TR-INT inserts sleep slots during task execution to lower operating temperature while R-DVFS does not. Lower temperature, especially in low-index bins, gives more reduction in leakage energy consumption.

As shown in Fig. 8.20(b), T-PG utilizes sleep slot insertions during task execution, yielding lower temperatures than T-DVFS. However, the amount of sleep slots is determined only by the ratio of WETC to time-to-deadline because T-PG uses only the fixed frequency, i.e., the critical speed. In addition, the locations of sleep slots are evenly distributed across bins without considering the runtime distribution. Because TR-INT inserts more sleep slots in lower-index bins, it yields lower temperatures and more energy savings than T-PG through aggressive sleep slot insertions.

In all the experiments above, it is assumed that the ambient temperature $T_{\mathrm{amb}}$ is constant. In order to evaluate the impact of ambient temperature change, different ambient temperature levels are applied to the methods. Figure 8.21 shows the energy consumption which is normalized with respect to T-DVFS. The figure shows that TR-INT yields more energy savings as the ambient temperature becomes higher. The reason is that, at high temperature, leakage power consumption becomes dominant and additional sleep state insertions by sacrificing switching power consumption becomes more effective for reducing the temperature and total energy consumption.

## 8.5 Conclusion

This chapter has extensively addressed system-level temperature-aware design methodologies, which is one of the most important challenges in 3-D ICs. The 3-D ICs need new power management techniques because of their different thermal characteristics (i.e., heterogeneous thermal coupling and cooling efficiency) compared with 2-D ICs. To maximize the performance of a temperature-constrained system based on the thermal characteristics of 3-D ICs, an analysis of instantaneous temperature and memory boundedness of applications is essential. To minimize temperature-induced power, the integration of DVFS and power gating has to be performed in a temperature and runtime distribution-aware manner. We hope that this chapter provides an overall understanding of thermal effects at the architectural level, and of the interaction of power, temperature, and performance in 3-D ICs.

**Fig. 8.21** Energy comparisons for H.264 decoder and ray tracing benchmark with different ambient temperature; energy consumptions presented in this figure are normalized against T-DVFS. (**a**) H.264 decoder when WETC $\geq D$. (**b**) H.264 decoder when WETC $< D$. (**c**) Ray tracing when WETC $\geq D$. (**d**) Ray tracing when WETC $< D$ [27]. © 2010 IEEE

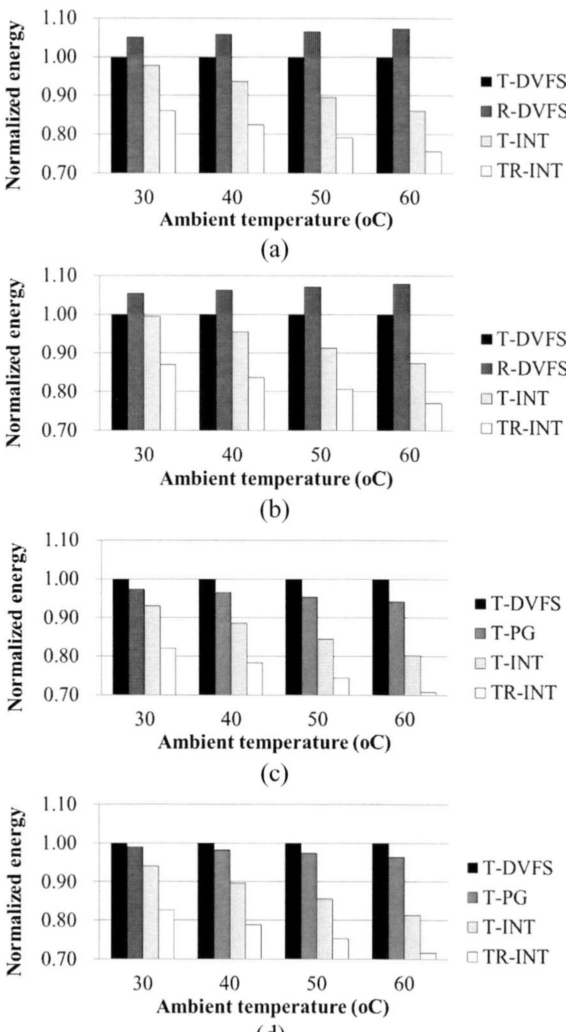

# References


1. Transistor counts on *Wikipedia* [Online]. Available http://en.wikipedia.org/wiki/Transistor_count
2. Benini, L., Micheli, G.D.: Networks on chips: a new SoC paradigm. Computer **35**(1), 70–78 (2002)
3. Xie, Y.: Processor architecture design using 3D integration technology. In: Proc. Int. Conf. on VLSI Design, pp. 446–451 (2010)
4. Joyner, J.W., Zarkesh-Ha, P., Meindl, J.D.: A stochastic global net-length distribution for a three-dimensional system-on-a-chip (3D-SoC). In: Proc. IEEE Int. ASIC/SOC Conf, pp. 147–151 (2001)
5. Loh, G., Xie, Y., Black, B.: Processor design in 3D die-stacking technologies. IEEE MICRO **27**(3), 31–48 (2007)

6. Zia, A., et al.: A 3-D cache with ultra-wide data bus for 3-D processor-memory integration. IEEE Trans. Very Large Scale Integr. **18**(6), 967–977 (2010)

7. Kumar, A., et al.: HybDTM: a coordinated hardware-software approach for dynamic thermal management. In: Proc. DAC, pp. 548–553 (2006)

8. Rao, R., et al.: An optimal analytical solution for processor speed control with thermal constraints. In: Proc. ISLPED, pp. 292–297 (2006)

9. Yang, J., et al.: Dynamic thermal management through task scheduling. In: Proc. Int. Symp. ISPASS, pp. 191–201 (2008)

10. Zhu, C., et al.: Three-dimensional chip-multiprocessor runtime thermal management. IEEE Trans. Comput.-Aided Des. Integr. Circuits Syst. **27**(8), 1479–1492 (2008)

11. Isci, C., et al.: An analysis of efficient multi-core global power management policies: maximizing performance for a given power budget. In: Proc. Int. Symp. Microarchitecture, Dec. 2006, pp. 347–358 (2006)

12. Kreith, F.: The CRC Handbook of Thermal Engineering, pp. 2.1–2.92. CRC Press, Boca Raton (2000)

13. Yang, Y., et al.: Adaptive multi-domain thermal modeling and analysis for integrated circuit synthesis and design. In: Proc. ICCAD, Nov. 2006, pp. 575–582 (2006)

14. Zhou, X., et al.: Thermal-aware task scheduling for 3D multi-core processors. IEEE Trans. Parallel Distrib. Syst. **21**(1), 60–71 (2010)

15. Li, T., John, L.K.: Runtime modeling and estimation of operating system power consumption. In: Proc. SIGMETRICS, pp. 160–171 (2003)

16. Chen, J.W., Dubois, M., Stenstrom, P.: Integrating complete-system and user-level performance/power simulators: the SimWattch approach. In: Proc. ISPASS, Mar. 2003, pp. 1–10 (2003)

17. Gomaa, M., Powell, M.D., Vijaykumar, T.N.: Heat-and-run: leveraging SMT and CMP to manage power density through the operating system. In: Proc. ASPLOS, Nov. 2004, pp. 260–270 (2004)

18. Performance Application Programming Interface [Online]. Available http://icl.cs.utk.edu/papi/

19. Corliss, G.: Which root does the bisection algorithm find? SIAM Rev. 325–327 (1977)

20. Lorch, J.R., Smith, A.J.: Improving dynamic voltage scaling algorithm with PACE. ACM SIGMETRICS Perform. Eval. Rev. **29**(1), 50–61 (2001)

21. Choi, K., Soma, R., Pedram, M.: Fine-grained dynamic voltage and frequency scaling for precise energy and performance tradeoff based on the ratio of off-chip access to on-chip computation times. IEEE Trans. Comput.-Aided Des. Integr. Circuits Syst. **24**(1), 18–28 (2005)

22. Everett, H. III: Generalized Lagrange multiplier method for solving problems of optimum allocation of resources. Oper. Res. **11**, 399–417 (1963)

23. Skadron, K., et al.: Temperature-aware microarchitecture: modeling and implementation. ACM Trans. Archit. Code Optim. **1**, 94–125 (2004)

24. Kang, K., Kim, J., Yoo, S., Kyung, C.-M.: Runtime power management of 3D multi-core architectures under peak power and temperature constraints. IEEE Trans. Comput.-Aided Design Integr. Circuits Syst. **30**(6), 905–918 (2011)

25. Huang, W., et al.: HotSpot: a compact thermal modeling method for CMOS VLSI systems. IEEE Trans. Very Large Scale Integr. **14**(5), 501–513 (2006)

26. Bao, M., et al.: Temperature-aware voltage selection for energy minimization. In: Proc. DATE, pp. 1083–1086 (2008)

27. Kang, K., Kim, J., Yoo, S., Kyung, C.-M.: Temperature-aware integrated DVFS and power gating for executing tasks with runtime distribution. IEEE Trans. Comput.-Aided Des. Integr. Circuits Syst. **29**(9), 1381–1394 (2010)

28. Yuan, L., et al.: Temperature-aware leakage minimization technique for real-time systems. In: Proc. ICCAD, pp. 761–764 (2006)

29. Krantz, S., Kress, S., Kress, R.: Jensen's Inequality. Birkhauser, Cambridge (1999)

30. Coskun, A.K., et al.: Dynamic thermal management in 3D multicore architectures. In: Proc. DATE, pp. 1410–1415 (2009)

# Chapter 9
# Low Power Mobile Storage: SSD Case Study

**Sungjoo Yoo and Chanik Park**

**Abstract** Solid state disks (SSDs) consisting of NAND flash memory are replacing conventional HDDs in mobile and server markets. The major reason for such a widespread replacement is the high performance and low power consumption enabled by an SSD. It especially gives better performance and power efficiency for random requests, because it does not require slow-moving and power-hungry mechanical operations, e.g., motor control. For the performance of sequential requests, multi-channel/way parallel architectures are adopted, and the speed of the host interface is increased. Even though an SSD has better power efficiency than an HDD, the SSD power consumption must be minimized due to the stringent power consumption constraints of mobile and server products. This chapter presents a case study of developing an SSD power model with an application to low power SSD design. The power model is based on real measurement data. For accurate modeling, one must take into account the internal parallelism as well as power states. The case study shows that the power model is useful in evaluating the designs of dynamic power management policy.

## 9.1 Power Consumption in Solid State Disk

Flash memory-based storage devices, e.g., solid state drives/disks (SSDs), USB sticks, etc., are becoming one of the major types of storage, both in personal usage and in industry.

Flash memory has the advantages of high performance, low power consumption, and high reliability compared with a conventional hard disk drive (HDD). High performance is the main reason that flash memory is adopted, especially for faster booting and application loading. The advantages come from the fact that flash memory

S. Yoo (✉)
POSTECH, Pohang, Republic of Korea
e-mail: sungjoo.yoo@postech.ac.kr

C. Park
Samsung Electronics, Hwasung-City, Republic of Korea
e-mail: ci.park@samsung.com

C.-M. Kyung, S. Yoo (eds.), *Energy-Aware System Design*,
DOI 10.1007/978-94-007-1679-7_9, © Springer Science+Business Media B.V. 2011

is based on electronic functions, e.g., program, read, and erase, whereas an HDD is based on mechanical ones, e.g., servo and spindle motors and arms. Driving those mechanical parts entails significant latency (on the order of milliseconds) and power consumption, especially when the HDD enters the active mode from a low power mode.

Among flash-based storage, SSD is becoming a major storage device, replacing the HDD in smart phones (e.g., iPhone 3GS with 32 GB flash memory) and net books as well as in notebook PCs and servers [1]. In this chapter, we present the issues of low power design for SSDs.

A SSD is inherently more power efficient (in terms of energy consumption per data transfer bandwidth) than an HDD because the HDD requires driving of both mechanical and electrical (SATA I/O, DRAM buffer, etc.) parts; whereas, the SSD consumes power only for electrical ones, including the flash memory.

One of the main reasons that SSD is favored over HDD is performance. The SSD offers higher performance than the HDD via parallel accesses; it utilizes relatively low speed flash devices in parallel. For instance, in order to achieve a throughput higher than 240 MB/s, we can utilize 8 flash devices with 33 MBps each in parallel. Recently, in order to obtain further performance improvement, high speed flash interface specifications have been presented, e.g., ONFI and toggle NAND. By adopting flash memories with high I/O bandwidth, the SSD performance can be improved significantly [2].

When the SSD is deployed in commercial products like notebook PCs, it is given a power consumption budget. Typically, the set makers allocate a similar or smaller power budget to an SSD compared with the traditional HDD. For instance, the SSD has a peak power budget of about 1 A and an average power consumption budget of about 1.2 W in typical notebook PCs [3] and is expected to have a much lower power budget in smart phones. Within this smaller power budget, the SSD must maximize performance. Further performance improvement requires more parallel operations in the SSD even though high speed I/O is adopted; in particular, the write performance is not improved significantly by high speed I/O. Thus, parallel writes are necessary to improve the write performance.

SSDs used in mobile devices, e.g., notebook PCs, do not have persistent traffic; instead, the traffic to the SSD depends on the user's usage patterns. Compared with SSDs used in high-end computing systems, e.g., servers, SSDs in mobile devices tend to have a significant amount of idle time. Thus, a reduction in idle power consumption is also important in an SSD.

Low power design for SSD has a huge design space: parameter sets in dynamic power management (DPM) (e.g., DPM policies including time-out parameters), flash translation layer (FTL) algorithms, and SSD controller architectural parameters (e.g., I/O frequency in flash devices and DRAM buffer size and caching/prefetch methods in the controller). Design space exploration can be performed based on two power estimation methods: real measurement and full system simulation with a power model.

We can obtain accurate information on power consumption from real measurement, which is practically used in SSD product designs. Real measurements have two drawbacks: long design cycle (e.g., hours of SSD execution is required for the evaluation of one design candidate) and changing battery characteristics over repeated runs [4]. Because of these two problems, design space exploration is performed in a set of limited choices.

The second option of cycle-level full system simulation with a power model can give detailed information on power consumption, e.g., peak power profile on a NAND flash command granularity. However, its simulation runtime is prohibitively long. For instance, assuming a simulation speed of $\sim$100 Kcycles/sec and a 200 MHz SSD controller, it may take 80 days to simulate 1 hour of real SSD execution.

In our work, we propose a real power measurement-based method of SSD power estimation. The proposed method tries to keep the advantages of real measurement, i.e., accuracy and the fact that both flash memory and controller are covered. In order to overcome its limitations of long execution time and lack of repeatability (and high cost), it performs a trace-based simulation of the SSD subsystem. The power estimation method takes as input real power measurements and SSD access trace and gives as output power profile (power consumption over time). In this chapter, we present two case studies where the power estimation method is utilized for the design of time-out-based DPM for the SSD.

## 9.2 Related Work

Several works have been published for low power HDD [5–11]. In [5], it is reported that the mechanical parts incur large overheads of power consumption, especially when the HDD starts to run the spindle and head. Also, the spin-up energy can vary between HDDs by an order of magnitude. A disk simulation model, DiskSim, is presented in [6]. In DiskSim, the HDD has four active-mode power states (seeking, rotation, reading, and writing) and two idle-mode ones. In [7], the authors show that fine-grained power states give further energy reduction in HDDs, because short idle periods can be exploited with fine-grained power states, thereby entering low power states more frequently.

Several adaptive methods for HDD power management have been presented. In [8], the authors present an adaptive DPM method based on the concept of a session, which represents the time period of frequent HDD accesses. The inter-session delay is predicted and exploited to make HDD enter into low power states. In [9], a time-out-based DPM policy is presented. In this work, if there is no new access during the time-out, a low power state is entered. The time-out is determined adaptively based on the accuracy of previous time-out predictions. In [10], the authors present a machine learning-based method to determine when to spin down the disk. In [11], an adaptive algorithm for time-out-based spin down is presented to calculate the time-out for disk spin down utilizing multiple time-out parameters and considering spin-up latency cost.

There are several studies on increasing HDD idle time proactively. In [12], prefetching is applied to prolong the idle time of HDDs, enabling them to stay in low power states for longer periods. In [13], a method of joint power management of memory and HDD is presented. In this work, the HDD idle time is prolonged by increasing the effective size of main memory, thereby reducing HDD accesses while also considering the overhead of energy consumption in main memory. In [14], the rotation speed of disk platters is adjusted for low power consumption.

Recently, a performance model of SSD has been presented and the performance of parallel SSD architectures has been explored [15]. Another flash-based memory simulator, called FlashSim, is presented in [16]. It is an event-based simulator and allows hybrid simulation between FlashSim and DiskSim. A power model of flash memory called FlashPower is presented in [17]. It is limited to the estimation of power consumption of single-level cell (SLC) NAND flash memory. In order to address the dynamic power consumption of flash memory, in [18] the authors present a low power coding scheme for multi-level cell (MLC) flash memory. They exploit the fact that MLC flash memory has value-dependent power characteristics. For instance, in a 2 bit MLC, the power consumptions of coding two 2 bit data, 00 and 01, are different from each other. This happens mainly because bit levels in MLC have different numbers of programming steps [19]. In [20] the authors present a low power solution for 3D-stacked flash memory. In that work, multiple flash dies share a common charge pump circuit which consumes the majority of standby power consumption in flash memory. In [21] the authors present three low power design methods: selective bit-line precharge (precharging only the bit line requiring a new programming pulse), advanced source-line program (dividing the source line into local ones), and intelligent interleaving (time-shifting program commands to avoid peak power problems). In [22] the authors present a method of low power buffer cache management for heterogeneous storage consisting of flash memory and HDD, exploiting the difference in their characteristics of power consumption and performance.

## 9.3 Flash Memory Operation and SSD Architecture

A flash memory device in package typically consists of vertically stacked silicon dies. The I/O signals of the flash memory package are shared by the flash memory dies in a time-multiplexed way, as will be explained later in this section. Figure 9.1 shows a simple block diagram of a single-level cell (SLC) flash memory die. It consists of blocks, each of which comprises 64 pages of 2 KB each.[1] The flash memory has an I/O buffer of page size. When a data read request is made by the controller which is connected to the flash memory device, the page containing the corresponding data is first fetched from the memory cell array to the I/O buffer. Then, the corresponding data is read by the controller via the byte-level data I/O. The byte-level data I/O of a memory die is shared in a time-multiplexed way by both commands/address and data transfer.

---

[1] To be specific, the page size is 2 KB + spare bits, i.e., 2111 bytes.

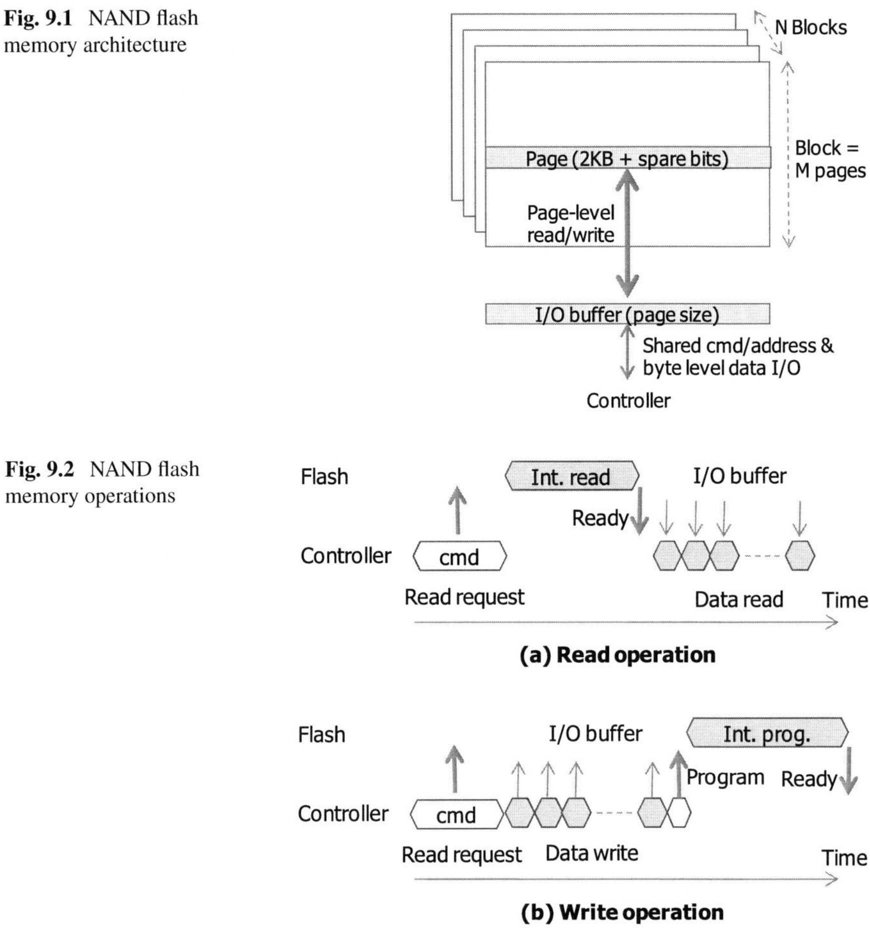

**Fig. 9.1** NAND flash memory architecture



**Fig. 9.2** NAND flash memory operations



Figure 9.2(a) illustrates a read operation to flash memory. The controller issues a read command including address (denoted "cmd" in the figure) to initiate flash internal read ("Int. read") to transfer the corresponding page from the memory cell array to the I/O buffer, which typically takes tens of microseconds, e.g., $t_R = 25\ \mu s$.[2] After the internal read is finished, the flash memory raises a ready pin to inform the controller that the data is ready. The controller then reads the corresponding data (even up to the entire page) via the data I/O. The I/O delay to read (or write) the entire 2 KB page from (to) the I/O buffer usually takes tens of microseconds, e.g., $t_{RC} = 63.33\ \mu s$ for a page of 2111 bytes at 33 MHz I/O frequency.

---

[2]The delay of internal read, program, and erase varies from manufacturer to manufacturer. Even for the same manufacturer, it depends on manufacturing conditions, e.g., process variations on the silicon die. During real usage, it also varies due to aging effects as well as environmental factors, e.g., temperature. For MLC memory, the program delay also depends on the value of data to be written, since different values require different program steps [19].

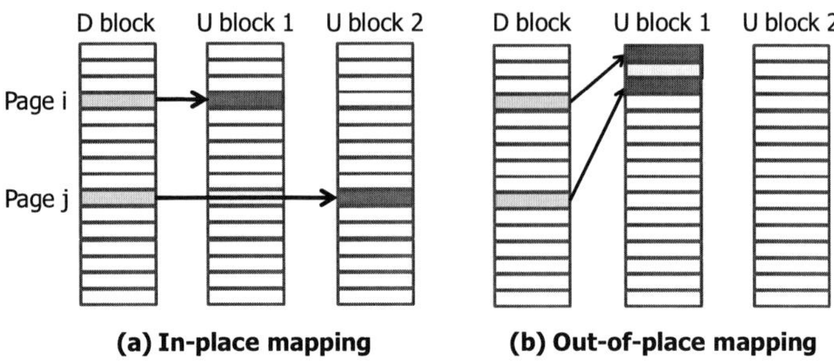



**Fig. 9.3** Logical-to-physical data mapping in flash memory

Figure 9.2(b) illustrates the write operation. In this case, after the write address is sent to the flash memory, the data is first written (sequentially or randomly) to the I/O buffer. Then, the controller initiates the flash internal program operation ("Int. prog." in the figure), which takes hundreds of microseconds: typically, $t_{PROG} = 300$ μs for SLC and 800 μs for MLC. In terms of power consumption, program consumes more power than read due to the flash memory characteristics of requiring charge pumping for program operation. For instance, sequential writes (writes to consecutive addresses) typically have about twice the power consumption of sequential reads.

One of the salient characteristics of flash memory is that no update is allowed on already-written data. When a memory cell needs an update, it first must be erased before writing a new data to it. We call such a constraint "erase before write." Erase is performed on a block basis and it takes milliseconds, e.g., $t_{ERASE} = 2$ ms. Thus, if we erase a block just to update a single datum on every write operation, the delay of the entire operation for a single write will be significant, e.g., $2$ ms $+ 63$ μs $+ 300$ μs $= 2.363$ ms per a single write. The latency corresponds to a write throughput of 423 B/sec, which is far less than the peak bandwidth of a single flash memory, 33 MBps at 33 MHz. In order to overcome the low performance due to "erase before write," log buffers (often called update blocks) are utilized; the new write data is written to the log buffers. Figure 9.3 illustrates two examples of log buffer schemes. Figure 9.3(a) is called an in-place scheme, where the page index is maintained in the log (update) blocks. The advantage of the in-place scheme is that page-level remapping does not need to be considered. However, it suffers from a low utilization of log blocks. As the figure shows, when the second write to page $j$ is performed, we must utilize a new log block (U block 2 in the figure), even though the first log block still has free space. In order to overcome the limitation of the in-place scheme, an out-of-place scheme, as shown in Fig. 9.3(b), is usually utilized. In this scheme, all the pages in a log block are utilized in a sequential manner. Thus, in Fig. 9.3(b), the second write to page $j$ is accommodated by the third page in U block 1 without utilizing another log block. The out-of-place scheme gives a good utilization of log block resource, but has an overhead of managing page-level mapping information. The flash translation layer (FTL) on the SSD controller maintains the address mapping between the logical and physical data in the log and data blocks

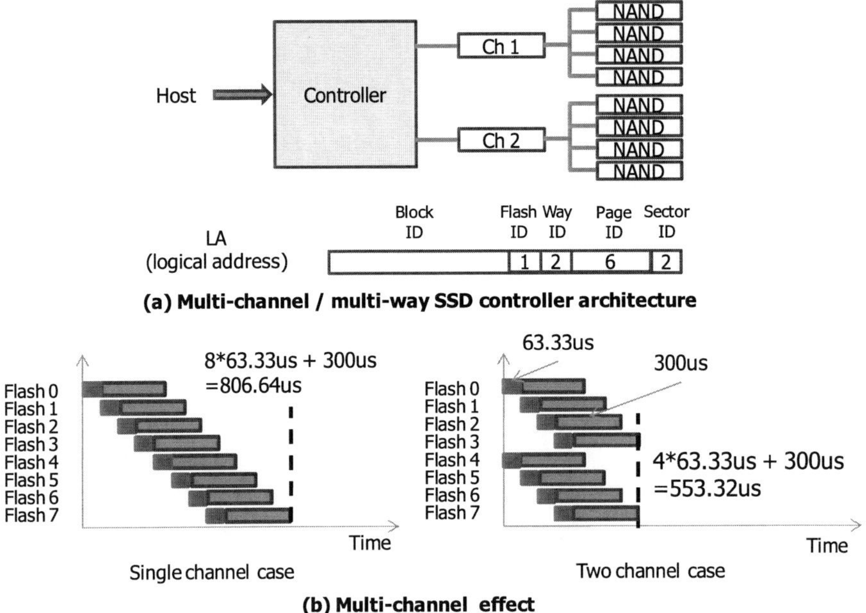



**(a) Multi-channel / multi-way SSD controller architecture**



**(b) Multi-channel effect**

**Fig. 9.4**  Multi-channel/multi-way SSD controller architecture

[23–25]. In reality, the controller (specifically FTL) determines the performance and power consumption of random reads/writes. Thus, the controller overhead must be taken into account in the power estimation of the SSD.

A single flash memory can support up to the data throughput of the data width * I/O frequency, e.g., 33 MBps at 33 MHz. In order to support higher bandwidths, typically, the SSD exploits two types of parallelism: channel and way. The SSD consists of multiple channels, each of which can be accessed independently (channel-level parallelism). A channel consists of one or multiple flash memory packages, each of which has multiple flash dies called ways. All the memory dies in a channel share the channel I/O signals in a time-multiplexed way. The execution of each die in the package can be overlapped (way-level parallelism) while considering the resource contention in the shared channel I/O signals.

Figure 9.4(a) illustrates an SSD controller architecture consisting of two channels and four ways per channel. The controller takes commands from the host (e.g., smart phone or notebook CPU) and performs its FTL algorithm to find the physical page address(es) to access. Then, it accesses the corresponding channels and ways, if needed, in parallel. On each channel, each way can run in parallel performing internal operations, e.g., internal read/program/erase. However, only one way can utilize the channel, i.e., the I/O signals of the package, at any time. Thus, the peak throughput is determined by the number of channels * I/O frequency.

Figure 9.4(b) illustrates how multiple channels can give higher performance. The figure depicts a case of sequential writes; the left-hand side shows the execution of sequential writes on a single channel consisting of eight ways. Flash *i* represents a flash memory die $i$ on the channel. The executions of flash memory dies overlap with each other. The controller writes a page to the I/O buffer of each flash memory die and launches the internal program operation, as shown in Fig. 9.2(b). Then, while the previously launched program is being performed, the controller starts writing the next page of data to the next flash memory die and launches the internal program operation after finishing the data write. The interleaving on a single channel is performed in this way. In the left-hand side of Fig. 9.4(b), the first small rectangle on each flash memory die represents the I/O buffer write operation and the second long rectangle the internal program operation.

The right-hand side of Fig. 9.4(b) shows the two-channel case where each channel consists of four ways. The controller needs to have two separate sets of I/O signals to be connected to the two channels. The controller can perform two I/O buffer write operations in parallel. After the two write operations are finished, it can launch two internal program operations which run concurrently. In this way, more flash memory dies execute their internal operations, thereby giving faster execution than in the single channel case.

## 9.4 SSD Power Estimation

Figure 9.5 shows the flow of the proposed SSD power estimation method. The power estimation flow takes the following as input:

- Performance and power consumption data from real measurements:
  - Information on the read/write latency for the data sizes of $1/2/4/8/16/32/64/128/256$ sectors (sector size $= 512$ B).
  - Information on the power consumption of active-mode operation represented by sequential reads/writes, power consumption per power state (idle, partial, and slumber), and power state transition delay values.
- Information of SSD architecture:
  - Number of channels, number of ways/channel.
  - Flash memory latency parameters ($t_R$, $t_{RC}$, $t_{PROG}$, and $t_{ERASE}$).
- SSD access trace obtained from real executions of host system (as shown on the left-hand side of Fig. 9.5).

The SSD power estimator consists of two parts: an event-driven simulation model and a power model for the SSD controller. As shown in Fig. 9.5, the power estimator takes a trace as input. It performs the event-driven simulation based on the information of read and write latency while considering architectural resource conflicts on channels and ways. It gives as output the power profile over time as well as the output execution trace (e.g., the status of each channel/way and latency of each SSD access).

The power estimator takes as input the parameters of power states, e.g., transition latency and power consumption per power state, and enables the simulation of dynamic power management (DPM) policy. In our experiments, we will present two case studies utilizing this feature.

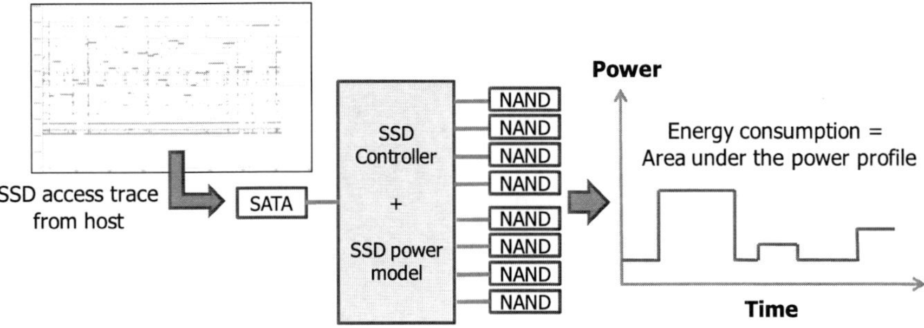



**Fig. 9.5**  SSD power estimation flow

## 9.4.1 Performance and Power Modeling

Figure 9.6 illustrates the simulation of a single write operation. First, we decompose the measured latency of the read and write operation into two parts: I/O and internal operation latency; see Fig. 9.6(a). In the figure, we assume a one page write operation to the SSD. The controller writes one page of data to the I/O buffer of the flash memory die in the time period between $t_0$ and $t_1$. The time period is determined by the I/O bandwidth of the target flash memory, e.g., 33 MBps. The internal operation part runs from $t_1$ and $t_2$, and it represents the actual program operation within the flash memory die and the operation of the FTL algorithm.

For the power modeling, we decompose the measured power consumption value into two parts: baseline and access-dependent parts. The baseline represents power consumption in the idle state. The power consumption of the idle state is measured

**Fig. 9.6**  Performance and power modeling: single write case

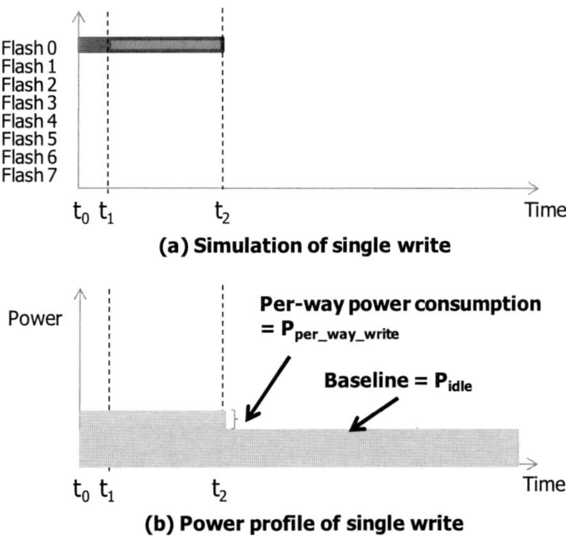

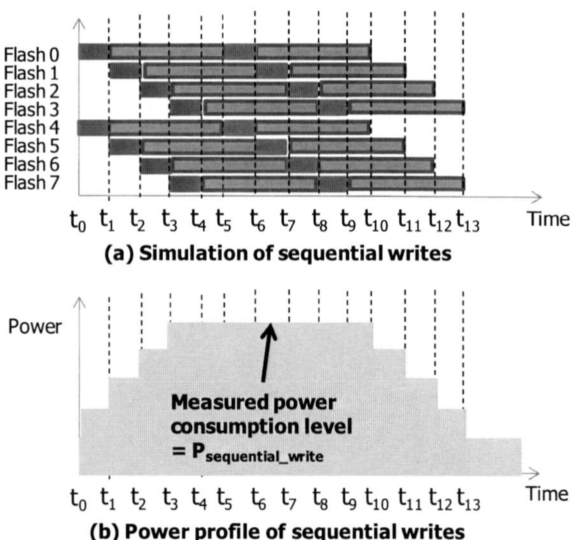



**Fig. 9.7** Performance and power modeling: sequential write case

separately. The access-dependent part is obtained by subtracting the baseline from the measured power consumption of the read and write operation.

Figure 9.6(b) illustrates the decomposition into baseline and access-dependent parts and shows the power profile for the case in Fig. 9.6(a). The SSD consumes the power consumption of the idle state until a transition to a low power state is made, as shown in Fig. 9.6(b). On top of the baseline level, we add the access-dependent part for a single write operation during the period of write operation between time $t_0$ and $t_2$. The power model presented in Fig. 9.6(b) is simple compared with real activities where the internal program operation has multiple current peaks due to its iterative program operation in MLC flash memory [19, 21]. For the purpose of exploring DPM policies in our case, the current power model is sufficient. However, for more sophisticated analyses, such as fine-grained peak power analysis, we will require more detailed power measurements enabling fine-grained decompositions down to the program iterations to give more accurate power estimation. This will be left for our future work.

Resource conflict modeling is the key function in the event-driven simulation. Figure 9.7 illustrates resource conflict and power modeling for the two-channel architecture shown in Fig. 9.4. Figures 9.7(a) and (b) illustrate the execution trace and the power profile, respectively, for the case of 16 page sequential writes to the SSD. Each of eight flash memory dies on the two channels receives one page of data from the controller at times $t_0$, $t_1$, $t_2$, and $t_3$. After the data transfer is finished, each die starts its internal program operation. After it is finished, each die receives a new page and performs another internal program operation.

Figure 9.7(b) shows the corresponding power profile. On top of the baseline level, the contribution of each flash memory die is added during the period while its (I/O or internal) operation is being performed. The figure shows a peak plateau from $t_3$ to $t_{10}$ when all the eight flash memory dies and controller are in active mode. Typically,

```
1 while (T_now < end of simulation) {
2    Advance time T_now to the next event
3    while (any event at time T_now) {
4       new_event = pop(event_list(T_now)) // pop the events of end of Flash operation first
5       If (new_event == host command)
6          Run FTL to find the corresponding PPAs
7          T_init = T_now
8          If (current state == low power state or transition to low power state)
9             PState = Transition2Active; P_total = P_idle; T_init = T_now + T_wakeup
10            Clear future events for transition to low power state
11            Schedule_events_for_Flash_operations(PPAs, T_init)
12         Else // current state == transition to active state
13            Schedule_events_for_Flash_operations(PPAs, T_init)
14      Else if (new_event==start or end of Flash operation)
15         If (new event == start of Flash operation), then
16            If (PState==Transition2Active), then
17               PState=Active
18            Add the power consumption of the newly started operation to P_total
19         Else, // new_event = end of Flash operation
20            Subtract the power consumption of the just finished operation from P_total
21            If there is no more future event for Flash operation, then
22               // Insert DPM policy here. The following is a TO-based DPM policy example
23               Schedule a TO event at T_now+TO
24      Else // power state transition event
25         // TO event for a power state transition in the DPM policy example
26         PState = LowPowerState
27         P_total = P_low_power_state
28         // If there is any lower power state, then schedule a TO event here
29   } // end of "any event at time T_now"
30 }
```

**Fig. 9.8**  Pseudo code of event-driven simulation method

the power consumption of sequential writes is measured while sequential writes are being performed continuously for a long period. Thus, the power consumption level at the plateau represents the power consumption of sequential writes. Thus, the per-way power consumption of the write operation ($P_{per\_way\_write}$) is calculated as follows:

$$P_{per\_way\_write}$$

$$= (P_{sequential\_write} - P_{idle})/\#\quad \text{active flash memory dies at the plateau}$$

We calculate the per-way power consumption of a read operation in the same way.

## 9.4.2 Event-Driven Simulation

Figure 9.8 shows the pseudo code of event-driven simulation. There are two types of events in the simulation: internal and external events. The simulation takes, as external events, host command arrivals (e.g., SATA read/write command) in the input trace. Based on the received external events and the status of the SSD model, the simulation schedules internal events of start/end of flash operations (I/O operation, and read/program/erase operation) and power state transition (e.g., transitioning to the active state on the command arrival during a low power state).

The simulation kernel advances the simulation time $T_{now}$ to the time point when the next event occurs (line 2 in Fig. 9.8). When the simulation time reaches the time point having any event (line 3), if there are multiple events scheduled at the time point, then the events are selected and processed in the order of end $\rightarrow$ start $\rightarrow$ host command $\rightarrow$ state transition, where "end" and "start" represent events for the end and start of flash operation, respectively (line 4). If a host command is selected, then the FTL algorithm runs to find the corresponding physical page addresses (PPAs) (line 6). Note that the overhead of running the FTL algorithm is taken into account by the access-dependent part of power consumption and runtime for each per-way operation.

If the selected event is a host command and the current state is a low power state, the SSD model must make a state transition from its current state to the active state. Since the state transition takes latency, a new event of state transition (in this example, a low power-to-active state transition) is scheduled at the time $T_{now} + T_{wakeup}$ where $T_{wakeup}$ is the state transition latency. The power state of the SSD model is changed to an intermediate state of state transition (PState = Transition2Active in line 9). The power consumption of the intermediate state is modeled as the idle state power consumption ($P_{total} = P_{idle}$ in line 9).

If the selected event is a host command and the current state is a transition state, depending on the direction of the state transition, events can be canceled or generated. To be specific, if the current state is the intermediate state in the transition to a low power state, then the future event of the state transition is canceled and a new state transition event as above is scheduled (line 10). If the current state is the intermediate one in the transition to an active state, then the selected command event is delayed and scheduled at the time point when the state transition to the active state is made (line 12).

The minimum and maximum size of read and write in the host command is 512 B (one sector) and 128 KB (256 sectors) and the page size is 2 KB in our model. Thus, a host command can create 2 (for the sizes of 512 B–2 KB) to 128 (for the size of 128 KB) event pairs for the start/end of flash operations. The created event pairs are scheduled by a function, Schedule_events_for_Flash_operations(PPAs, $T_{init}$), where PPAs is a list of physical page addresses obtained from the FTL algorithm run (line 6). A single call to the function can schedule up to 128 events. An ASAP (as soon as possible) scheduling algorithm is utilized to schedule the events. Thus, a start event is scheduled to the time point when the earliest idle time starts at the corresponding flash channel for flash I/O operation. The corresponding end event is also scheduled at a future time point with the time gap of the latency of internal operation.

If the new event (selected on line 4) is a start event and the current power state is a transition state (line 16), then the power state is set to the active state (line 17). Then, the total power consumption increases by the amount of the power consumption of newly started operation (line 18). If the new event is an end event (line 19), then the total power consumption decreases by the amount of the corresponding power consumption (line 20). After serving any end event, it needs to be checked whether there remains any pending event for internal flash operation or not. If there is no future event for flash operation, then an idle period starts. We can insert the policy of DPM (under design) into this part of the event-driven simulation algorithm. In Fig. 9.8, a time-out ($TO$)-based policy is simulated. Thus, an event for power state transition, a $TO$ event, is scheduled at $T_{\mathrm{now}} + TO$ (line 23).

If the new event (selected in line 4) is a power state transition event, it is an event for the transition to a low power state, i.e., a $TO$ event in the case of simulating the $TO$-based policy, since events for the transitions to active state have been handled previously. In this case, i.e., of a $TO$ event occurs, the power state is set to the low power state (line 26). The total power is set to that of the low power state (line 27). If there is any lower power state, then another $TO$ event can be scheduled (at line 28). The entire event-driven simulation continues until all the input external events, i.e., SSD commands, are served.

Figure 9.9 illustrates the event-driven simulation of a $TO$-based DPM policy. In the figure, at time $t_{13}$, an idle period starts and a $TO$ timer starts to count down. A $TO$ event is scheduled at time $t_{14}$ (line 23 in Fig. 9.8) in this case. At $t_{14}$, the $TO$ timer expires and a low power state is entered (lines 24–28). The figure shows that the total power consumption drops down to the level of the entered low power state (line 27). At $t_{15}$, the controller sends a host read command for 8 pages. A state transition event is scheduled at $t_{16}$ ($= t_{15} + T_{\mathrm{wakeup}}$).

## 9.5  Time-Out-Based SSD Dynamic Power Management

For the reduction in idle-mode power consumption, commercial SSD products adopt a time-out ($TO$)-based policy and device initiated power management (DIPM). Thus, if an idle period is detected via the timer event by the SSD, then the SSD asks the host to grant a transition to low power states [26, 27]. In the $TO$-based policy, the selection of $TO$ value per power state transition is an important design problem.

Figure 9.10 illustrates the $TO$-based policy assuming the ideal condition that $T_{\mathrm{wakeup}}$ is zero. In the figure, vertical arrows represent incoming host commands. Shaded rectangles represent the useful operation of SSD, e.g., I/O and internal flash operations. The horizontal arrows represent the amount of $TO$ period. Thus, in the figure, the low power state is entered in the third idle period since the first two idle periods are shorter than the $TO$. The blank rectangles represent the time periods at a low power state. Thus, their area represents the amount of saved energy obtained by entering the low power state.

Figure 9.11 shows the case when $T_{\mathrm{wakeup}}$ is considered in Fig. 9.10. The transition overhead of wakeup latency (dark rectangles) incurs performance loss as shown in Fig. 9.11. If the selected $TO$ is too big, the low power state is rarely entered, thus

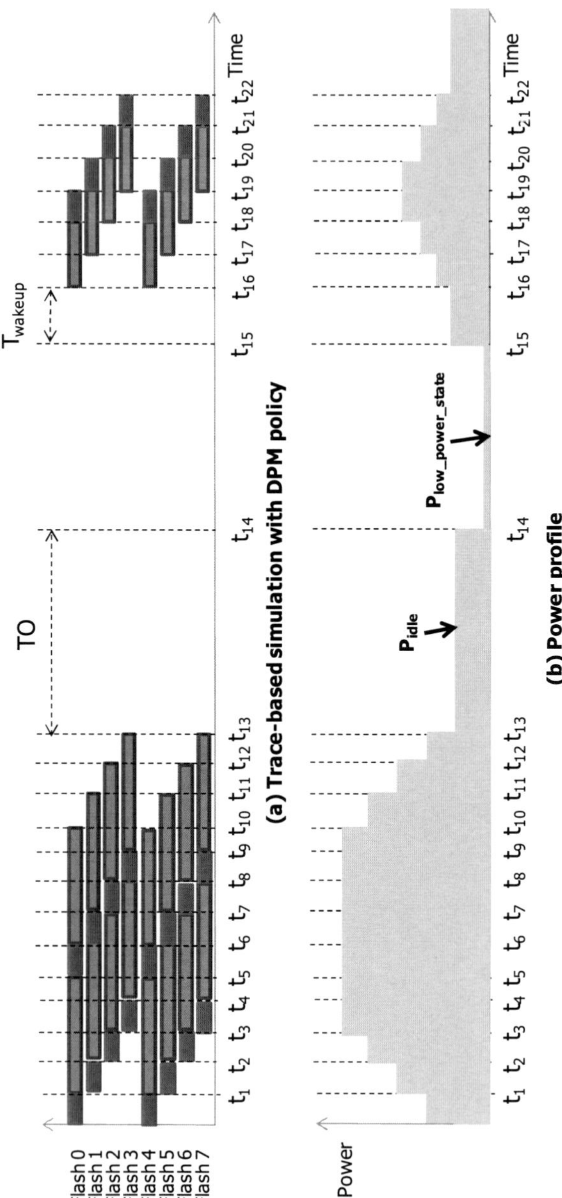


**Fig. 9.9** Example of event-driven simulation

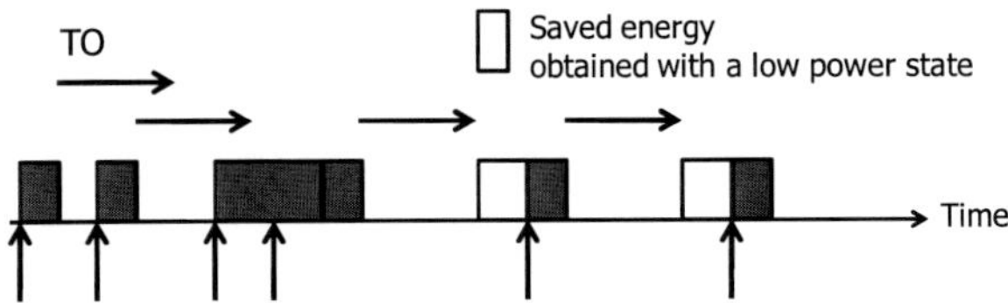



**Fig. 9.10**  *TO*-based policy for SSD power management

**Fig. 9.11**  Performance loss in the *TO*-based policy

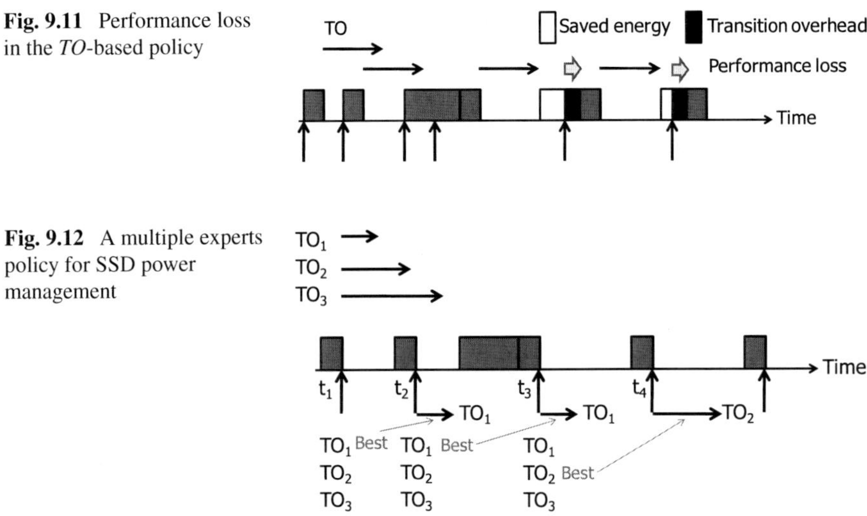



**Fig. 9.12**  A multiple experts policy for SSD power management



giving small energy reduction and performance loss. If it is set to a small value, then the energy reduction can be large while the performance loss also increases. Thus, the *TO* selection must take into account both energy reduction and performance loss.

The SSD access behavior changes over time depending on the programs running on the host and program phases even for a single program run. Thus, a single fixed *TO* may not be effective. Instead, an adaptive *TO* selection is favored. As mentioned in Sect. 9.2, several studies on adaptive *TO* selection based on the previous history of HDD accesses have been presented [7–11].

In this section, we present a case study of applying a method of multiple experts to the dynamic *TO* selection. In the multiple experts policy, multiple *TO* values are evaluated, and the one whose recent performance is the best is selected.

Figure 9.12 illustrates a multiple experts policy. In this example, three *TO* values, $TO_1$, $TO_2$, and $TO_3$, are managed. At time $t_1$, $TO_1$ is assumed to give the best score and is selected. Thus, at time $t_2$ when a new idle period starts, $TO_1$ is applied. During the time period from $t_3$ to $t_4$, $TO_2$ gives the best score. $TO_2$ is now selected to be applied at time $t_4$. As shown in Fig. 9.12, one of multiple *TO* values can be selected depending on their previous scores. Thus, dynamic behavior change in SSD accesses can be tracked.

**Fig. 9.13** A score function in multiple experts policy

```
1  For each expert i,
2     energy_loss[i] = TO[i]>T_idle ? 1 : TO/T_idle;
3     perf_penalty[i] = TO[i]>T_idle ? 1 : T_wakeup/T_idle;
4     ratio[i] = α * energy_loss[i] + (1- α) * perf_penalty[i]
5     score[i] = prev_score[i] * β^ratio[i]
```

The score function plays an important role in the multiple experts policy. Figure 9.13 shows a pseudo code of the score function used in our experiments. The score consists of two parts: energy loss and performance penalty (both are normalized). The ratio $\alpha$ represents their relative importance to be set by the designer (line 4 in Fig. 9.13). The energy loss is set to its maximum value of 1 when the $TO$ value is greater than $T_{idle}$ (line 2). In this case, the $TO$ is larger than the idle time period. Thus, if the SSD took this $TO$ value, it could not utilize the idle period for energy reduction and would lose this opportunity. Thus, we set the energy loss to the maximum value in this case. If the $TO$ value is smaller than $T_{idle}$, then the smaller the $TO$, the larger the energy gain that can be obtained. Thus, we set the energy loss to the ratio of $TO$ to $T_{idle}$ in this case (line 2).

Performance penalty is calculated in a similar way to energy loss. If the $TO$ value is larger than $T_{idle}$, if the SSD utilized this $TO$ value, it would have lost performance due to the wakeup without obtaining any reduction in energy consumption. Thus, we set the performance loss to the maximum value of 1 in this case. If the $TO$ value is smaller than $T_{idle}$, then the performance loss of $T_{wakeup}$ is considered to be amortized over $T_{idle}$. Thus, the larger $T_{idle}$, the smaller performance loss is obtained, as shown in line 3 of Fig. 9.13.

For the calculation of the final score for each $TO$ value, both energy loss and performance penalty are combined to give a single value called ratio (line 4 in Fig. 9.13). As mentioned before, the parameter $\alpha$ can be adjusted depending on the design goal. For instance, if energy reduction is more urgent than performance improvement (e.g., if the battery power is running out), a high value of $\alpha$ can be set. The final score is calculated utilizing both the score history, i.e., the previous score value, and an exponential weighting function. The parameter $\beta$ is also specified by the designer depending on the decision of how much of the previous score is considered in calculating the current score. The selection of two parameters $\alpha$ and $\beta$ is an important step in designing the multiple experts policy. In our experiments, we will report the impact of different parameter settings on performance and energy consumption.

## 9.6 Experiments

Table 9.1 details our experimental setup. The power estimator was designed in Matlab. We used the measured SSD power consumption obtained by running sequential reads/writes. We also measured the power consumption of the idle state and two low power states called *partial* and *slumber*.

**Table 9.1** Experimental setup

|  | Features |
| --- | --- |
| SSD product | A Samsung SSD (2.5", 128 GB, and SATA2) [28] |
| Host OS | Windows VISTA |
| Measured latency and power in the active state | Read/write commands for the data sizes of 1, 2, 4, 8, 16, 32, 64, 128, and 256 sectors |
| Low power states | Partial and slumber |
| Traces to the SSD input | Obtained from a Samsung notebook PC (Sens X360) running MobileMark2007 (Reader, Productivity, and DVD) [29] and PCMark05 [30] |

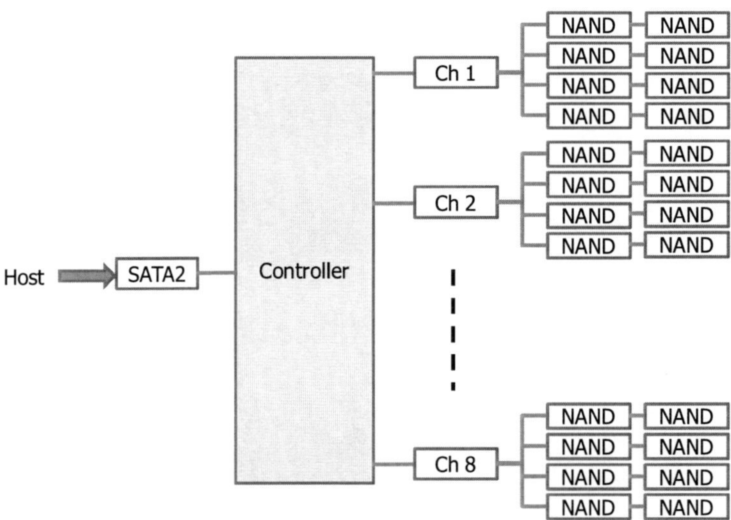


**Fig. 9.14** Multi-channel and multi-way SSD architecture used in the experiments

Figure 9.14 shows the SSD architecture used in the experiments. From the product specification of peak performance and capacity in [28], we assumed MLC, 4 KB/page, 33 MHz I/O frequency, 8 channels, and 8 ways/channel.

Figure 9.15 shows a trace of SSD accesses of the MobileMark 2007 Reader scenario. The $x$ and $y$ axes represent time and logical block address (LBA). As shown in the figure, the access trace shows both randomness and temporal locality. Spatial locality does exist but is not shown in the figure due to the coarse granularity of the $y$ axis (for all the address space in the figure).

Figure 9.16 shows the statistics of SSD accesses for the scenarios Productivity and DVD in MobileMark 2007. The figure shows that small size accesses (less than or equal to 8 sectors) are dominant in terms of access frequency and they tend to be random compared with large size accesses, which justifies that several FTL

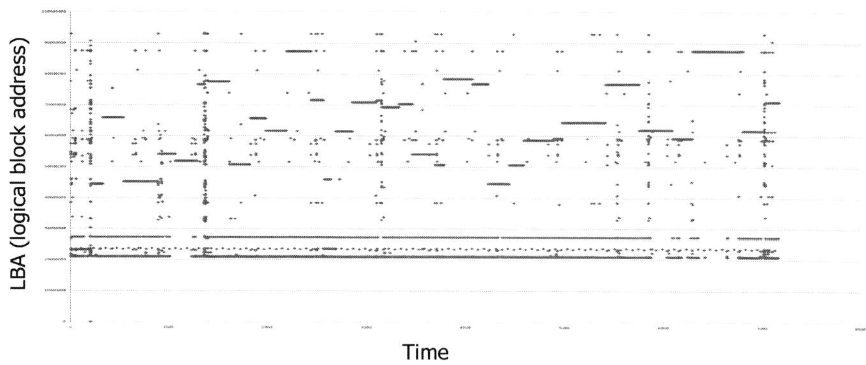

**Fig. 9.15** SSD access trace obtained from running MobileMark 2007 Reader

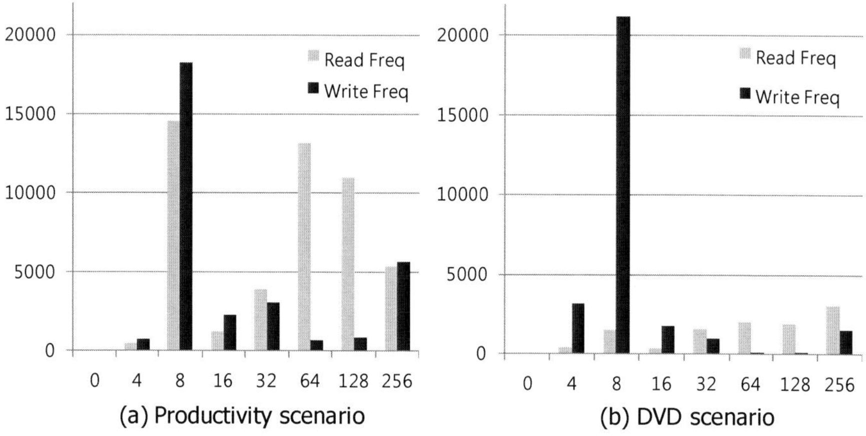

**Fig. 9.16** Statistics of SSD accesses for two MobileMark 2007 scenarios

algorithms, e.g., LAST [25], classify SSD accesses into short random and long sequential ones.

We evaluated the accuracy of performance/power estimation by comparing the estimation results obtained from our SSD power estimator and the corresponding measurement data. The comparison showed that the SSD power estimator gives the same estimation results as the measurement data in both power consumption of sequential reads/writes and latency of all the read/write data sizes. Note that the accuracy check was performed only for the training set, i.e., the measured data. When compared with other measured data than the training set, we expect estimation errors, which will be analyzed and reduced in our future work.

We applied the event-driven simulation to the two *TO*-based policies presented in Sect. 9.5. As mentioned before, the *TO*-based policies require selecting a suitable set of parameters, e.g., a single *TO* or a set of *TO*s, parameters $\alpha$ and $\beta$, etc. Figure 9.17 shows the trend of energy reduction and performance loss in the case of the single

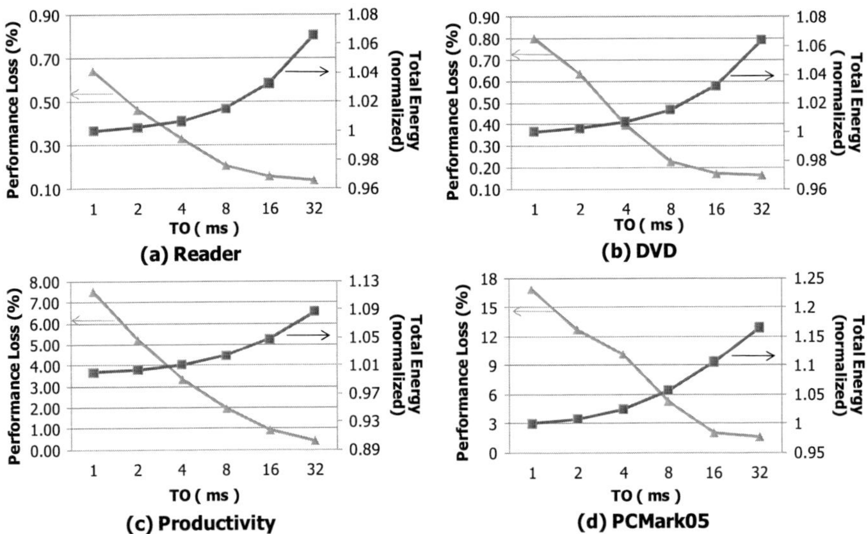

**Fig. 9.17**  Results for MobileMark 2007 (**a**)–(**c**) and PCMark05 (**d**)

*TO*-based policy. In the experiments, we sweep the *TO* values from 1 ms to 32 ms. We utilize the same value for the two *TO* parameters of the two low power states, partial and slumber. It takes 1–20 minutes to run the event-driven simulation with a given *TO* value. The runtime varies depending on the number of accesses and *TO* values. Although the simulation runtime is on the order of minutes, it is 6–100+ times faster than real SSD runs.[3]

Figure 9.17 shows that as *TO* increases, energy consumption increases since we can exploit fewer idle periods for energy reduction. In this case, the performance penalty (due to accumulated wakeup delay) also decreases since there are fewer wakeups from low power states. The general trends look similar in all the four scenarios in Fig. 9.17. However, the amount of energy reduction and performance penalty is significantly different between scenarios. Comparing the PCMark05 and Reader scenarios, at the *TO* value of 1 ms, PCMark05 gives more than 15% reduction in energy reduction and a performance penalty of 17% compared with the case when the *TO* value is 32 ms. The Reader scenario gives an energy reduction of about 6% and less than 1% of performance penalty in this case. We analyze that this difference results from the fact that idle periods dominate in the MobileMark scenarios. According to the traces, we found that about 95% of the total runtime is idle in the MobileMark scenarios. Thus, the performance loss of aggressive policy, e.g., *TO* = 1 ms is not easily visible, especially with the two MobileMark scenarios, Reader and DVD.

---

[3]We expect a faster event-driven simulation to be obtained when the algorithm is implemented in C/C++ rather than in Matlab.

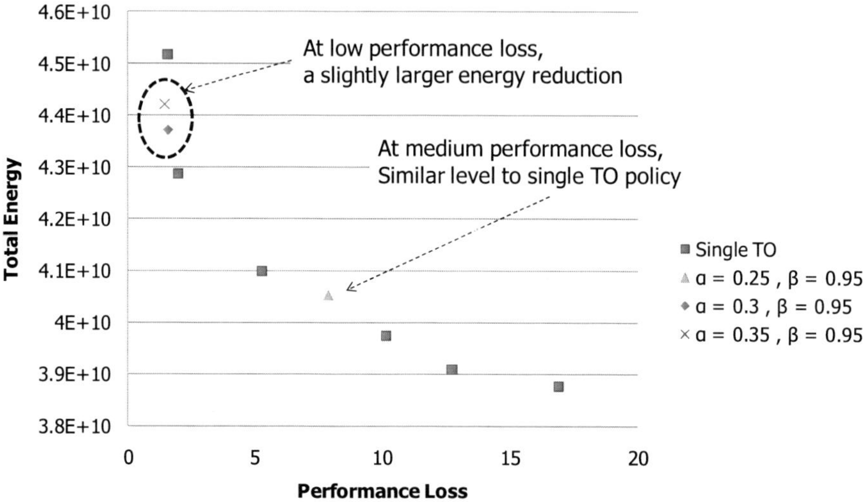


**Fig. 9.18** Comparison of multiple experts and single *TO* policies: PCMark05

PCMark05 is a benchmark to evaluate PC performance. Thus, it runs continuously, thereby generating SSD accesses without long idle periods. Instead, it tends to have many short idle periods. Thus, the performance of PCMark05 is more susceptible to the aggressive policy of $TO = 1$ ms.

Figure 9.18 shows the result of the second policy utilizing multiple experts described in Sect. 9.5. We performed an experiment where the parameter space of the multiple experts policy was explored extensively. As *TO* values, we used 11 values ranging from 2 ms to 32 ms. In Fig. 9.18, we show the results obtained from three representative sets of parameters $\alpha$ and $\beta$. Overall, compared with the result of the single *TO* policy, the multiple experts policy does not give a significant improvement in the relationship of energy reduction and performance loss. As parameter $\alpha$ becomes larger in line 4 of Fig. 9.13, the energy consumption increases and the performance penalty decreases. Compared with the single *TO* policy, the multiple experts policy gives a slight improvement in energy consumption when performance loss is small.

The comparison between the multiple experts and single *TO* policies in Fig. 9.18 hints that the idle time pattern in the SSD is not easily utilizable by the *TO*-based policies, and Fig. 9.19 supports this. The figure shows the cumulative idle time. Figure 9.19(a) shows a large-scale view covering up to 1000 ms; Fig. 9.19(b) shows a small-scale view focusing on small idle periods (0–100 ms). Figure 9.19(a) also shows that the conventional *TO*-based policy for HDDs covers large idle periods whose sizes are typically larger than 1 second. For SSDs, much smaller idle periods are utilized, mainly because the latency of SSD access, especially random access, is much shorter than that of HDD. Figure 9.19(b) shows the range of *TO* values (1–32 ms) used in our experiments.

As shown in Fig. 9.19, the DPM for SSD Must handle very short idle periods whose behavior changes very quickly. Figure 9.20 gives a detailed view (histogram)

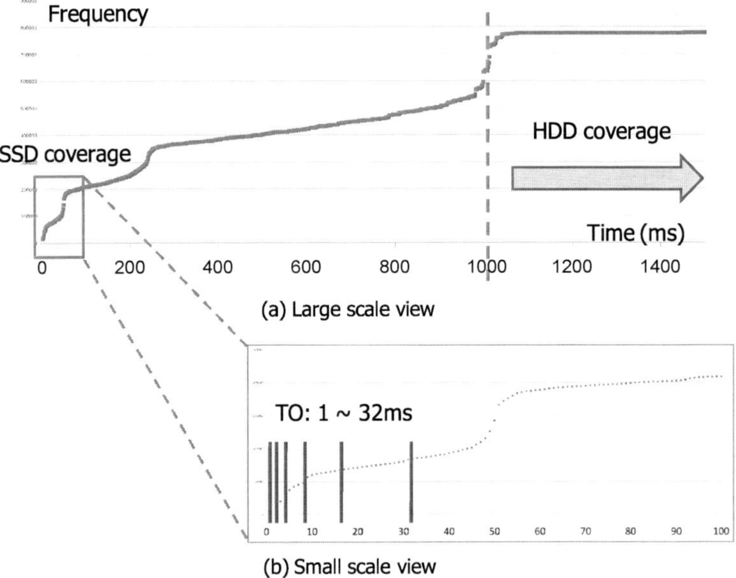



**Fig. 9.19**  Cumulative idle time of SSD accesses

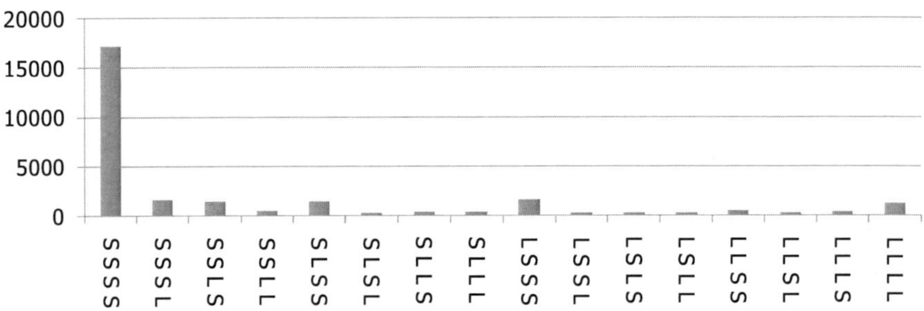



**Fig. 9.20**  SSD access patterns: PCMark05

of short idle period patterns. We classified idle periods into short and long ones with a threshold of 10 ms and collected the frequency of 16 patterns of four consecutive idle periods. In the figure, $S$ and $L$ represent short and long idle periods, respectively. Thus, *SSSS* represents the pattern where four consecutive idle periods are short ones. As shown in the figure, the pattern *SSSS* dominates the number of patterns. In addition, short idle periods are included in most of the next most dominant patterns, e.g., *SSSL*, *SSLS*, *SLSS*, and *LSSS*.

DPM in SSDs is different from that in HDDs since SSD is characterized by many short idle periods, as shown in Figures 9.19 and 9.20. Considering the overhead of wakeup from low power states to the active state, short idle periods are not easy to exploit. However, if they can be exploited, their potential in reducing SSD energy

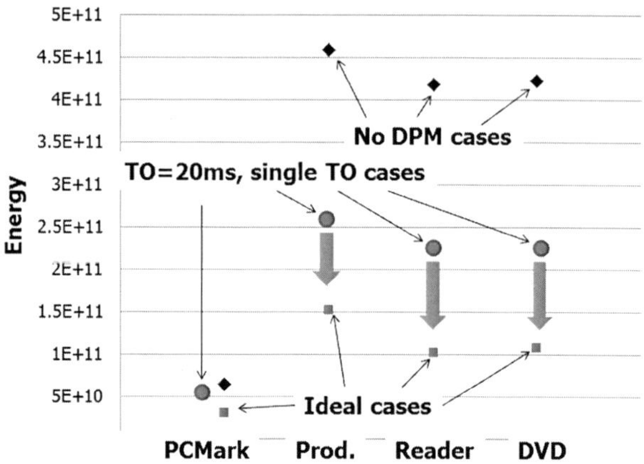



**Fig. 9.21** Room for improvement in SSD power management

consumption will be huge. Figure 9.21 shows their potential. In the cases of Mo-bileMark scenarios, compared with the cases of no DPM, the single *TO* policy of $TO = 20$ ms gives significant reductions in energy consumption. However, as shown in the figure, there is still much room for improvement in energy consumption between the ideal cases ($TO = T_{\text{wakeup}} = 0$) and the real cases of $TO = 20$ ms and a positive $T_{\text{wakeup}}$ value.

Closing the gap in Fig. 9.21 is one of the most important problems in DPM for SSD. As shown in Figures 9.19 and 9.20, there are many short idle periods in SSD accesses. Conventional reactive approaches, which make the SSD enter low power states only after idle periods start, can suffer from low efficiency, as in the case of multiple experts policy due to high overhead caused by frequent wakeups. We think a proactive and holistic approach can tackle this problem more efficiently. One possibility will be to have both the operating system (or device driver) and the SSD work together in order to transform short idle periods into long idle ones. For instance, existing approaches which perform prefetches or enlarge buffer cache can be applied to the combination of OS and SSD. In such a solution, prefetches and larger buffer cache will reduce infrequent read (e.g., via prefetching the infrequently accessed data) and write accesses (e.g., via holding the write data for a longer period in the enlarged buffer cache). The coverage of the holistic approach can be extended to the application software by making it aware of the existence of the underlying SSD and transforming its SSD accesses by minimizing infrequent isolated accesses. Such holistic approaches will be very interesting and practical research topics.

## 9.7 Summary

In this chapter, we introduced a method of power estimation and performance modeling for SSDs. Our power estimator performs an event-driven simulation. It takes as input an SSD access trace and the measurement data of performance and power consumption of SSD accesses. The output is the power profile (power consumption over time) including a trace of internal flash operations. Our method gives accurate power estimations since it is based on real measurement data. It also gives a fast simulation by applying an event-driven simulation with traces. We presented two case studies in which the method was applied to designing a DPM policy. As future work, we expect that holistic approaches, in which both software (application and/or operating system) and SSD collaborate in order to transform short idle periods into longer ones for better efficiency of DPM policies, will be practically required.

# References


1. Kim, B.: Design space surrounding flash memory. In: International Workshop on Software Support for Portable Storage, IWSSPS (2008)
2. Open NAND Flash Interface. http://onfi.org
3. Creasey, J.: Hybrid hard drives with non-volatile flash and longhorn. In: Windows Hardware Engineering Conference, WinHEC. MicroSoft, Redmond (2005)
4. Communications with Samsung engineers
5. Hylick, A., Rice, A., Jones, B., Sohan, R.: Hard drive power consumption uncovered. ACM SIGMETRICS Perform. Eval. Rev. **35**(3), 54–55 (2007)
6. Zedlewski, J., Sobti, S., Garg, N., Zheng, F., Krishnamurthy, A., Wang, R.: Modeling hard-disk power consumption. In: The USENIX Conference on File and Storage Technologies, FAST, pp. 217–230. USENIX Association, Berkeley (2003)
7. IBM.: Adaptive power management for mobile hard drives. In: IBM. http://www.almaden. ibm.com/almaden/mobile_hard_drives.html (1999)
8. Lu, Y., De Micheli, G.: Comparing system-level power management policies. IEEE Des. Test Comput. **18**(2), 10–19 (2001)
9. Douglis, F., Krishnam, P., Bershad, B.: Adaptive disk spin-down policies for mobile computers. In: 2nd Symposium on Mobile and Location-Independent Computing, pp. 121–137. USENIX Association, Berkeley (1995)
10. Helmbold, D., Long, D., Sconyers, T., Sherrod, B.: Adaptive disk spin-down for mobile computers. Mob. Netw. Appl. **5**(4), 285–297 (2000)
11. Bisson, T., Brandt, S.: Adaptive disk spin-down algorithms in practice. In: The USENIX Conference on File and Storage Technologies, FAST. USENIX Association, Berkeley (2004)
12. Papathanasiou, A.E.: Scott. M. L.: Energy efficient prefetching and caching. In: The USENIX Conference on File and Storage Technologies, FAST. USENIX Association, Berkeley (2004)
13. Cai, L., Lu, Y.H.: Joint power management of memory and disk. In: Design Automation and Test in Europe (2005)
14. Gurumurthi, S., et al.: DRPM: dynamic speed control for power management in server class disks. In: International Symposium on Computer Architecture (2003)
15. Dirik, C., Jacob, B.: The performance of PC solid-state disks (SSDs) as a function of bandwidth, concurrency, device architecture, and system organization. In: International Symposium on Computer Architecture, pp. 279–289. ACM, New York (2009)
16. Kim, Y., et al.: FlashSim: a simulator for NAND flash-based solid-state drives. In: The International Conference on Advances in System Simulation, SIMUL (2009)
17. Mohan, V., Gurumurthi, S., Stan, M.R.: FlashPower: a detailed power model for NAND flash memory. In: Design Automation and Test in Europe (2010)
18. Joo, Y., Cho, Y., Shin, D., Chang, N.: Energy-aware data compression for multi-level cell (MLC) flash memory. In: Design Automation Conference, pp. 716–719. ACM, New York (2007)
19. Suh, K.D., et al.: A 3.3 V 32 Mb NAND flash memory with incremental step pulse programming scheme. IEEE J. Solid-State Circuits **30**(11), 1149–1156 (1995)

20. Ishida, K., Yasufuku, T., Miyamoto, S., Nakai, H., Takamiya, M., Sakurai, T., Takeuchi, K.: A 1.8 V 30 nJ adaptive program-voltage (20 V) generator for 3D-integrated NAND flash SSD. In: International Solid-State Circuits Conference, pp. 238–239. IEEE Press, New York (2009)
21. Takeuchi, K.: Novel co-design of NAND flash memory and NAND flash controller circuits for sub-30nm low-power high-speed solid-state drives (SSDs). In: Symposium on VLSI Circuits (2008)
22. Kang, H., et al.: Low-power buffer cache management for heterogeneous storage devices in mobile systems. In: The Fourth International Workshop on Software Support for Portable Storage (2009)
23. Lee, S., Park, D., Jung, T., Lee, D., Park, S., Song, H.: A log buffer-based flash translation layer using fully-associative sector translation. ACM Trans. Embed. Comput. Syst. **6**(3) (2007)
24. Kang, J., Cho, H., Kim, J., Lee, J.: A superblock-based flash translation layer for NAND flash memory. In: The 6th ACM & IEEE International Conference on Embedded Software, EMSOFT, pp. 161–170. ACM, New York (2006)
25. Lee, S., Shin, D., Kim, Y., Kim, J.: LAST: locality-aware sector translation for NAND flash memory-based storage systems. ACM SIGOPS Oper. Syst. Rev. **42**(6) (2008)
26. Intel, Co.: X25-M and X18-M mainstream SATA solid-state drives. Intel. http://www.intel.com/design/flash/nand/mainstream/index.htm (2009)
27. Intel, Co.: Designing energy efficient SATA devices. Intel. http://download.intel.com/technology/EEP/Designing_energy_efficient_SATA_devices.pdf (2010)
28. Samsung SSD. Samsung. Available at http://www.samsung.com/global/business/semicondu-ctor/products/flash/ssd/2008/product/pc.html (2009)
29. Business Applications Performance Corporation: MobileMark 2007. In: BAPCo. http://www.bapco.com/products/mobilemark2007/ (2007)
30. Futuremark, co.: PCMark05. In: Futuremark. Available at http://www.futuremark.com/products/pcmark05/ (2009)

# Chapter 10
# Energy-Aware Surveillance Camera

**Sangkwon Na and Chong-Min Kyung**

**Abstract** In this chapter, we introduce an application example of a wireless surveillance camera (WSC) consisting of image sensor, event detector, video encoder, flash memory, wireless transmitter, and battery. The battery- and flash-constrained WSC records images when significant events, such as suspicious pedestrians or vehicles, are detected, based on a hierarchical event detection method to avoid wasting energy on insignificant events. In an energy-aware sense, the recorded images are stored in non-volatile (flash) memory or transmitted to the base station according to the urgency of the event. Balancing the usage of all resources including battery and flash is critical in prolonging the lifetime of a WSC, because a shortage of either battery charge or flash capacity could lead to a complete loss of events, or a significant loss of quality in the recorded image of events. We assume that the resources of the WSC, i.e., the battery and flash, are refreshed every system maintenance period (SMP). The proposed method controls the bit rate of encoded videos and sampling rate, e.g., resolution and frame rate, to prolong the lifetime of the WSC until the next SMP. Experimental results show that the proposed method prolongs the lifetime of the WSC by up to 88.41% compared with an existing bit-rate allocation method that does not consider resource usage balancing.

## 10.1 Overview

The recent progress in low-power video and non-volatile memory technology along with a growing interest in safety has led to an increased demand for the wireless video surveillance camera, which captures events of interest for possible offline interpretation. The main function of a wireless surveillance camera (WSC), which consists of sensor, event detector, video encoder, flash, and battery, is to detect suspicious objects in its camera scope, and to store the images with recognizable quality.

S. Na (✉)
Samsung Electronics, Seoul, Republic of Korea
e-mail: sangkwon.na@gmail.com

C.-M. Kyung
KAIST, Daejeon, Republic of Korea
e-mail: kyung@ee.kaist.ac.kr

C.-M. Kyung, S. Yoo (eds.), *Energy-Aware System Design*,
DOI 10.1007/978-94-007-1679-7_10, © Springer Science+Business Media B.V. 2011

As the cost of non-volatile memory has been falling and the demand for prolonged operating time of the WSC is growing, non-volatile memory is used in the WSC for the disposal of gathered data because of the lower energy consumption for flash writing than for wireless transmission. However, the gathered data needs to be transmitted to the base station for immediate interpretation according to the urgency of the recorded event. When a WSC detects and captures an event, it must decide on the storage location, i.e., internal or external, according to the wireless channel condition, the flash capacity, and the event urgency. Because a WSC is powered by battery and has limited flash capacity, it is crucial to evenly wear out these two resources and, therefore, to balance the usage of battery charge and of flash storage in its operation, while satisfying the specified image quality requirement and interpretation accuracy. To avoid wasting the battery energy and flash for recording uncritical events, most WSCs operate in an event-driven manner, where all functional blocks, except the event detector, are normally power-gated and only wake up when the event detector detects an event [1–7].

The video encoding complexity can be scaled by adjusting encoding parameters, e.g., frame size and type, number of reference frames, block partition size, motion search method, etc. [8]. An encoding configuration that gives a low compression efficiency and, therefore, a high bit rate for encoded frames generally has a low computational complexity [8, 9]. We can reduce the energy consumption of video encoding by increasing the bit rate, which, however, increases the energy consumption for writing the encoded data into flash or transmitting to the base station. In addition, the energy consumption of video encoding is also scaled by the sampling rate, i.e., the number of pixels per second.

Various methods for obtaining a trade-off between the energy consumption of video encoding and flash writing/wireless transmission have been proposed based on finding the optimal video encoding configuration in terms of the total energy consumption [10–14]. In this chapter, we present an application example of energy-aware system design utilizing a method of finding the video encoding configuration to minimize the energy consumption of a WSC and maximize the utilization of the flash, while satisfying the given image quality requirement by choosing either an internal or external storage destination. This method finds the optimal output bit rate that is expected to evenly wear out the battery and flash. We also present a method of controlling the sampling rate so that the WSC survives the system maintenance period (SMP) for replenishing the resources, i.e., battery and flash.

This chapter is organized as follows. Section 10.2 introduces a target system architecture. Section 10.3 explains the energy-rate-distortion relationship in the target system. Section 10.4 presents the proposed solution. Section 10.5 reports experimental results.

## 10.2 Target System Architecture

The target system consists of image sensor, hierarchical event detector, video encoder, control logic, local storage (flash), wireless transceiver, and battery, as de-

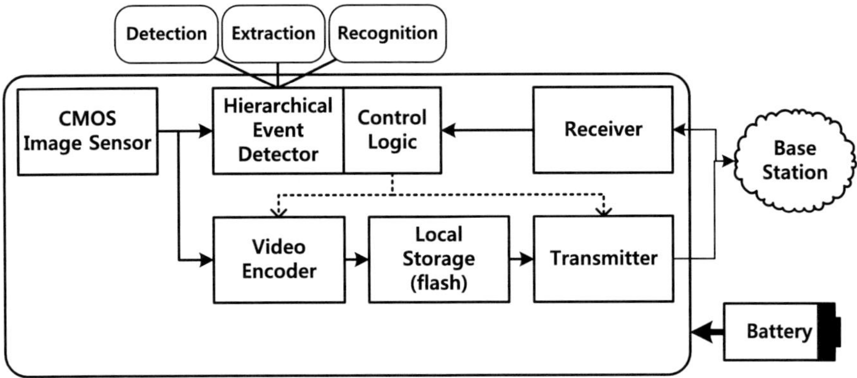


**Fig. 10.1** Target system architecture of our WSC

picted in Fig. 10.1. An image sensor captures an image with the given image resolution (width × height) and frame rate (frames per second, *fps*); the energy consumption of the image sensor is scaled by the number of pixels per second. An event detector determines the occurrence of an event and evaluates its criticality based on hierarchical event detection. The evaluation basis is customized according to the target application. When the event detector judges the occurrence of an event, the video encoder compresses the bits representing captured images with the video quality, bit rate, and sampling rate determined by the control logic. The control logic monitors the remaining battery charge to determine the sampling rate and output bit rate which maximizes the usage of the WSC. Then, the control logic also determines the storage location of encoded videos according to the wireless channel condition, flash space, and battery charge. The distance between the WSC and the base station can vary from time to time and is given statistically.

### 10.2.1 Event Model

The WSC captures an image of the scene to be observed at a fixed position. When a suspicious object appears in the scene, the WSC detects that an event has happened. The degree of the event urgency, defined as the *event criticality*, is classified into two levels: low and high. We build an event model that consists of three parameters, (1) the number of total events ($N_{\text{event}}$) during the system maintenance period ($T_{\text{SMP}}$), (2) probabilities of low-criticality ($\alpha$) and high-criticality ($\beta$) events, and (3) average time duration per event whose criticality is low ($t_{\text{event}}^{\text{l}}$) and high ($t_{\text{event}}^{\text{h}}$). We assume that the event time duration follows

$$t_{\text{event}} \sim N\left(\mu_{\text{event}}, \sigma_{\text{event}}^2\right) \qquad (10.1)$$

where $\mu_{\text{event}}$ and $\sigma_{\text{event}}^2$ are the average and standard deviation of the event time duration, respectively.

                                                        

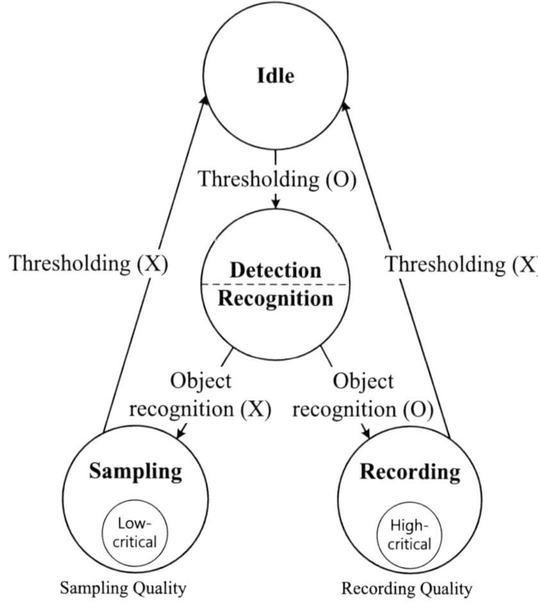


**Fig. 10.2** WSC operation states corresponding to each event criticality

## 10.2.2 Operation States

According to the event criticality, we define four operation states of the WSN: *Idle* (*IS*), *Detecting & Recognition* (*DS*), *Sampling* (*SS*), and *Recording* (*RS*). The higher the event criticality, the more important the image of that event. Thus, the image of high-criticality events must be recorded with very little distortion for future interpretation. To prolong the lifetime of the WSC, the energy that it consumes to record the image of low-criticality events must be reduced. In *Idle* state (*IS*), a *thresholding*-based event detector observes the captured image and judges the occurrence of an event by checking whether the sum of absolute differences (SAD) between pixel values of the current frame and background frame exceeds the given threshold or not. Therefore, other components except the image sensor and event detector are power-gated. In *Detection & Recognition* state (*DS*), an *edge*-based object detector searches for moving objects that belong to a certain range in terms of size and aspect ratio. Once the detector finds any moving object in the scene, the *pattern recognition*-based object detector recognizes the object(s). We classify the scene where a human body or face appears as a high-criticality event. In *Sampling* state (*SS*), the video encoder compresses captured images with sampling to allow recognition of the objects in the compressed images. In *Recording* state (*RS*), the video encoder compresses images with the best quality. The thresholding-based event detector still runs in *SS* and *RS* to judge the termination of the event.

The WSC operation state changes according to event occurrence and criticality. Initially, WSC runs in *IS*, and judges the event occurrence. Once the event occurrence is detected, the state of the WSC is changed to *DS*, i.e., the *Thresholding (O)* edge in Fig. 10.2. In *DS*, when the moving object has a certain size and aspect ratio and is recognized as a human body, the state of the WSC is changed to *RS*, i.e., the *Object recognition (O)* edge. If it fails to find a human body in the image, the WSC operating state changes to *SS*, i.e., the *Object recognition (X)* edge in Fig. 10.2. In *SS* and *RS*, the WSC detects the persistence of an event using the *thresholding*-based event detector and changes the state to *IS* when an event disappears, i.e., *Thresholding (X)* in Fig. 10.2.

When the remaining battery charge and/or flash space is not enough to record a couple of events in the future, the WSC transmits an alert signal to the base station (BS), notifying of the scarcity of the WSC resources. Then, an operator comes to the WSC to replace the battery and dump the data stored in the flash memory. Actually, we assume that periodic maintenance, i.e., the system maintenance period (SMP), is carried out to replenish the resources in the WSC.

## 10.3 Energy-Rate-Distortion Relationship of Target System

The WSC carries out its given task in an energy-aware manner, i.e., with the minimum energy required to handle it. Thus, it must find the appropriate configuration in which it can operate as long as possible without failure of capturing critical events. Conventional imaging systems have been characterized in terms of rate and distortion, in other words, to try to find how many bits are needed to represent the source with the given distortion. However, it seems necessary to introduce another term, energy, to design an energy-aware WSC and characterize the energy-rate-distortion relationship of the target system. Three major components that consume the largest portion of the battery charge are the event detector, the video encoder, and the wireless transmitter. In the following section, we analyze the relationship among energy, rate, and distortion in order to find the appropriate operating configuration in an energy-aware manner.

### *10.3.1 Event Detection*

There are many ways to detect suspicious objects in an image and judge the occurrence of an event. Simple event detection algorithms consume low energy but frequently detect meaningless events; sophisticated event detection algorithms consume high energy due to the increased computational complexity. Depending on the algorithm accuracy, some significant events can be missed. Thus, appropriate event detection algorithms need to be utilized to trade off energy consumption and detection accuracy for the given target application. Based on the computational complexity and detection accuracy, we classify event detection into three levels: *thresholding*-based event detection, *edge*-based moving object detection, and *pattern recognition*-based object detection. We have compared the accuracy and complexity of three event detection levels implemented with OpenCV library [15] in Table 10.1. The detection rate denotes the proportion of correctly detected events to all actual

**Table 10.1** Comparison of accuracy (detection rate and false alarm rate) and complexity (clock cycles per frame) with respect to three event detection levels for three test sequences (*PetsC1D2*, *ThreePerson_Circles*, and *IndoorGTTest2*)

| Detection algorithm | Accuracy | | Complexity |
| --- | --- | --- | --- |
| | Average detection rate (%) | Average false alarm rate (%) | Average clock cycles per frame |
| *Thresholding* | 99.18% | 41.37% | 3,201,347 |
| *Edge* | 72.94% | 23.55% | 14,048,533 |
| *Pattern recognition* | 78.32% | 6.17% | 133,235,221 |

events, and the false alarm rate denotes the proportion of false detected events to all detected events. The results of Table 10.1 show the tendency between detection accuracy and complexity with respect to the detection algorithm. We briefly describe the three levels of event detection in the following paragraphs.

### 10.3.1.1 Thresholding-Based Event Detection

The event occurrence is detected using the sum of absolute differences (SAD) between pixel values of the current frame and a background frame generated using the image without any object in the scene [4–7]. When the value of SAD is larger than a threshold value, the event detector acknowledges the occurrence of an event. Because this method considers the difference of pixel values, environmental changes such as fluttering leaves and daylight changes can erroneously be detected as an event. As shown in Table 10.1, this level has a high false alarm rate, i.e., it frequently detects meaningless events.

### 10.3.1.2 Edge-Based Moving Object Detection

To detect moving objects such as human bodies and cars, an *edge*-based detection algorithm can be utilized. Edge detection is based on a gradient of pixel values. The edge operator is applied to the difference image in successive frames for a noise-robust edge detection [16]. In our work, an aspect ratio and an area of the minimum bounding box of the detected moving object are used to judge human bodies. When an aspect ratio lies within a certain range, the event detector judges the target object that appears in the scene, as shown in Figs. 10.3(b) and (e) where the white lines and regions denote the parts of moving objects. With this increased complexity, *edge*-based moving object detection shows a lower false alarm rate than *thresholding*-based event detection, as shown in Table 10.1.

### 10.3.1.3 Pattern Recognition-Based Object Detection

Pattern recognition algorithms classify and detect an object based on predefined features and their statistics acquired from a training set. Face recognition is one of the

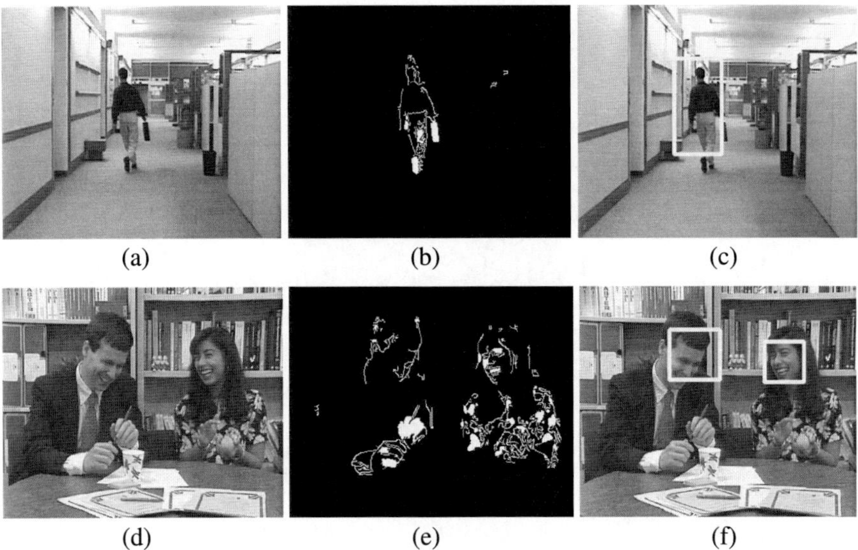

**Fig. 10.3** An example of event detection: (**a**), (**d**) original images, (**b**), (**e**) edge-based moving object detection, and (**c**), (**f**) pattern recognition-based object detection for (**a**)–(**c**) *hall* and (**d**)–(**f**) *paris* sequence

most popular pattern recognition areas. In this work, Haar wavelet-based classification is utilized to recognize the human body [17, 18]. The wavelet representation is used to capture the structural similarity between various instances of the class. Haar coefficients are compared to the trained wavelet template within a given template window, i.e., $30 \times 30$. Figures 10.3(c) and (f) show the result of *pattern recognition*-based object detection for body and face detection, respectively. The yellow rectangles in Figs. 10.3(c) and (f) denote the region that contains detected objects. Although it requires the largest computational complexity among the three event detection levels, the false alarm rate is minimized, as shown in Table 10.1. With a state-of-the-art algorithm, the detection rate of the pattern recognition can be improved up to about 100%. The accuracy of pattern recognition can be improved through learning.

## *10.3.2 Video Encoding*

### 10.3.2.1 Power-Rate-Distortion Model of Video Encoding

A power-rate-distortion ($P$-$R$-$D$) model [13] was proposed by integrating the classical Laplacian-based rate-distortion model and the computational complexity model, which is represented as a function of the power consumption of the *software-based* video codec. This paper deals with the variation of achievable rate and/or dis-

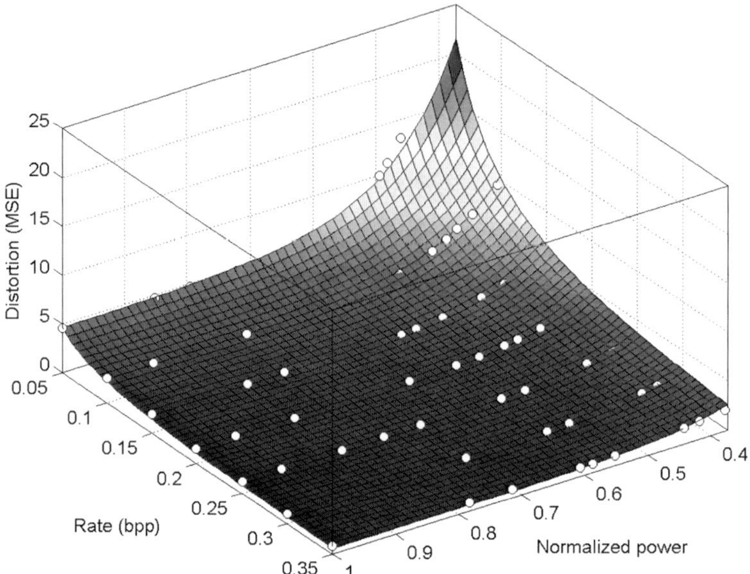

**Fig. 10.4** 3D surface plot of $P$-$R$-$D$ for *akiyo* sequence; the power is normalized to the maximum power consumption; the rate is represented as *bits per pixel*; the distortion is represented as mean squared error (MSE); white circles denote the actual $P$-$R$-$D$ points obtained from experiments [19]

tortion according to the power consumption based on the power model of the energy-aware *hardware codec*-based system using the Cauchy-density-based rate-distortion model. We prepared an operational framework for the $P$-$R$-$D$ analysis [19] as follows.

Based on the encoder power model obtained through the gate-level simulation of the target video encoder, we developed a power-distortion ($P$-$D$) model. Inputs of the modeling process are encoder power model, reference frame buffer power model, input video sequence, and target bit rate. The total power consumption of video encoding is given as the sum of the video encoding power and the reference frame buffer access power. The reference frame buffer storage power including refresh is assumed as negligible.

Based on our $P$-$D$ model and the Cauchy-density-based rate-distortion model [20], we proposed the $P$-$R$-$D$ model of the target video encoder [19] as follows:

$$D = c_0\left(c_1 e^{-c_2 P} + 1\right) R^{-\gamma} \tag{10.2}$$

where $c_0$, $c_1$, and $c_2$ are fitting parameters and $P$ is the encoding power consumption normalized by its maximum value. Figure 10.4 shows a 3D surface plot of the proposed $P$-$R$-$D$ model, where the white circles denote the actual $P$-$R$-$D$ points obtained from experiments. $R^2$ of the curve fitting was about 0.98 with $c_0 = 3.055$, $c_1 = 5.086$, and $c_2 = 11.15$.

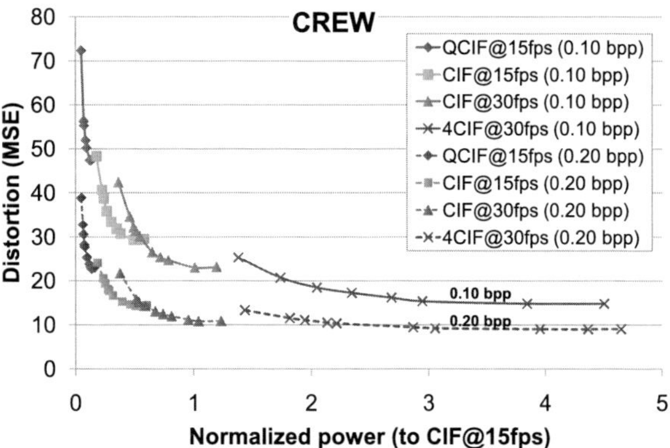



**Fig. 10.5** $P$-$D$ curves with respect to bit rate and sampling rate given as the combination of resolution and frame rate for *crew* sequence

## 10.3.2.2 Sampling Rate

In addition to bit rate and distortion of the encoded video, the overall power and bit rate can be controlled by changing the resolution and/or frame rate, i.e., the *sampling rate*. In this work, we define the *sampling rate*, $\psi$, as the product of the image resolution (the product of the image width and height in pixels) and the frame rate; $\psi = n_{\mathrm{w}} n_{\mathrm{h}} r_f$, where $n_w$ and $n_h$ are the image width and height, respectively, and $r_f$ is the frame rate (frames per second, hereafter *fps*).

Figure 10.5 shows the $P$-$D$ relationship for various resolutions and frame rates as obtained from the proposed $P$-$R$-$D$ analysis framework. We used four sampling rates: QCIF ($176 \times 144$)@15 fps, CIF ($352 \times 288$)@15 fps, CIF@30 fps, and 4CIF ($704 \times 576$)@30 fps. In Fig. 10.5, the solid lines and dashed lines denote $P$-$D$ curves for the bit rates of 0.10 and 0.20 bpp (bits per pixel), respectively. The fitting parameters $c_0$, $c_1$, and $c_2$ in (10.2) are separately adjusted according to the given sampling rate. Lowering the resolution and frame rate leads to reduced power consumption of the video encoder at the increase of the distortion. With limited resources (battery and flash), it is required to determine the sampling rate to maximize the WSC lifetime under the distortion requirement corresponding to the given sampling rate.

## 10.3.2.3 Energy Model of Video Encoding

After the event detector detects the occurrence of an event, the video encoder compresses the number of bits representing captured frames. We used the proposed $P$-$R$-$D$ model for the video encoder in (10.2) to calculate the energy consumption of video encoding, $E_{\mathrm{enc}}$, given by

$$E_{\mathrm{enc}} = P_{\mathrm{enc}}^{\psi}(R, D) t_{\mathrm{event}} \tag{10.3}$$

where $P_{\mathrm{enc}}^{\psi}$ is the power consumption of video encoding for the given sampling rate ($\psi$); $D$ is the distortion represented as the *mean squared error* (MSE); $R$ is the bit rate represented as *bits per pixel*; and $t_{event}$ is the total event time duration. The power consumption of the video encoder can be derived from (10.2) as follows:

$$P_{enc}^{\psi}(R, D) = P_{max} \frac{1}{c_2^{\psi}} \ln\left( \frac{c_0^{\psi} c_1^{\psi}}{D R^{\gamma} - c_0^{\psi}} \right) \tag{10.4}$$

where $c_0^{\psi}$, $c_1^{\psi}$, and $c_2^{\psi}$ are the fitting parameters for the given sampling rate $\psi$, and $P_{max}$ denotes the maximum power consumption of video encoding.

### 10.3.3 Storage and Transmission

The flash-constrained WSC needs to decide the storage location of encoded videos in terms of energy consumption to fully utilize flash space as well as battery charge. Because the energy consumption of both flash access and wireless transmission is mainly proportional to the output bit rate of the encoded videos, we need to find the optimal bit rate in terms of the total energy consumption, i.e., the sum of the video encoding, flash access, and wireless transmission energy.

There are four major factors related to the storage location decision: the wireless channel condition, flash space, battery charge, and the urgency (criticality level) of the corresponding event. The energy consumption of the flash access, $E_{acc}$, can be modeled as a function of bit rate and writing energy per bit as follows:

$$E_{acc} = P_{acc}(R) t_{event} \tag{10.5}$$

$$P_{acc}(R) = e_{acc} \psi R \tag{10.6}$$

where $e_{acc}$ denotes the energy consumption of the flash per writing a bit, modeled in [21], and $\psi$ represents the given sampling rate.

The energy consumed to transmit encoded data depends on the transmission distance, path loss exponent, and modulation scheme (e.g., BPSK, QPSK, QAM), etc. Based on the power model in [22], we modeled the energy consumed by transmitting encoded data ($E_{tx}$) as follows:

$$E_{tx} = P_{tx}\left( \frac{t_{event} \psi R}{BW_{ch}} \right) \tag{10.7}$$

In (10.7), $P_{tx}$ denotes the power consumption of the transmitter; $\psi$ is the sampling rate (= image width $\times$ image height $\times$ frame rate); $R$ is the bit rate of the encoded frames; $BW_{ch}$ is the channel bandwidth. The total number of bits of encoded videos is calculated as the product of the total event time duration ($t_{event}$) and the number of bits per second ($\psi R$). Because encoded videos contain frames encoded in *SS* or *RS*, $R$ can be the bit rate in either *SS* or *RS*.

If the wireless channel condition does not change rapidly and the WSC is mobile, the major variable to be predicted in the wireless transmission energy is the distance between the WSC and the base station (BS). Based on the statistics on the distance between the WSC and BS, the WSC decides the storage location of encoded videos, either the WSC itself or the BS.

### *10.3.4 Problem Definition*

Given the distortion requirement ($D^{SS}$ and $D^{RS}$) and the minimum event detection rate ($\Phi_{ed}$), the objective of the energy-aware system design is to determine the top event detection level ($L_{ed}$) and the video encoding configurations of $SS$ and $RS$, so as to minimize the total energy consumption of the WSC under the flash space constraint ($C_{fla}$). The event detection rate is the ratio of actual *high*-criticality events to all detected events. The problem can be formulated as follows:

$$\text{minimize} \quad E_{\text{tot}} = E_{ed} + E_{enc}^{SS} + E_{enc}^{RS} + E_{tx}^{SS} + E_{tx}^{RS} + E_{acc}^{SS} + E_{acc}^{RS} \tag{10.8}$$

$$\text{such that} \quad N_{bit}^{SS} + N_{bit}^{RS} \le C_{fla,0} \tag{10.9}$$

$$\Phi_{ed}^{h} \ge \Phi_{ed} \tag{10.10}$$

where $\Phi_{ed}^{h}$ is a user-defined event detection rate for *high*-criticality events, $E_{\square}^{SS}$ and $E_{\square}^{RS}$ denote the energy consumption of $SS$ and $RS$, respectively, $N_{bit}^{SS/RS}$ is the total number of bits in $SS$ and $RS$, respectively, and $C_{fla,0}$ is the initial flash space. As $L_{ed}$ goes up, $E_{ed}$, the event detection energy, increases exponentially. On the other hand, the false alarm rate decreases drastically as $L_{ed}$ goes up. Consequently, the energy consumption of the other components decreases. In addition, the sampling rate also scales the total energy consumption. $E_{enc}^{SS/RS}$ is affected by $R^{SS/RS}$ and $D^{SS/RS}$, respectively. $E_{acc}^{SS/RS}$ and $E_{tx}^{SS/RS}$ are modeled as a function of $R^{SS/RS}$, respectively.

## 10.4 Lifetime Maximization of Wireless Surveillance Camera

In this section, we introduce a method to maximize the WSC lifetime, i.e., the usage of both battery charge and flash space, by finding the energy-optimal operating configuration of the event detector, video encoding, and storage location. As depicted in Fig. 10.6, the proposed lifetime maximization engine finds the optimal bit rate and sampling rate for the estimated event parameters and the remaining battery charge and flash, given the $P$-$R$-$D$ model parameters, video encoder power model, wireless channel condition, and the statistics on the distance between the WSC and BS.

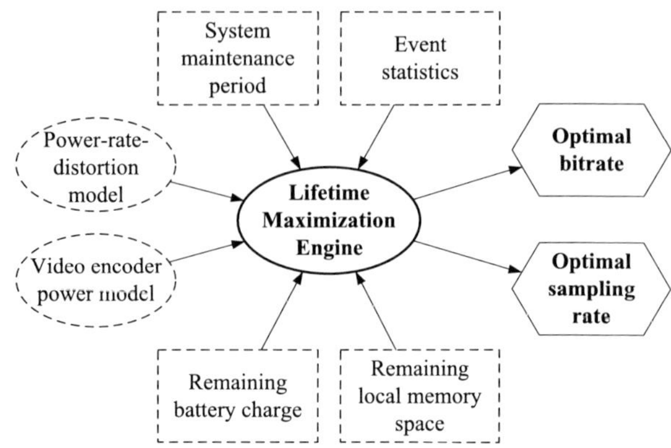

**Fig. 10.6** Lifetime maximization engine

## *10.4.1 Definition*

To maximize the lifetime of the resource-constrained WSC, it is necessary to utilize the given resources effectively while monitoring the remaining resources. The remaining lifetime of the WSC at time $i$, $f_i$, can be defined by

$$f_i = \min_{\kappa \in \{\text{bat}, \text{fla}\}} g_i^\kappa, \quad \text{for } \forall i = 0, 1, \ldots \tag{10.11}$$

where $g_i^\kappa$ is called the *resource lifetime* of the WSC at time $i$ (launch of the WSC is assumed to occur at $i = 0$) as defined by the exhaustion of resource $\kappa$, i.e., either battery (bat) or flash (fla). The battery lifetime is affected by the energy consumption of the WSC, and the flash lifetime is affected by the bit rate of the encoded video. As the WSC lifetime is defined as the time to exhaustion of whichever of the two resources wears out first, it is important to evenly wear out both resources. For example, if the flash is filled with encoded bits while there is some remaining battery charge, the remaining lifetime is still zero. The WSC lifetime also depends on event characteristics such as the probability of events and statistics of the event time duration.

The energy consumption of the WSC's major energy consumers, i.e., the video encoder, flash, and wireless transmitter, can be controlled by the bit rate of the encoded video. As shown in Fig. 10.7, the energy consumption of the video encoder decreases as the bit rate of the encoded output increases. On the other hand, the energy consumption of flash access and wireless transmission is proportional to the bit rate. Eventually, the battery lifetime becomes a function of bit rate and distortion, while the flash lifetime is a function of bit rate. The storage location also affects both the battery and flash lifetime. For example, when a large amount of encoded videos are stored externally, i.e., at the BS, the battery wears out faster than the flash. Thus, there is a trade-off point in terms of energy consumption. Because the energy consumption of the WSC and the amount of encoded bits can be expressed as a function of $R_i$, the bit rate at time $i$, the battery lifetime $g_i^{\text{bat}}$ and the flash lifetime $g_i^{\text{fla}}$ at time

**Fig. 10.7** Total energy consumption of WSC: (**a**) $E_{enc} + E_{acc}$, (the sum of the video encoder, $E_{enc}$, and flash access energy, $E_{acc}$) and (**b**) $E_{enc} + E_{acc}$ (the sum of the video encoder and wireless transmission energy, $E_{tx}$) with respect to bit rate for *news* sequence where $R_{int}^{opt}$ and $R_{ext}^{opt}$ denote, respectively, the optimal bit rate of output video stored internally and externally in terms of the energy consumption of WSC [23]. © 2011 IEEE

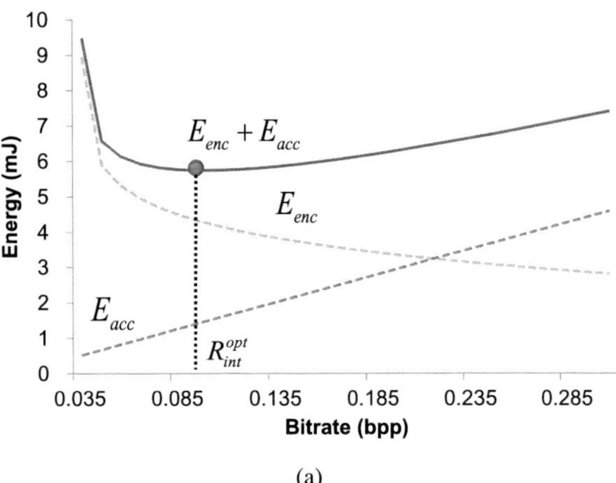


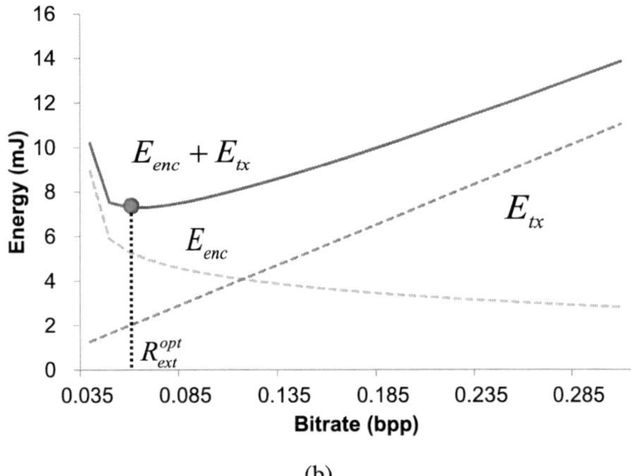


$i$ are defined by

$$g_i^{bat} = \frac{E_{bat,i}}{[\omega_i(P_{enc}^{\psi_i}(R_i^{int}, D_i) + P_{acc}(R_i^{int})) + (1 - \omega_i)(P_{enc}^{\psi_i}(R_i^{ext}, D_i) + P_{tx}(R_i^{ext}))]\hat{\pi}_i}$$

$$\sim \frac{1}{E_{tot}}$$

$$g_i^{fla} = \frac{C_{fla,i}}{\omega_i \psi_i R_i^{int} \hat{\pi}_i}$$

$$(10.12)$$

where $E_{bat,i}$ and $C_{fla,i}$ are the remaining battery energy and remaining flash capacity at time $i$, respectively; $R_i^{int}$, $R_i^{ext}$, and $D_i$ denote the bit rate stored internally and externally and distortion at time $i$, respectively; $\hat{\pi}_i$ is the estimated duty cycle of the WSC at time $i$ based on prior event statistics; $\omega_i$ denotes the portion of the time du-

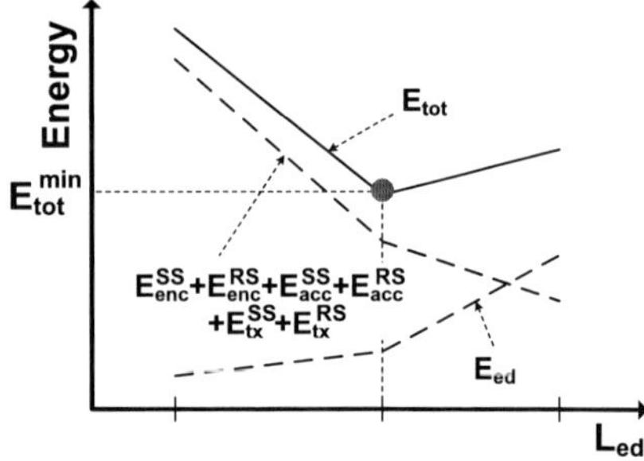



**Fig. 10.8** Energy-optimal event detection level allocation [24]. © 2010 IEEE

ration of an internally stored event. In (10.12), $P_{enc}^{\psi_i}$ denotes the power consumption of the video encoder at the current sampling rate, $\psi_i$. The battery lifetime, $g_i^{bat}$, is an inverse function of the average energy consumption of the WSC, $E_{tot}$.

## 10.4.2 Hierarchical Event Detection

An event detection level ($L_{ed}$) affects the energy consumption of an event detector ($E_{ed}$) as well as the duty cycle of the WSC ($\pi_i$). Figure 10.8 shows the energy consumption of the WSC as the event detection level ($L_{ed}$) changes. As $L_{ed}$ increases, the energy consumption of the event detector ($E_{ed}$) increases, while that of the video encoder, flash, and wireless transmitter ($E_{enc}^{SS} + E_{enc}^{RS} + E_{acc}^{SS} + E_{acc}^{RS} + E_{tx}^{SS} + E_{tx}^{RS}$) drastically decreases due to the reduced $\pi_i$. Because of the trade-off relationship between $E_{ed}$ and $E_{enc}^{SS} + E_{enc}^{RS} + E_{acc}^{SS} + E_{acc}^{RS} + E_{tx}^{SS} + E_{tx}^{RS}$, the curve of the total energy consumption ($E_{tot}$) has convexity with respect to $L_{ed}$; the energy-optimal event detection level is marked as the circle in Fig. 10.8. When the detection rate of the obtained event detection level, i.e., $\Phi_{ed}^{h}$, is lower than the detection rate requirement ($\Phi_{ed}$), a suboptimal event detection level is utilized. Because the event detection level is discrete and the operation varies little for the given sequence, it is easy to find a feasible solution. On the other hand, it is necessary to measure the detection rate for the given test sequences.

Figure 10.9 shows the energy consumption (solid lines with left $y$-axis) of our WSC with respect to the event detection level for three test sequences. The detection rate and false alarm rate are depicted in Fig. 10.9 (bars with right $y$-axis). As shown in Figs. 10.9(a) and (b), the energy consumption of the video encoder, flash, and wireless transmitter ($E_{enc}^{SS} + E_{enc}^{RS} + E_{acc}^{SS} + E_{acc}^{RS} + E_{tx}^{SS} + E_{tx}^{RS}$) decreases drastically when sophisticated event detectors are used due to the reduced $\pi_i$ in the case of frequent occurrence of *low-criticality* events. Although $E_{ed}$ is stable for $\beta$ corresponding to the test sequence, $E_{enc}^{SS} + E_{enc}^{RS} + E_{acc}^{SS} + E_{acc}^{RS} + E_{tx}^{SS} + E_{tx}^{RS}$ is sensitive to $\beta$. For *PetsC1D2* and *ThreePerson_Circles*, all of the three event detection levels show a sufficient detection rate, i.e., $\Phi_{ed}^{h} > \Phi_{ed}$. The *thresholding*-based event detector does not miss any single significant event, but it mistakes meaningless events for *high-criticality* events. As a result, the energy-optimal event detectors for *PetsC1D2* and *ThreePerson_Circles* test sequence are an *edge*-based and a

**Fig. 10.9** Comparison of energy consumption (*solid lines with left y-axis*) of $E_{ed}$ (event detector), $E_{enc}^{SS} + E_{enc}^{RS} + E_{acc}^{SS} + E_{acc}^{RS} + E_{tx}^{SS} + E_{tx}^{RS}$ (video encoder, flash, and wireless transmitter), and $E_{tot}$ (total energy), and detection and false alarm rate (*bars with right y-axis*) with respect to event detection level for three sequences: (**a**) *PetsC1D2* ($\alpha = 0.63$, $\beta = 0.37$), (**b**) *ThreePerson_Circles* ($\alpha = 0.44$, $\beta = 0.56$), (**c**) *IndoorGTTest2* ($\alpha = 0.26$, $\beta = 0.74$) where a *dashed line* denotes the required detection rate ($\Phi_{ed}$) [24]. © 2010 IEEE

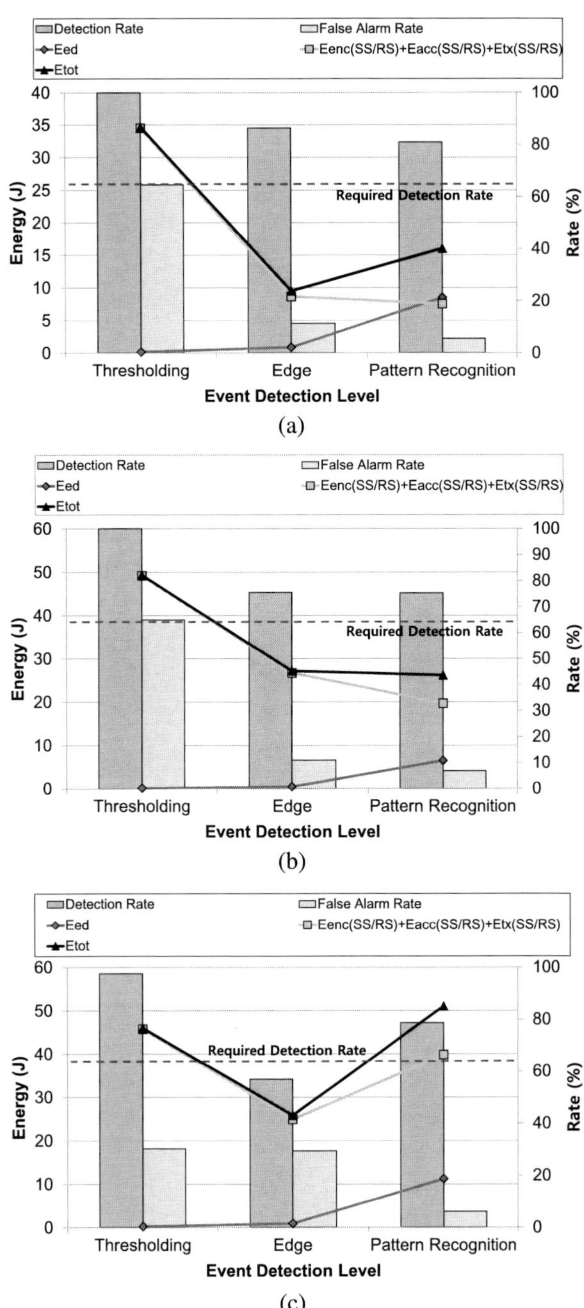


*pattern recognition*-based event detector, respectively. When $\beta$ is relatively high, $E_{enc}^{SS} + E_{enc}^{RS} + E_{acc}^{SS} + E_{acc}^{RS} + E_{tx}^{SS} + E_{tx}^{RS}$ dominates $E_{ed}$ even with the high computational complexity of sophisticated event detectors, as shown in Fig. 10.9(c). The

**Fig. 10.10** Sum of power consumption of the video encoder and wireless transmitter according to the distance between WSC and BS

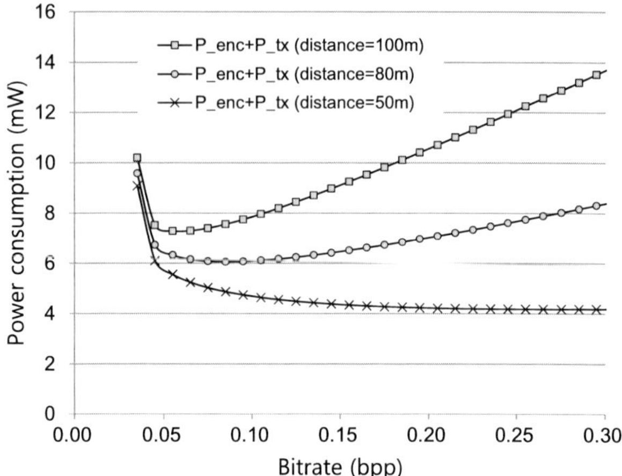


*edge*-based event detector shows an abnormally reduced total energy consumption in Fig. 10.9(c), because of the low detection rate, i.e., 56.9%. Thus, the *edge*-based event detector is excluded from the event detector candidates due to detection rate constraint (56.9% < $\Phi_{ed}$). In this case, the *thresholding*-based event detector is chosen as a suboptimal event detection level in terms of energy consumption. The proposed method provides 73.4% energy savings on average compared to the method of [13] with the *thresholding*-based event detector; i.e., it prolongs the lifetime of the WSC up to 3.76 times compared to [13].

## 10.4.3 Storage Location Decision

The factors that affect the energy consumption of the wireless transmitter are distance, bit rate, and channel condition. The wireless channel condition is uncontrollable, but the distance and bit rate can be determined in terms of the WSC lifetime. Especially, if the WSC is mobile, the increase of the distance between WSC and BS drastically increases the energy consumption of the wireless transmitter, as shown in Fig. 10.10. The optimal bit rate of the output video is determined for both internal and external cases (see the discussion in Sect. 10.4.4).

When the distance is given statistically, the problem of determining the storage location can be rewritten as finding the distance threshold ($d$) which maximizes the battery lifetime redefined as a function of bit rate and the probability density function (pdf) of the distance. When the distance between WSC and BS is less than the obtained distance threshold, the encoded video is transmitted to BS. Otherwise, the encoded video is written in internal (flash) storage.

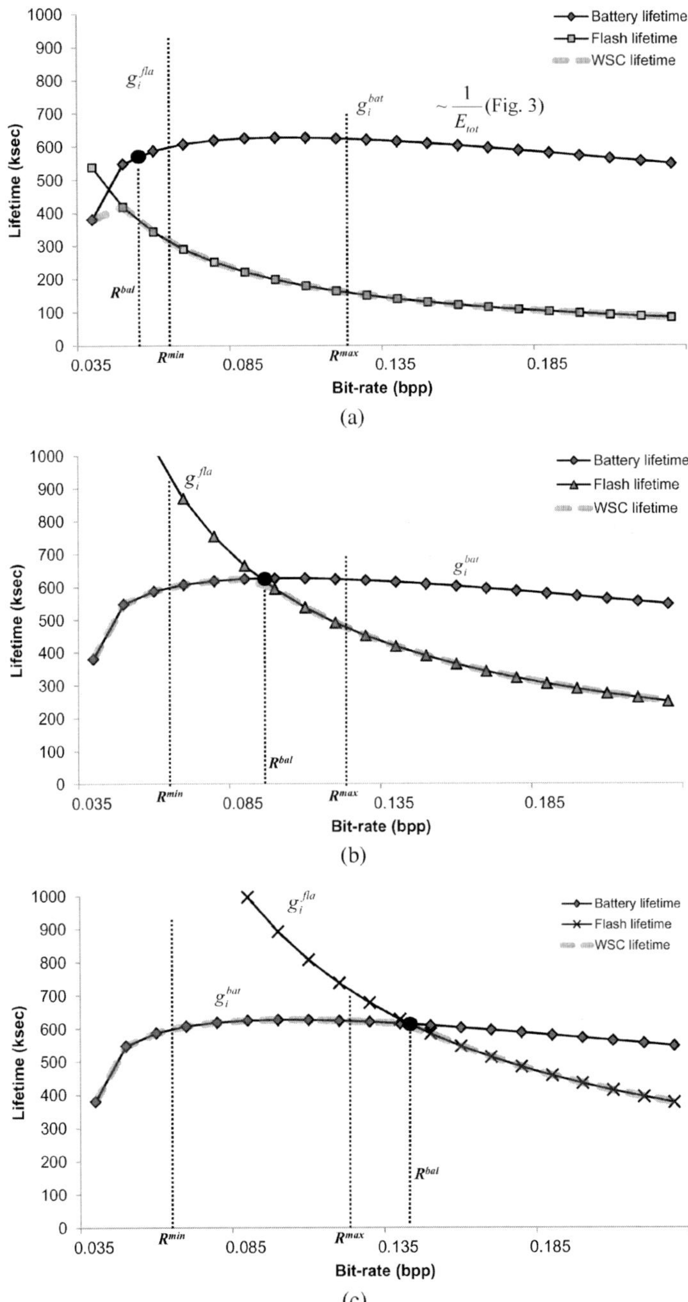







**Fig. 10.11** Lifetime curves of battery and flash ($g_i^{\mathrm{bat}}$ and $g_i^{\mathrm{fla}}$) with respect to bit rate where *solid lines* denote the resource lifetimes and a *dashed line* denotes the WSC (Wireless Surveillance Camera) lifetime in case of (**a**) $R^{\mathrm{bal}} < R^{\mathrm{min}}$, (**b**) $R^{\mathrm{min}} < R^{\mathrm{bal}} < R^{\mathrm{max}}$, and (**c**) $R^{\mathrm{max}} < R^{\mathrm{bal}}$ for *news* sequence [23]. © 2011 IEEE

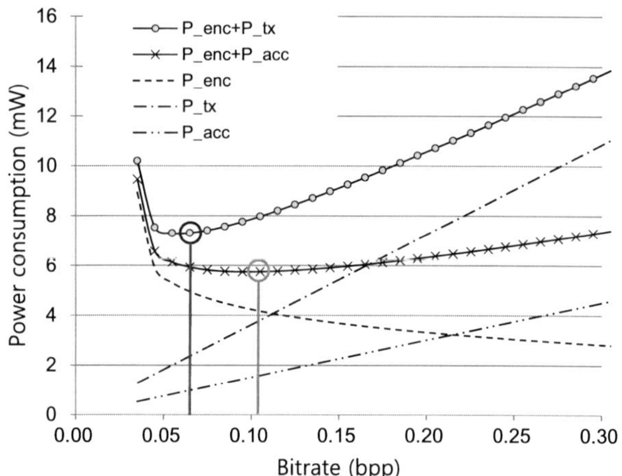

**Fig. 10.12** Power consumption of the video encoding, flash access, and wireless transmission with respect to bit rate and energy-optimal bit rate for internal (*red*) and external (*blue*) storage cases

## 10.4.4 Bit Rate Decision

Maximizing the WSC lifetime is achieved by balancing the lifetime of its two resources: the battery and flash. Figure 10.11 shows the tendencies of the battery lifetime and the flash lifetime according to the bit rate; thereby the WSC lifetime is maximized at the bit rate shown as $R^{\mathrm{bal}}$ (if there are multiple intersection points between the two tendency curves, the one with the larger lifetime is simply chosen as $R^{\mathrm{bal}}$). Actually, the optimal bit rates for the internal and external storage cases in terms of the energy consumption are different, as shown in Fig. 10.12. In this section, $R^{\mathrm{bal}}$ denotes the bit rate of the output video stored internally (the bit rate for external storage case can be obtained with the same method). The problem is formulated as follows:

$$R_i^{\mathrm{bal}} = \arg\min_{R_i} \left| g_i^{\mathrm{bat}} - g_i^{\mathrm{mem}} \right| \quad \text{such that } D_i \leq D^*(\psi_i) \tag{10.13}$$

where $R_i^{\mathrm{bal}}$ denotes the bit rate where the battery lifetime and flash lifetime are balanced, and $D^*(\psi_i)$ is the required distortion level for the given sampling rate, $\psi_i$.

When $R_i^{\mathrm{bal}}$ lies outside the bit rate range supported by the video encoder, it means that the WSC lifetime is dominated by one of the two resources. When $R_i^{\mathrm{bal}}$ is lower than $R^{\mathrm{min}}$, i.e., the lower bound of the feasible bit rate in the video encoder as shown in Fig. 10.11(a), the flash determines the overall lifetime of the WSC. In this case, $R_i^{\mathrm{bal}}$ is given by

$$R_i^{\mathrm{bal}} = R^{\mathrm{min}} \tag{10.14}$$

because the flash lifetime is monotonically decreasing. When the battery lifetime determines the WSC lifetime, i.e., $R_i^{\mathrm{bal}}$ is higher than $R^{\mathrm{max}}$ (the upper bound of the feasible bit rate in the video encoder) as shown in Fig. 10.11(c), $R_i^{\mathrm{bal}}$ is determined to maximize the battery lifetime as follows:

$$R_i^{\text{bal}} = \arg \max_{R^{\min} \le R_i \le R^{\max}} g_i^{\text{bat}} \qquad (10.15)$$

This is because the $g_i^{\text{bat}}(\sim \frac{1}{E_{\text{tot}}})$ is concave ($E_{\text{tot}}$ is convex, as shown in Fig. 10.11).

### 10.4.5 Sampling Rate Decision

Even if the WSC lifetime is maximized by dynamically balancing the resources, the expected lifetime may be shorter than the system maintenance period (SMP). To ensure that the WSC survives through the SMP, we propose to control another feature of the video, the *sampling rate*. In other words, the sampling rate (resolution $*$ frame rate) is adaptively determined to prolong the lifetime of the WSC until the SMP. The optimal sampling rate at time $i$, $\psi_i^{\text{opt}}$, is given by

$$\psi_i^{\text{opt}} = \arg \min_{\{\psi_i \mid \psi_i \in \Psi\}} \left( \frac{f_i}{\tau_i} \,\middle|\, \frac{f_i}{\tau_i} \ge \mu \right) \quad \text{such that } D_i \le D^*(\psi_i) \qquad (10.16)$$

where $D^*(\psi_i)$ is the required distortion level for the given sampling rate, $\psi_i$; $f_i$ is the remaining lifetime of the WSC at time $i$; $\tau_i$ is the remaining time until the SMP at time $i$; $\mu(\ge 1)$ is the lifetime fitting ratio which determines the degree of the lifetime margin; and $\Psi$ is the set of the sampling rate candidates. The lower the value of $\psi_i$, the longer $f_i$. Thus, (10.16) finds the highest sampling rate, $\psi_i$, that enables the WSC to survive through the SMP.

We define the situation $f_i < \tau_i$ as a *violation*. To avoid frequently changing the sampling rate, the violation check and $\psi$ control are adaptively performed using the difference between $f_i$ and $\tau_i$. As shown in Fig. 10.13, the next violation check point, $T_{\text{check}}$, is set at

$$T_{\text{check}} = i + \frac{|\tau_i - f_i|}{2} \qquad (10.17)$$

regardless of whether $f_i$ is shorter than $\tau_i$ or not.

The bit rate control process explained in Sect. 10.4.4 is performed again after a change of $\psi_i$ because the rates of increase of $g_i^{\text{bat}}$ and $g_i^{\text{fla}}$ according to $\psi_i$ are not the same.

## 10.5  Experimental Results

### 10.5.1  Configuration

For the simulation environment and assumptions we used (1) event detector: OpenCV [15], (2) video encoder: our power-scalable H.264/AVC video encoder

---

[1] Weighted mean bit rate obtained with training sequences that have similar event characteristics.

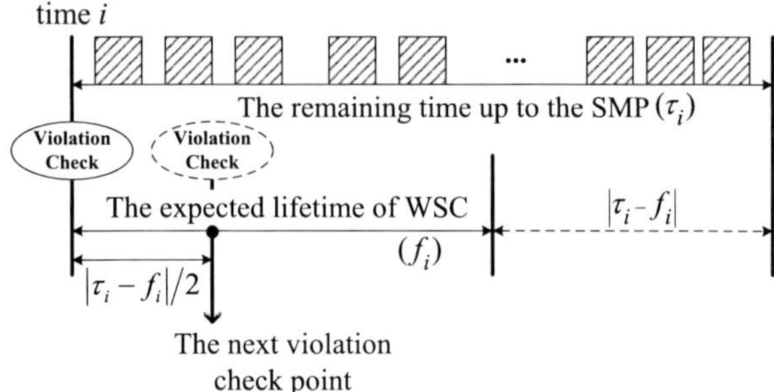


**Fig. 10.13** The next violation check point according to the remaining time up to the SMP and the expected lifetime of BSN

**Table 10.2** Comparison of BORA[1] without flash [25] and with flash [13], with the proposed BRC and BRC + SRC schemes in terms of the lifetime and the remaining resources when the system maintenance period ($T_{\text{SMP}}$) is 120 days, the initial battery charge is 3600 mWh, and the initial memory capacity is 100 GB

| | | BORA w/o flash[a] [25] | BORA w/o flash [13] | Proposed (BRC) | Proposed (BRC + SRC) |
|---|---|---|---|---|---|
| Lifetime | (hours) | 929 hours | 1657 hours | 2195 hours | 3,124 hours |
| | (days) | 38 days | 69 days | 91 days | 130 days |
| | % improvement vs. BORA w/o flash | 0.00% | +81.57 % | +139.47% | +242.10% |
| | % improvement vs. BORA w/o flash | – | 0.00 % | +31.88% | +88.41% |
| Remaining battery (%) | | 0.00% | 35.94% | 0.00% | 0.00% |
| Remaining flash (%) | | – | 0.000% | 0.000% | 0.001% |

[a]The encoded data are transferred to the base station immediately

with baseline profile, (3) flash: NAND flash memory [21], (4) the system maintenance period: 120 days, (5) the initial battery charge: 3600 mWh, and (6) the initial flash space: 100 gigabytes (GB). We used three test sequences captured for the purpose of surveillance: *PetsD1C1*, *ThreePersons_Circles*, and *IndoorGTTest2* [26]. Eventually, we characterized these three sequences to obtain the event parameters used in our simulator.

Our simulator, named *Energy-AWES³OME* (Energy-Aware WirelEss Smart Surveillance Sensor netwOrk ModEl), was built to evaluate the WSC lifetime. *Energy-AWES³OME* generates the event trace from the characterized event parameters and models the energy consumption of the WSC. It takes an event from the event trace, calculates the total energy consumed to process the event, and updates the remaining battery and flash capacity. This process is repeated until any one of the two resources is exhausted.

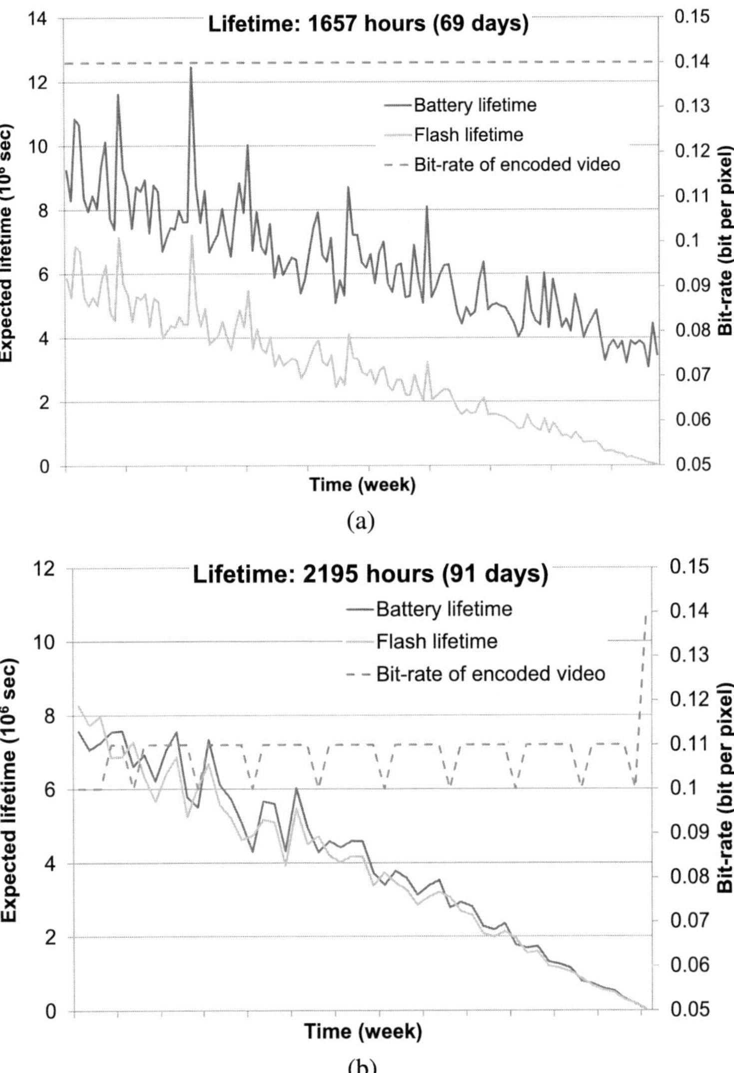



**Fig. 10.14** Comparison of (**a**) BORA with flash [13], (**b**) BRC, and (**c**) BRC + SRC in terms of the expected lifetime (left *y*-axis) and the bit rate (right *y*-axis) where the initial battery capacity is 3600 mWh and the initial flash capacity is 100 GB [23]. © 2011 IEEE

## 10.5.2  Bit Rate and Sampling Rate Control

Table 10.2 compares the proposed bit-rate control (BRC) and proposed BRC + SRC (sampling rate control) schemes vs. the battery-oriented bit rate allocation (BORA) without flash [25] and with flash [13] in terms of lifetime and remaining resource capacities. The BORA scheme decides the bit rate in terms of the energy consumption

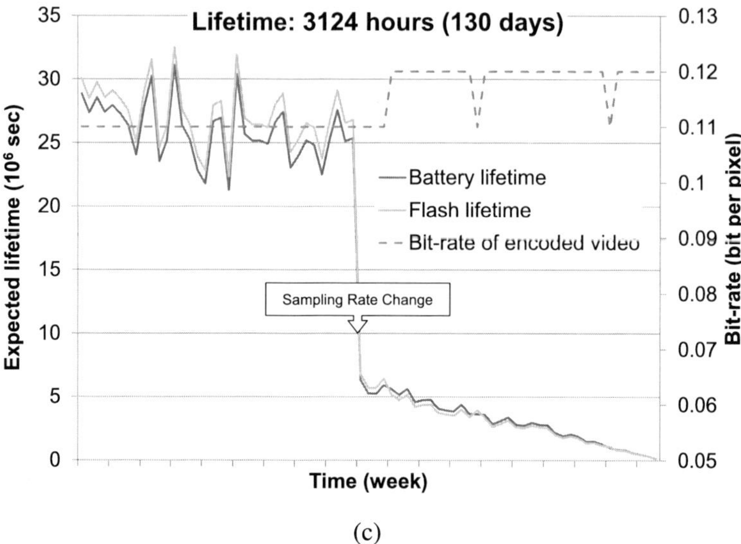


(c)

**Fig. 10.14**   (Continued)

considering only the battery. In particular, the encoded data are transmitted immediately to the base station (BS) for the BORA scheme without flash [25]; the energy consumption of the wireless transmitter is modeled based on [22]. Because the energy consumption per bit disposal of the transmitter is higher than that of the flash write, the BORA scheme with flash has a lifetime 81.57% longer than the BORA scheme without flash.

For the BORA scheme with flash, the bit rate obtained from the optimization of the energy consumption of the WSC belongs to the flash-bound region, which means that the flash dominates the WSC lifetime. Accordingly, the BORA scheme with flash achieves a relatively short lifetime compared with the proposed BRC and BRC + SRC schemes, and 35.95% of battery charge remains when the flash is full. The BRC scheme achieved 139.47% and 31.88% increased lifetime compared with the BORA scheme without flash and with flash, respectively, by evenly wearing out all the resources. Although the proposed BRC scheme prolongs the lifetime of the WSC further than the BORA scheme with flash, the WSC still cannot survive the SMP. On the other hand, the BRC + SRC scheme can survive the SMP by adaptively changing the sampling rate; i.e., the lifetime of the BRC + SRC scheme exceeds 120 days. Eventually, the BRC + SRC scheme achieved 242.10% and 88.41% increased lifetime compared with the BORA scheme without flash and with flash, respectively.

Figure 10.14 shows the expected lifetime (left $y$-axis) and the bit rate (right $y$-axis). The expected lifetime fluctuates due to the estimation error of the event duty cycle obtained by a scalar Kalman filter. In this work, the step of the bit-rate control (BRC) is 0.01 bpp (bits per pixel). For the BORA scheme with flash in Fig. 10.14(a), the weighted mean bit rate [13] obtained with training sequences that

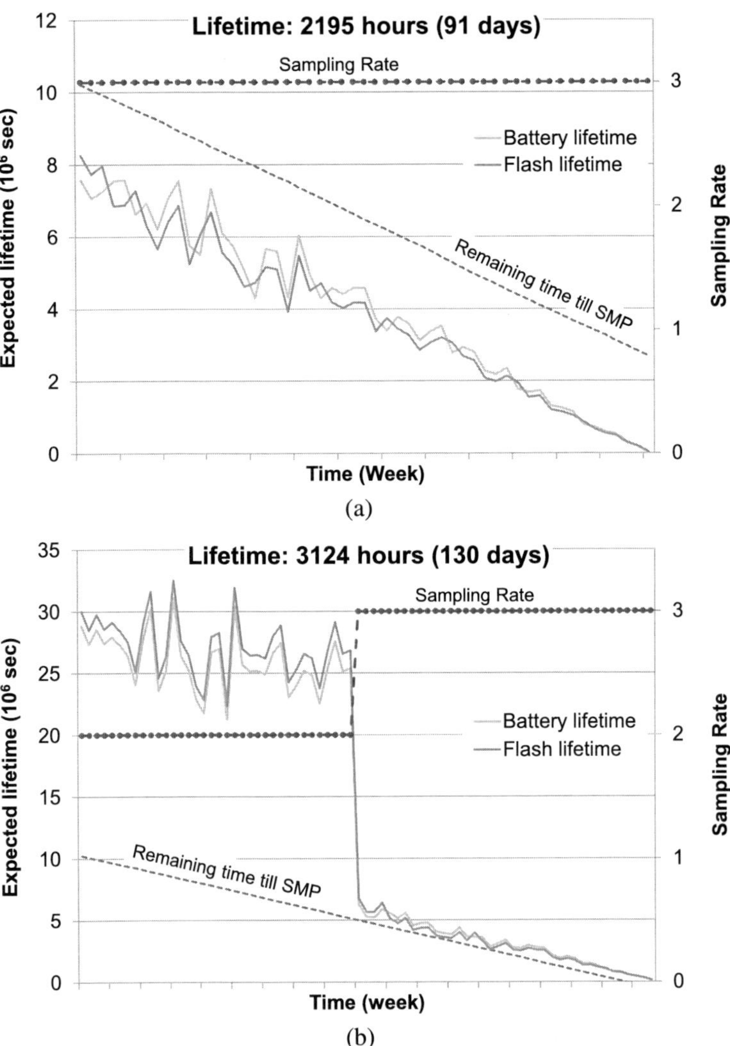



(a)

(b)

**Fig. 10.15** Comparison of (**a**) BRC and (**b**) BRC + SRC in terms of the expected lifetime, the remaining time until the SMP (left $y$-axis), and the sampling rate (right $y$-axis, $3 = 4$CIF@30 fps, $2 =$ CIF@30 fps, $1 =$ CIF@15 fps, and $0 =$ QCIF@15f ps) where the initial battery capacity is 3600 mWh and the initial flash capacity is 100 GB [23]. © 2011 IEEE

have similar event characteristics is applied to the video encoder. Because the allocated bit rate in Fig. 10.14(a) is battery-oriented, i.e., it does not consider the flash lifetime, the expected flash lifetime is different from the battery lifetime. This is in contrast with the BRC and BRC + SRC schemes, where the given resources are evenly utilized.

Figure 10.15 shows the expected lifetime (left $y$-axis) of the resources until the death of the WSC, and the sampling rate (right $y$-axis) for the BRC and BRC + SRC schemes. The initial sampling rate was set to 4CIF(704 × 576)@30 fps. For the BRC scheme, the expected lifetime of both resources is always shorter than the remaining time until the SMP. On the other hand, the BRC + SRC scheme changed the sampling rate to CIF(352 × 288)@30 fps by (10.16) at the first violation check point because the expected lifetime of the WSC with 4CIF(704 × 576)@30 fps, $f_i$, is shorter than the remaining time until the SMP, $\tau_i$. On the 62nd day, the sampling rate was switched back to 4CIF@30 fps. As a result, the WSC survived until the SMP with the BRC + SRC scheme, as shown in Fig. 10.15.

## 10.6 Conclusion

In this chapter, we presented an application example of energy-aware design: a wireless surveillance camera (WSC) that controls the storage location, bit rate, and sampling rate of the output video in terms of the WSC lifetime. Based on an event-driven method, the WSC reduces the energy wasted on capturing and storing uncritical events. With limited resources, i.e., battery and flash, the demonstrated method maximizes the lifetime of the WSC, which is defined as the time until one of the two resources wears out.

## References


1. Valera, M., Velastin, S.: Intelligent distributed surveillance systems: a review. IEE Proc., Vis. Image Signal Process. **152**(2) (2005)
2. Talukder, A., et al.: Optimal sensor scheduling and power management in sensor networks. In: Proc. SPIE (2005)
3. Talukder, A., et al.: Autonomous adaptive resource management in sensor network systems for environmental monitoring. In: Proc. IEEE Aerospace Conference (2008)
4. Hengstler, S., et al.: MeshEye: a hybrid-resolution smart camera mote for applications in distributed intelligent surveillance. In: Proc. IPSN (2007)
5. He, T., et al.: Energy-efficient surveillance system using wireless sensor networks. In: Proc. MobiSys (2004)
6. Feng, W., et al.: Panoptes: scalable low-power video sensor networking technologies. ACM Trans. Multimed. Comput. Commun. Appl. **1**(2) (2005)
7. Hampapur, A., et al.: Smart video surveillance: exploring the concept of multiscale spatiotemporal tracking. IEEE Signal Process. Mag. **22**(2) (2005)
8. Vanam, R., et al.: Distortion-complexity optimization of the H.264/MPEG-4 AVC encoder using GBFOS algorithm. In: Proc. IEEE Data Compression Conference (2007)
9. He, Z., et al.: Power-rate-distortion analysis for wireless video communication under energy constraints. IEEE Trans. Circuits Syst. Video Technol. **15**(5) (2005)
10. Zhang, Q., et al.: Power-minimized bit allocation for video communication over wireless channels. IEEE Trans. Circuits Syst. Video Technol. **12**(6) (2002)
11. Liang, Y., et al.: Joint power and distortion control in video coding. In: Proc. SPIE (2005)
12. He, Z., Wu, D.: Resource allocation and performance analysis of wireless video sensors. IEEE Trans. Circuits Syst. Video Technol. **18**(5) (2006)

13. He, Z., et al.: Energy minimization of portable video communication devices based on power-rate-distortion optimization. IEEE Trans. Circuits Syst. Video Technol. **18**(5) (2008)
14. Su, L., et al.: Complexity-constrained, H. 264 video encoding. IEEE Trans. Circuits Syst. Video Technol. **19**(4) (2009)
15. Open Source Computer Vision Library (OpenCV). http://sourceforge.net/projects/opencv/
16. Kim, C., Hwang, J.: Video object extraction for object-oriented applications. J. VLSI Signal Process. **29**, 7–21 (2001)
17. Oren, M., Papageorgiou, C., Sinha, P., Osuna, E., Poggio, T.: Pedestrian detection using wavelet templates. In: Proceedings of IEEE Conference on Computer Vision and Pattern Recognition, CVPR '97, p. 193 (1997)
18. Papageorgiou, C.P., Oren, M., Poggio, T.: A general framework for object detection. In: Proc. IEEE 6th International Conference on Computer Vision, p. 555 (1998)
19. Kim, J.: Analysis and architecture design of power-aware H.264/AVC encoder using power-rate-distortion optimization. Ph.D. dissertation, Korea Advanced Institute of Science and Technology (KAIST) (2010)
20. Kamaci, N., Altunbasak, Y., Mersereau, R.M.: Frame bit allocation for the H.264/AVC video coder via Cauchy-density-based rate and distortion models. IEEE Trans. Circuits Syst. Video Technol. **15**(8) (2005)
21. Chouhan, S., et al.: Ultra-low power data storage for sensor network. In: Proc. IPSN (2006)
22. Mathur, G., et al.: A framework for energy-consumption-based design space exploration for wireless sensor nodes. IEEE Trans. Comput. Aided Des. Integr. Circuits Syst. **28**(7) (2009)
23. Na, S., Kim, G., Kyung, C.-M.: Lifetime maximization of video blackbox surveillance camera. In: Proc. of IEEE International Conference of Multimedia and Expo (2011)
24. Na, S., Kim, J., Kim, J., Kim, G., Kyung, C.-M.: Design of energy-aware video codec-based system. In: Proc. of IEEE International Conference on Green Circuits and Systems, ICGCS, pp. 201–206 (2010)
25. Lu, X., Erkip, E., Wang, Y., Goodman, D.: Power efficient multimedia communication over wireless channels. IEEE J. Sel. Areas Commun. **21**(10), 1738–1751 (2003)
26. Performance evaluation of surveillance systems. http://www.research.ibm.com/peoplevision/performanceevaluation.html

# Chapter 11
# Low Power Design Challenge in Biomedical Implantable Electronics

Sung June Kim

**Abstract** The quality of life for patients with sensory loss can be drastically improved by the use of biomedical implantable electronics. The implantable electronics can replace the functions of sensory organs that are congenitally defective or damaged by accidents. In the field of neural prosthesis, these implantable devices are often the most important constituents of the whole system. Two neural prosthetic devices are commercially available at present; the *cochlear implant* for the severely hearing impaired, and the *deep brain stimulator* for Parkinson's disease patients. Under development is a device for the vision impaired, known as *artificial vision* or a *retinal implant*.

A neural prosthetic device consists of an internal implant, an external processor, and a telemetry module connecting the two. The external processor determines electrical stimulation parameters for a sensory input, encodes them, and transmits the encoded data to the implant unit. Often power for the implant electronics is also delivered from outside by telemetry. The implanted electronics then receives data, decodes them, and generates stimulation waveforms. These waveforms are sent to the electrode array to deliver charges to stimulate target neurons.

Unlike the ICs used in consumer electronics, safety is the number one concern for implantable devices in humans. Not only must the materials be biocompatible, but also any signals presented to the body must be proven safe. Long-term reliability of the neural interface is another concern. Due to glial tissue encapsulation, the impedance of the electrode-tissue interface can change over time.

In this chapter we consider two representative IC design examples of biomedical implantable electronics: cochlear and retinal implants. Low power design is critical in these applications. For example, in a cochlear implant, the typical power consumption of the current products is on the order of 100 milliwatts. The power depends on the number of channels or stimulation electrodes. While the number stays on the order of 10 for effective stimulation in cochlear implants, the number of required channels, or pixels, is expected to be on the order of 1000 for reasonable visual acuity in retinal implants. Thus low power design is essential for future biomedical implant systems. Here we discuss some of the design examples that have

S.J. Kim (✉)
Seoul National University, Seoul, Republic of Korea
e-mail: kimsj@snu.ac.kr

C.-M. Kyung, S. Yoo (eds.), *Energy-Aware System Design*,
DOI 10.1007/978-94-007-1679-7_11, © Springer Science+Business Media B.V. 2011

been implemented so far, but further innovative ideas are critically needed to ensure the success of future products.

## 11.1 Cochlear Implant

### *11.1.1 Background of Cochlear Implant*

A cochlear implant is a neural prosthetic device that restores hearing for persons with moderate-to-severe sensorineural hearing loss [1–3]. According to the World Health Organization (WHO), 250 million people worldwide have a moderate-to-severe hearing loss, many of whom can be helped by this advanced technology.

Several cochlear implant devices have been developed. All of them have the following components in common: a microphone that picks up the sound, a speech processor that converts the sound into electrical signals and encodes them, a transmission system that sends the electrical signals on a radio frequency (RF) carrier signal and transmits them wirelessly to the implanted electronics, a receiver/stimulator that receives and decodes the received signals to form the stimulation waveforms, and an intracochlear electrode array (consisting of multiple electrodes) that is inserted into the cochlea by a surgeon [4, 5].

Cochlear implants have been successfully used in hundreds of thousands of adults and children worldwide. Significant open-set recognition of speech can be achieved with commercially available multichannel cochlear systems. A majority of users can converse over the telephone for everyday communication. At present, there are three major cochlear implant manufacturers: Cochlear (Australia) [6], Advanced Bionics (USA) [7], and Med-El (Austria) [8]. Recently a "low cost but effective" cochlear implant concept has emerged, and a product was developed in Korea by Nurobiosys Corporation in collaboration with the author's team at Seoul National University [9]. Figure 11.1 shows the conceptual view of a typical cochlear implant system, consisting of the components mentioned above.

### *11.1.2 Cochlear Implant System Design*

The external speech processor of the cochlear implant system consists of a microphone, an analog preprocessor, digital signal processor (DSP) hardware, and a power amplifier. The implantable unit consists of a receiver/stimulator integrated circuit and an intracochlear array of electrodes. The inductive telemetry link consists of circuits and coils for forward transmission of power and data, and backward transmission of indicator signals of selected chip and electrode functions. These are shown in Fig. 11.2 in block diagram format.

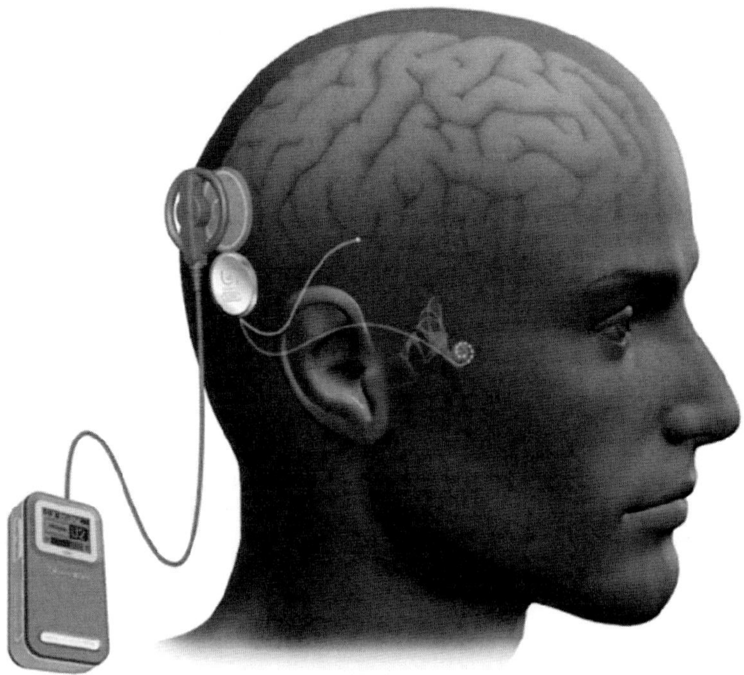

**Fig. 11.1** Conceptual view of the cochlear implant (shown with permission from Nurobiosys Corporation). The cochlear implant consists of a microphone, a speech processor, a transmission system, a receiver/stimulator, and an intracochlear electrode array

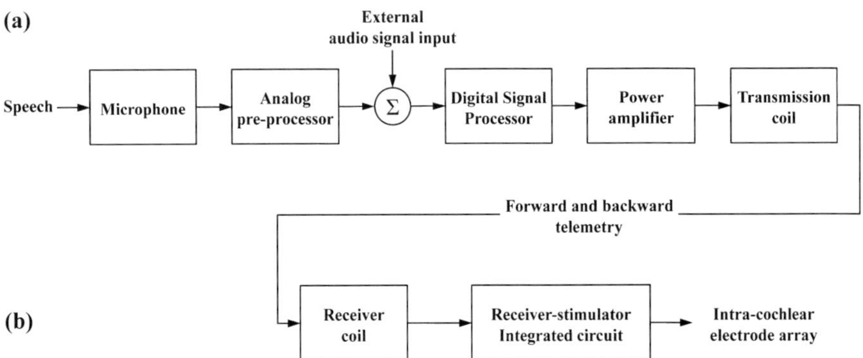


**Fig. 11.2** Block diagram of the cochlear implant system [9]. It consists of two parts: (**a**) an external speech processor and (**b**) an implantable unit. The inductive telemetry link consists of circuits and coils for forward transmission of power and data, and backward transmission of indicator signals of selected chip and electrode functions. © 2007 IEEE

### 11.1.2.1 Communication

A pulse width modulation (PWM) encoding combined with amplitude shift keying (ASK) modulation and demodulation based on pulse counting has been adopted as an effective yet simple method of communication between the external speech processor and the implantable unit [10].

In PWM, three bits are encoded as follows: a logical "one" and "zero" are encoded to have a duty cycle of 75% and 25%, respectively, and an end-of-frame (EOF) bit has a 50% duty cycle. Such an encoding method allows easier synchronization and decoding because each bit has a uniform rising edge at its beginning. According to this bit coding scheme, every bit has a high state, which is advantageous for uniform RF energy transmission.

Each data frame for forward telemetry contains several types of information for controlling the implant. This includes pulse duration, stimulation mode setting, and stimulation electrode and pulse amplitude setting for each stimulating channel. Information for the feedback telemetry includes voltages at critical internal nodes and impedance values of the electrode array. For efficient transcutaneous transmission of PWM encoded data, a class-E tuned power amplifier is used with ASK modulation [11].

### 11.1.2.2 Speech Signal Processing

Figure 11.3 shows a functional block diagram of the signal processing performed within the DSP hardware. After digitizing the speech signal, the DSP performs frequency analysis using a fast Fourier transform (FFT). The chip then computes the average power of each channel by simple summation and averaging, according to predetermined channel-frequency allocation. Band-pass filters for the channels are formed by integration across FFT bins. As an alternative to the FFT method, digital filters can be implemented using an infinite impulse response (IIR) or a finite impulse response (FIR) filter with envelope detection. However, the FFT approach is more efficient, as the total number of computations is substantially lower than in the IIR or FIR approaches. The FFT option also provides flexibility because many functions can be implemented by simple arithmetic operations [12].

## *11.1.3 Circuit Design for Cochlear Implant*

### 11.1.3.1 The Receiver/Stimulator Chip

A custom designed chip for the implanted part of the cochlear implant system is described with the block diagram shown in Fig. 11.4. The chip consists of a forward data decoder, a stimulation controller, and a current stimulator for electrical stimulation of auditory neurons. It also has a voltage sampling circuit and a backward

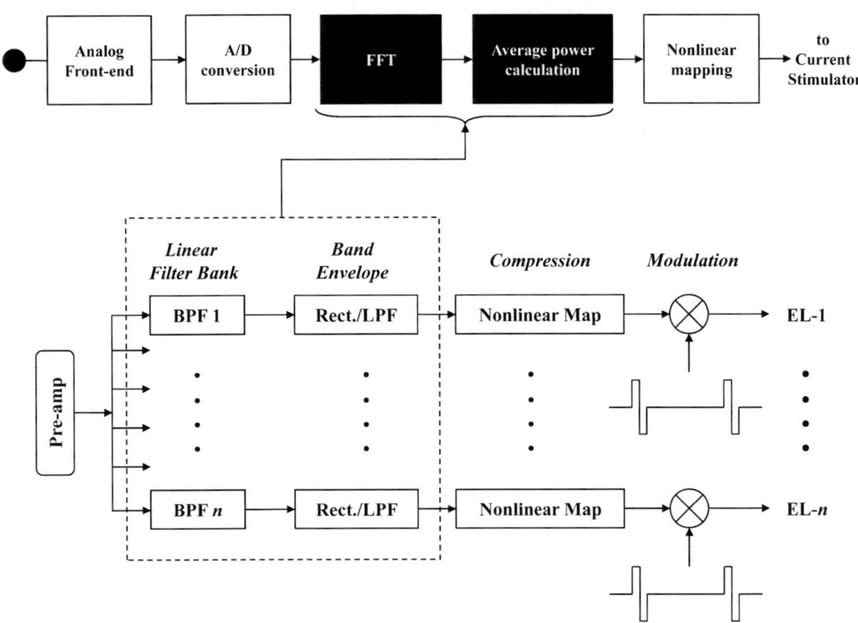



**Fig. 11.3** Block diagram of the speech signal processing. After digitizing the speech signal, the DSP performs frequency analysis using FFT. The chip computes the average power of each channel by simple summation and averaging, according to predetermined channel-frequency allocation. Band-pass filters for the channels are formed by integration across FFT bins

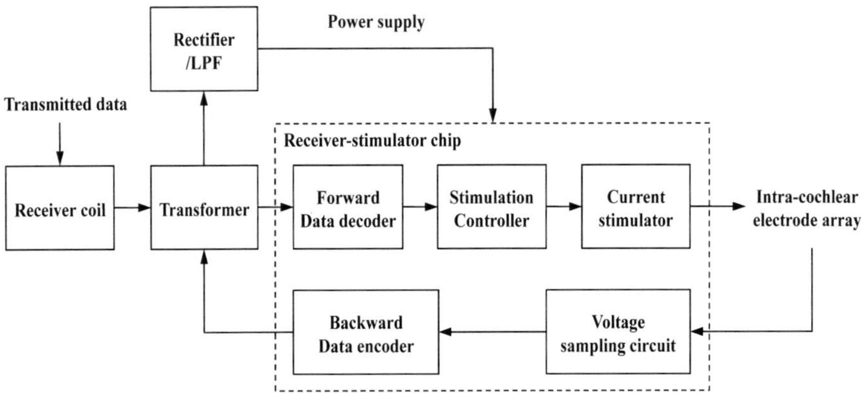



**Fig. 11.4** Block diagram of the receiver/stimulator chip for cochlear implant with its peripheral circuit [9]. The chip consists of a forward data decoder, a stimulation controller, and a current stimulator for electrical stimulation. It also has a voltage sampling circuit and a backward data encoder for backward telemetry. © 2007 IEEE

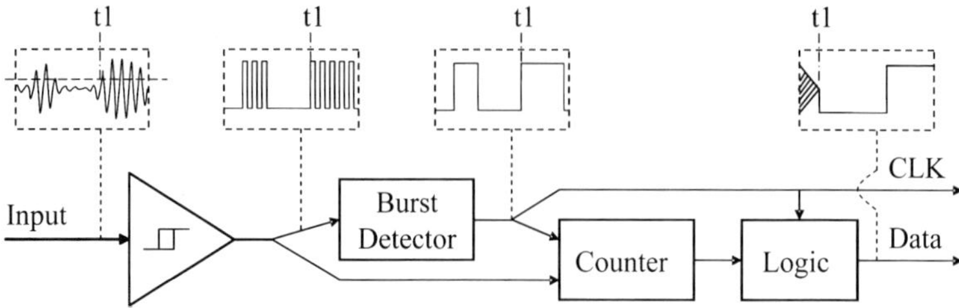


**Fig. 11.5** Functional block diagram of forward data decoder. A Schmitt trigger circuit is used to count the number of pulses to measure the duration of the received data to determine the received bits. The forward data decoder performs both data and clock recovery

data encoder for backward telemetry of internal information. In this configuration, the power and the data are both received by a pair of coils [13]. The implanted coil has a small number of turns compared to the external transmitting coil so that the induced voltage can be lower than a few volts, for safety reason. This small voltage is then stepped up internally, using a small transformer, and a regulated supply of power is obtained.

### 11.1.3.2 Forward Data Decoder

The transmitted signal in the RF telemetry has a 2.5 MHz carrier frequency. The absorption of skin is relatively low at that frequency [14]. As stated earlier, a PWM encoding is used with an ASK modulation and demodulation. A Schmitt trigger circuit is used to count the number of pulses to measure the duration of the received data to determine the received bits. The forward data decoder also performs clock recovery. Figure 11.5 shows the block diagram of the forward data decoder.

### 11.1.3.3 Current Stimulator

For safety reasons, the electrical stimulation is delivered in a biphasic configuration. Net accumulation of charge can result in lowering of the pH level, which can adversely affect the viability of cells [15]. The current stimulation circuit shown in Fig. 11.6 is designed to generate current pulses with equal amplitude in the anodic and cathodic directions. When the current flows from the channel electrode (CH) to the reference electrode (REF), switches PM1 and NM2 are closed; for the current flow in the opposite direction, switches PM2 and NM1 are closed. After stimulation, both the channel electrodes and the reference electrode are grounded. This ensures safety against passing of any unwanted charge to the body.

### 11.1.3.4 Multichannel Current Stimulator

In the circuit shown in Fig. 11.7, Elec1–Elec16 represent active electrodes in the intracochlear electrode array, and Reference represents the extracochlear reference

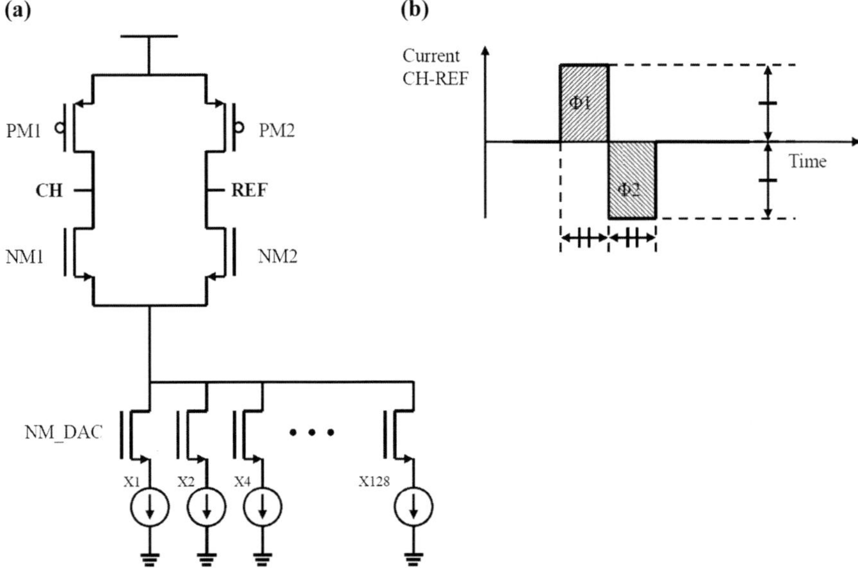


**Fig. 11.6** Diagram of the current stimulator (**a**) and output current pulse (**b**). When the current flows from the CH to the REF, switches PM1 and NM2 are closed, while for the current flow in the opposite direction, switches PM2 and NM1 are closed. The electrical stimulation is delivered in a charge-balanced biphasic configuration

electrode. By controlling the on/off status and duration of the switches, we can easily determine the shape of stimulation pulses and the stimulation mode (monopolar or bipolar). This circuit is designed so that the biphasic pulse is formed by switching from a single set of current sources to ensure charge balance. Passage of any dc current due to the remaining residual charge at the electrodes is precluded with the use of blocking capacitors between the (switch) current source and all electrodes.

### 11.1.3.5  Backward Telemetry

In the backward telemetry system, pieces of information of the implanted unit such as electrode impedances, power supply voltages, and communication errors are fed back to the DSP chip using a technique called load modulation [13]. For example, to send a value of electrode impedance, the receiver/stimulator samples the voltage difference between an active and the reference electrode. The sampled voltage is then sent to the external speech processor via the backward telemetry link. In the circuit shown in Fig. 11.8, the voltage is converted to a proportional pulse duration, during which the quality factor of the receiver resonant circuit is reduced. The external processor then measures the duration of the pulse for which the amplitude is reduced, to measure the voltage which is representative of the impedance. Using the load modulation method, we only need one set of coils to cover the bidirectional communication.

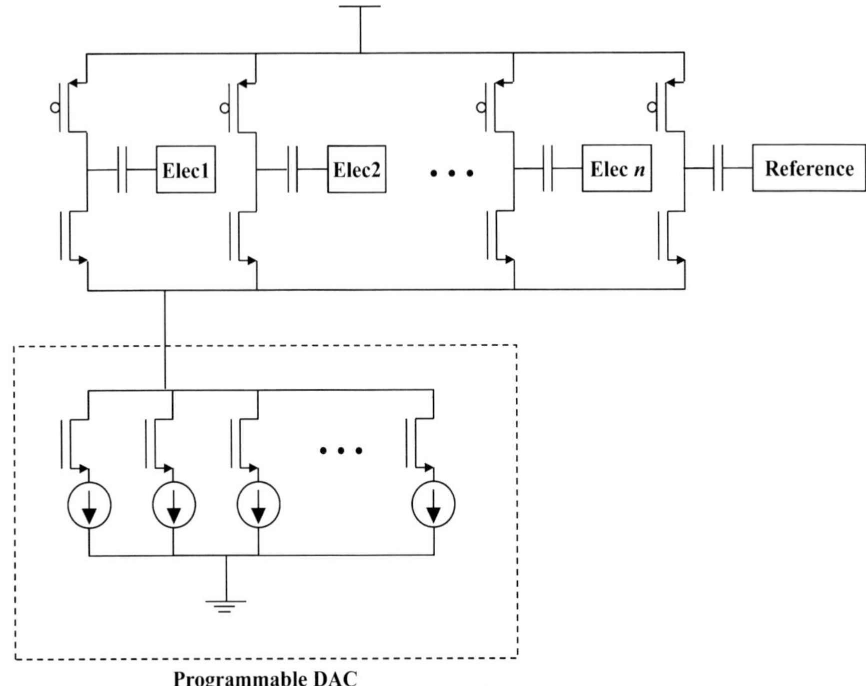

Programmable DAC

**Fig. 11.7** Diagram of the multichannel current stimulator [9]. By controlling the on/off status and duration of the switches, we can easily determine the shape of stimulation pulses and the stimulation mode (monopolar or bipolar). This circuit is designed so that the biphasic pulse is formed by switching from a single set of current sources to ensure charge balance. © 2007 IEEE

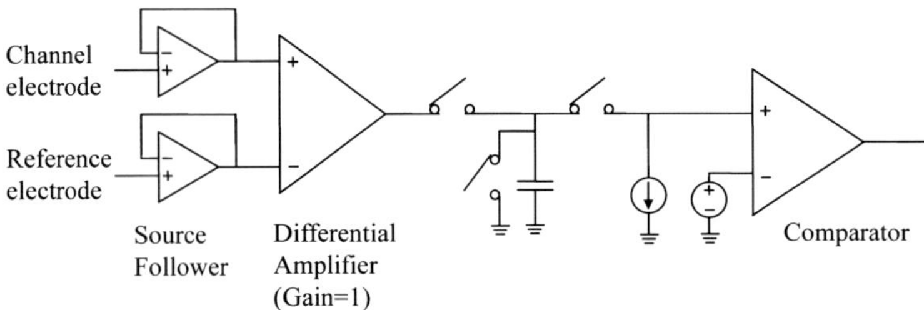

**Fig. 11.8** Backward telemetry system using load modulation technique. The voltage from the electrodes is converted to a proportional pulse duration, during which the quality factor of the receiver resonant circuit is reduced. The external processor then measures the duration of the pulse for which the amplitude is reduced, to measure the voltage which is representative of the impedance

### 11.1.3.6 Fabricated Cochlear Implant Chip

The receiver/stimulator chip design described above has been fabricated using a 0.8-μm high-voltage CMOS technology (Austria Micro Systems; AMS, Austria). Figure 11.9 shows a microphotograph of the fabricated chip die. The regulator, data

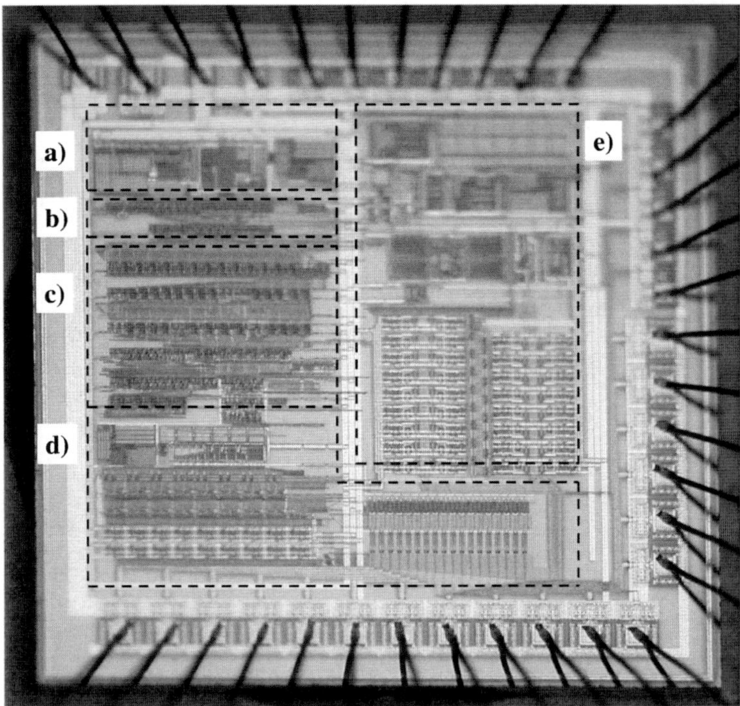


**Fig. 11.9**  Chip die photograph of the receiver/stimulator chip. (**a**), (**b**), (**c**), (**d**), and (**e**) are the circuit blocks for regulator, data receiver, control logic, current source, and voltage-to-time converter, respectively

**Table 11.1**  Specifications of the fabricated receiver/stimulator chip

| Technology | HV CMOS 0.8 μm |
|---|---|
| Die size | 3.5 mm × 3.5 mm |
| Carrier frequency | 2–5 MHz (typical 2.5 MHz) |
| Data rate | 100–250 kbps (typical 125 kbps) |
| Maximum pulse rate | 6.4–16 Mpps (typical 8 Mpps) |
| Number of electrodes | 18 (2 reference electrodes) |
| Current amplitude | Maximum 1.86 mA in 7.3 μA steps |
| Power supply range | 6–16 V |
| Sampling voltage range | 0.5–4.5 V |
| Current consumption | 0.8 mA (@ excluding stimulation current) |

receiver, control logic, current source, and voltage-to-time converter are shown in Figs. 11.9(a), (b), (c), (d), and (e), respectively. Some details of the chip specifications are shown in Table 11.1.

## 11.2  Retinal Implant

### 11.2.1  Background of Retinal Implant

#### 11.2.1.1  Retinal Degeneration and Artificial Retina

Photoreceptor loss due to retinal degenerative diseases such as age-related macular degeneration (AMD) and retinitis pigmentosa (RP) is a leading cause of adult blindness. In those patients, macular photoreceptors are almost totally lost. Fortunately, however, the inner nuclear and ganglion layer cells in the macula often survive at a fairly high rate. The artificial retina, or retinal implant, is designed to alleviate the disabilities produced by these diseases [16]. It has been known clinically that electrical stimulation of the remaining retinal neurons is effective in restoring some useful vision. Perception of small spots of light called phosphenes, is evoked by application of controlled electrical signals to a small area of the retina of a blind volunteer via a microelectrode. Thus, the majority of the development efforts in the retinal prosthesis is carried out by means of electrical stimulation [16–22].

#### 11.2.1.2  Three Methods of Retinal Implants

Depending on the location of the implant within the retina, the retinal implantation has been developed in three different directions as follows:

1. Epiretinal implant [18]: The electrode array is inserted into the vitreous cavity and attached to the retinal surface using tacks. The surgery is almost the same as a routine vitrectomy. However, this method requires extra devices (tacks), and the stimulation may involve axon fibers originating from other locations, resulting in unwanted crosstalk.
2. Subretinal implant [19]: The electrode array is implanted in a space between the retina and the choroidal layer. Locating the electrodes here can be advantageous in that it is the same location as that of the damaged photoreceptors, thus we can take advantage of the natural signal processing of neural networks. However, the surgery is difficult and may involve massive bleeding in the choroidal layer where the electrode needs to penetrate.
3. Suprachoroidal implant [20]: The electrode array is inserted into a space between sclera and choroid through a simple sclera incision. This method is the safest in terms of surgery, but, since the electrode is placed farther away from the target ganglia, a higher threshold stimulation current may be needed.

Similar to the cochlear implant, the retinal implant consists of a capture device (a camera), a signal (image) processor, telemetry, an implantable receiver/stimulator, and an electrode array. Figure 11.10 shows a conceptual view of the retinal implant system showing these elements and the three methods of implantation described above.

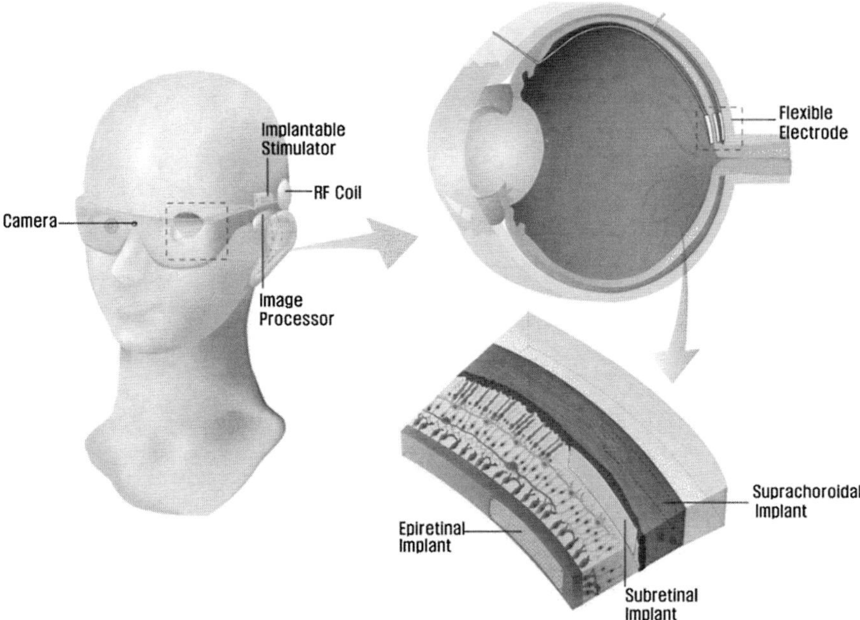



**Fig. 11.10** Conceptual view of the retinal implant and the three methods of implantation (epiretinal, subretinal, and suprachoroidal). The retinal implant is divided into three methods according to the inserting location of the retinal electrode. The retinal implant consists of a camera, an image processor, telemetry, an implantable receiver/stimulator, and an electrode array

## *11.2.2  Retinal Implant System Design*

### 11.2.2.1  Electrical Stimulation Pulse

An electrical stimulation waveform for a retinal implant is characterized by four parameters: amplitude, width, interphase delay, and frequency (see Fig. 11.11) [21]. A biphasic waveform with charge balancing between anodic and cathodic pulse is preferred for safety. The purpose of using the interphase delay is to separate the two pulses so that the second pulse does not reverse the physiological effect of the first pulse.

### 11.2.2.2  Components of Retinal Implant System

We have designed an implantable retinal implant system consisting of an external unit (for stimulus control), a telemetry unit, an implantable receiver/stimulator, and a rechargeable battery (see Fig. 11.12) [22]. The rechargeable battery is added so that this device can be easily applied to animal experiments. It can also be useful in the future for building a totally implantable system where the implant is required to be equipped with a rechargeable battery. The selected stimulation data is modulated,

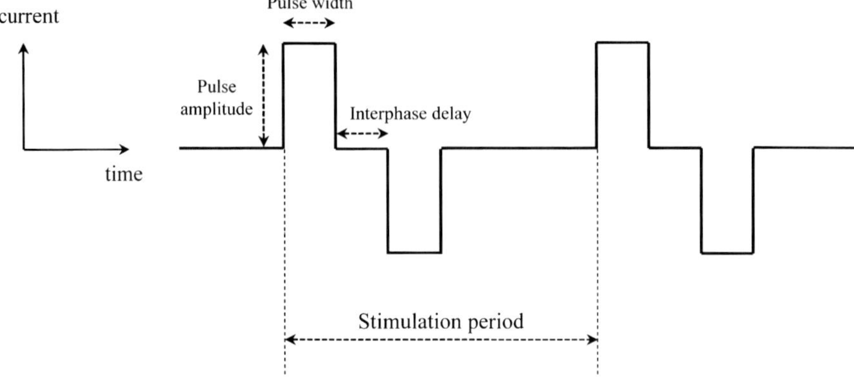



**Fig. 11.11** Electrical current stimulation waveform for retinal implant

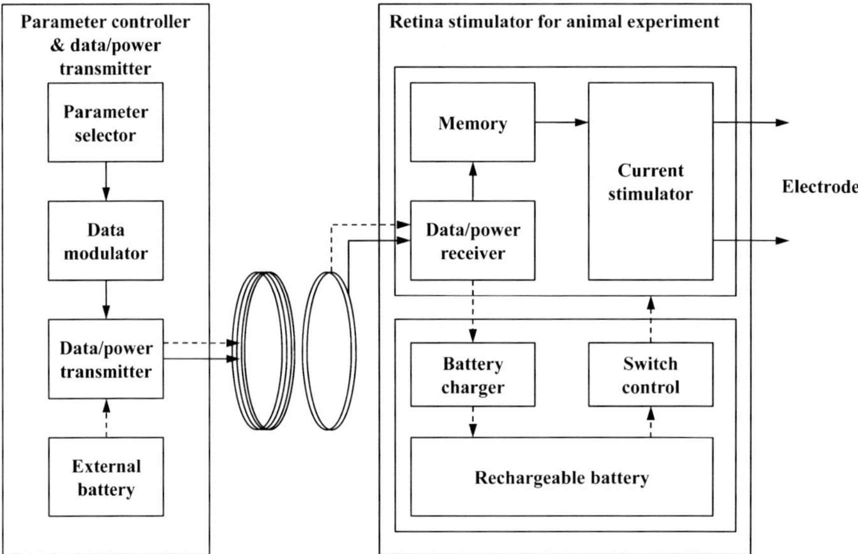



**Fig. 11.12** Block diagram of the retinal implant system [22]. It consists of an external unit (for stimulus control), a telemetry unit, an implantable receiver/stimulator, and a rechargeable battery

and the modulated carrier signal is transmitted to the implantable unit through a pair of RF coils.

### 11.2.2.3 Communication

The communication method of the retinal implant is similar to that of the cochlear implant. A pulse width modulation (PWM) encoding combined with amplitude shift keying (ASK) modulation and demodulation based on pulse counting has been

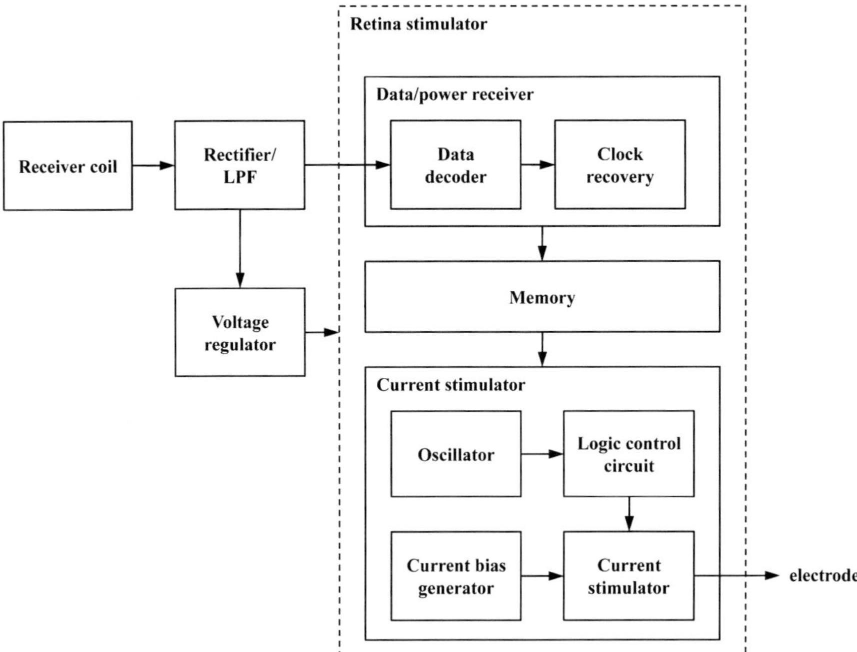


**Fig. 11.13** Implanted part of the retinal implant. Enclosed by the *dotted line* is the IC implementation of the receiver/stimulator. The IC consists of circuits for a data/power receiver, a current stimulator, and a parameter memory

adopted as an effective yet simple method of communication between the external parameter controller and data/power transmitter and the implantable retinal stimulator. The external unit is used to select stimulation parameters and to generate a parameter data frame. For efficient transcutaneous transmission of PWM encoded data, a class-E tuned power amplifier is used with ASK modulation. We designed the system so that the telemetry system is operated only when changes of the stimulus parameters are required, or during the charging of the rechargeable battery.

## 11.2.3 Circuit Design for Retinal Implant

### 11.2.3.1 Retinal Stimulator

The implantable retinal stimulator (receiver/stimulator) is shown in Fig. 11.13 in a block diagram format. When the transmitted data are received by the internal coil, their envelopes are extracted using a half-wave rectifier and a low-pass filter. The data decoder in the data/power receiver then recovers the parameter data and saves them in the parameter memory. The voltage regulator uses the same envelope signal to generate power. The IC consists of circuits for a data/power receiver, a current stimulator, and a parameter memory.

### 11.2.3.2 Operating Modes

The retinal stimulator has two modes of operation: stimulation and battery recharging. In the stimulation mode, the parameter data stored in the parameter memory are sent to the current stimulator. The current stimulator consists of multiple (seven) current sources and a timing logic circuitry. The current generator circuitry has a current bias circuitry and an 8-bit binary current-weighted digital-to-analog converter. The timing logic circuitry has a 2.5 MIIz oscillator and switch control logic circuitry for controlling the current stimulation waveform. The battery charging mode is described in the following section.

### 11.2.3.3 Battery Charging

The system is based on a single inductive link for both data transmission and battery charging. Simultaneous transmission of the stimulation parameter and charging power is difficult, because the battery charging circuit influences the precisely designed load value of the data/power receiving circuit and can cause failure in data reception. To separate the stimulation mode and battery charging mode, a switch circuit is positioned between the voltage regulator of the data/power receiver circuit and the battery charge circuit. The switch consists of two PMOS transistors, one capacitor, and one resistor (see Fig. 11.14). The resistor and capacitor form a parallel connection with an RC time constant of 100 milliseconds, which is much longer than the clock period of the data/power receiver chip. Therefore, the voltage of the "a" node in Fig. 11.14 is higher than the threshold of the Q2 switch when a data signal (PWM) is applied, causing Q2 to turn off. The data decoding can then be successfully carried out without a loading effect. In the battery charging mode, only sinusoidal waveforms are present in the receiver coil. In this case, the level of CLK is logically high, Q1 is turned off, the voltage of node "a" is made logically low, and Q2 is turned on, enabling the charging of the battery.

### 11.2.3.4 Stimulator Array with Distributed Sensor Network

For a reasonable human visual acuity, a retinal implant is expected to bring an image with at least 1000 pixels. No other neural prostheses demand such a large array. However, conventional neural prosthetic implant devices, in which the stimulus electrodes are directly connected to the telemetry chip by wires, may be restricted in meeting such a large pinout requirement. This problem is illustrated in Fig. 11.15 (top). The maximum number of I/O pads or connection wires limits the maximum number of stimulus electrodes. Researchers in NAIST, Japan, proposed a distributed network method [23–25]. In their implementation, schematically shown in Fig. 11.15 (bottom), the conventional micromachined stimulus electrodes have been replaced with microsized CMOS devices that consist of an on-chip electrode, a serial interface circuit, and a photosensor. Because each micronode is linked through the single-wire serial bus, the restriction on the number of the stimulus electrodes is resolved.

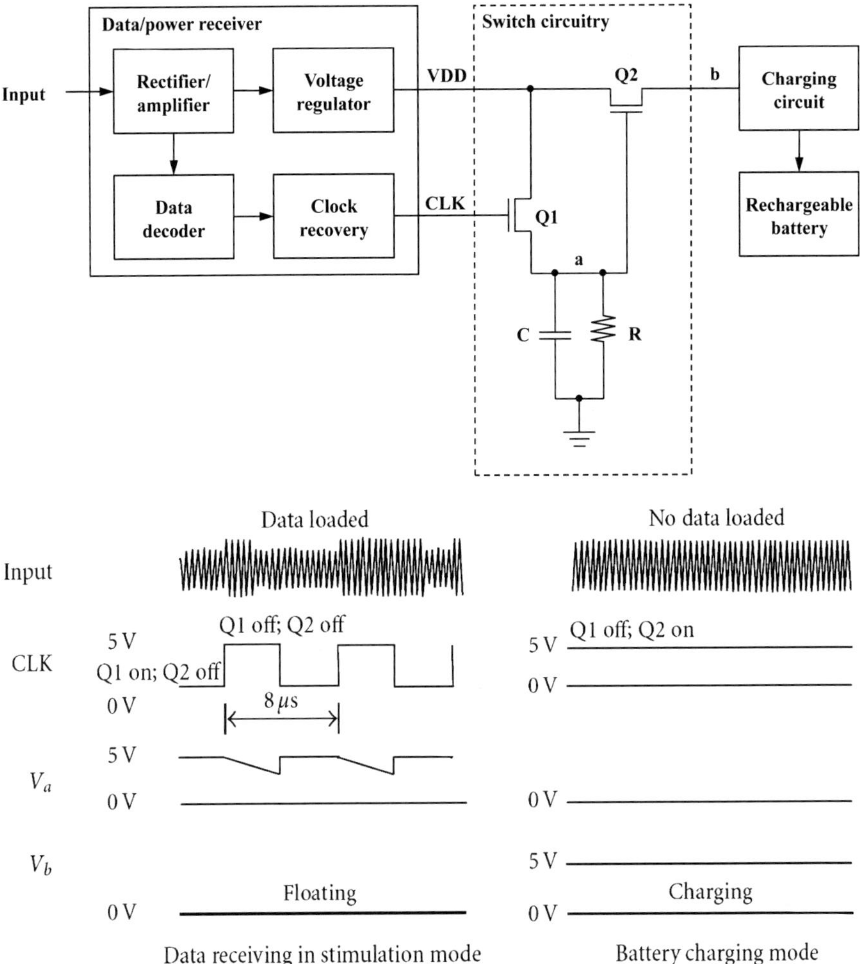


**Fig. 11.14** Operation of switch circuitry in stimulation and battery charging modes: (*top*) circuit design and (*bottom*) output waveform at each node. When no data are loaded on the carrier frequency, CLK is logic high, therefore switch Q1 is turned off, and the voltage of "a" node is logic low, so switch Q2 would be turned on (battery recharging mode). If any data are loaded on the carrier frequency, the CLK recovers and the voltage of node "a" nerve goes below the threshold of switch Q2, so Q2 would be turned off

## 11.3 Conclusion

We have shown IC design examples for a couple of successful areas of biomedical implantable electronics; however, they still involve great challenges. For example, although the majority of users can enjoy everyday conversation using cochlear implants, it is not certain why the device does not give the same level of benefit to other patients. Further development is also necessary if the device is to deliver finer information such as music. We need to study not just the auditory system, but also

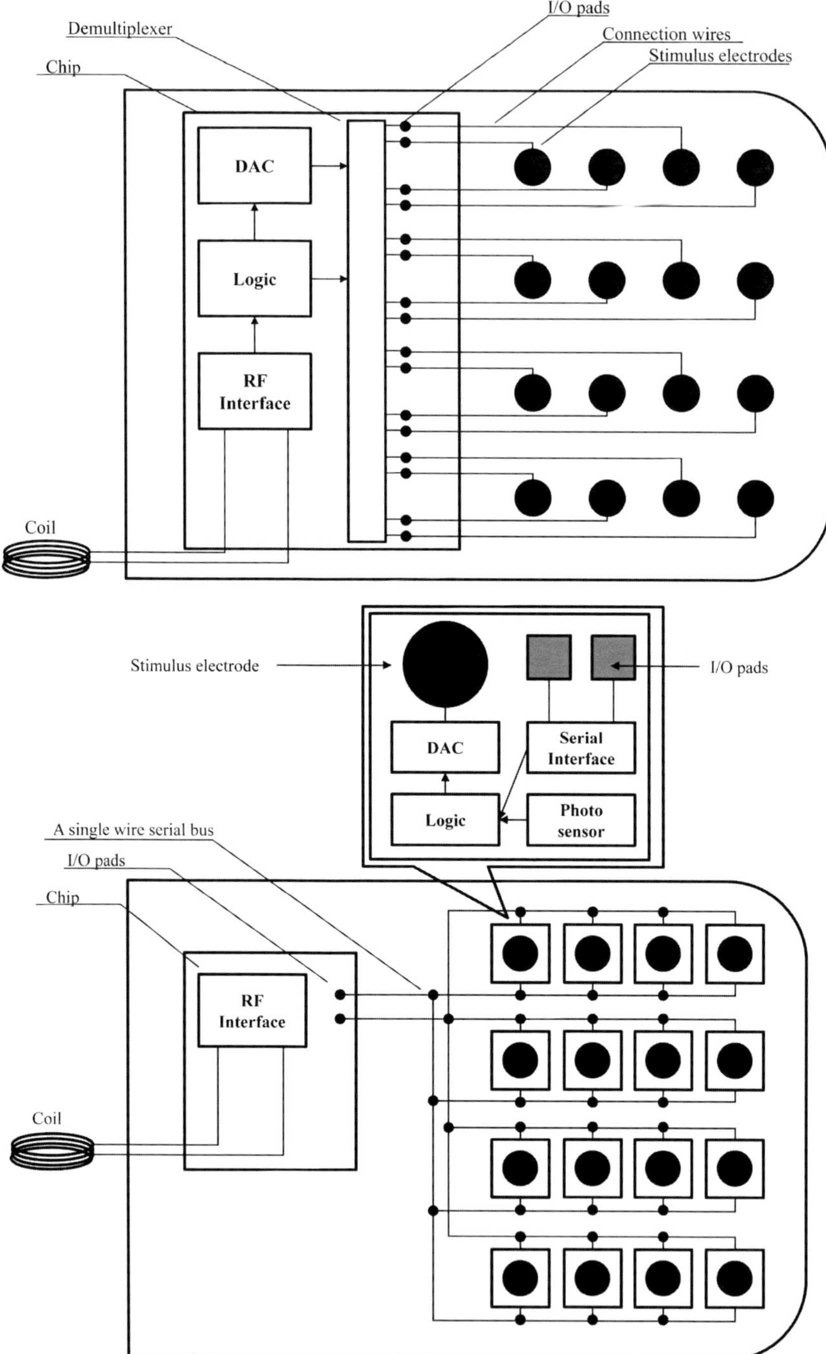



**Fig. 11.15** Diagram of retinal prosthetic devices: (*top*) conventional device, (*bottom*) proposed device (reproduced with permission from authors of [23])

how the brain perceives artificial hearing [26]. Cost is another important issue. Two thirds of hearing loss patients live in developing countries [27] and cannot afford the current price of the cochlear implant.

For the retinal implant, the state-of-the-art device has an electrode array of less than 100 channels, whereas an array size on the order of 1000 is required to obtain a practically useful vision. This presents a big challenge for IC design, not only in terms of the pin connections to the electrodes, but also in terms of power management. The device also needs improvement so that the electrodes can be placed closer to the ganglia for threshold reduction and for higher resolution. Minimization of crosstalk for increased array size is also critical to achieve the desired level of spatial resolution.

The battery, disposable or rechargeable, can last about a day or two for average users of the cochlear implant with channels as few as 20. What would be the size of the battery for the retinal implant when the channel count reaches 1000? Obviously, we need a battery with a much larger capacity or, even better, ICs with reduced power consumption.

As stated earlier, safety is the highest priority in the implantable electronics, even while designing for efficiency and low power. Double-checking to ensure safety through redundant design is a must. For example, shorting all the electrodes to ground at the end of a stimulation cycle can be done combined with the use of decoupling capacitors to prevent any dangerous charges from reaching body tissues. An "ac logic" concept has been developed to satisfy both safety and power saving concerns [28]. The ac logic is designed so that the implantable electronics runs on ac power instead of dc power.

To optimize power use, we can switch the ac power so that it is supplied only during a selected portion of a cycle. For example, by operating at the positive phase of the ac power, the circuit can have a resting state at the negative phase. Since all the pins of the implanted IC chip, including the power pins, have charge-balanced ac waveforms, this method can be used to enhance the safety of the implant.

# References


1. Wilson, B.S., Finley, C.C., Lawson, D.T., Wolford, R.D., Eddington, D.K., Rabinowitz, W.M.: Better speech recognition with cochlear implants. Nature **352**, 236–238 (1991)
2. Kessler, D.K., Loeb, G.E., Barker, M.J.: Distribution of speech recognition results with the Clarion cochlear prosthesis. Ann. Otol. Rhinol. Laryngol. **104**(Suppl. 166), 283–285 (1995)
3. Skinner, M.W., Clark, G.M., Whitford, L.A., Seligman, P.M., Staller, S.J., Shipp, D.B., Shallop, J.K.: Evaluation of a new spectral peak coding strategy for the nucleus 22 channel cochlear implant system. Am. J. Otol. **15**(Suppl. 2), 15–27 (1994)
4. Dorman, M., Loizou, P., Rainey, D.: Speech intelligibility as a function of the number of channels of stimulation for signal processor using sine-wave and noise-band outputs. J. Acoust. Soc. Am. **102**, 2403–2411 (1997)
5. Wilson, B.S., Finley, C.C., Lawson, D.T., Wolford, R.D., Zerbi, M.: Design and evaluation of a continuous interleaved sampling (CIS) processing strategy for multichannel cochlear implants. J. Rehabil. Res. Dev. **30**, 110–116 (1993)
6. http://www.cochlear.com/

7. http://www.advancedbionics.com/

8. http://www.medel.com/

9. An, S.K., Park, S.I., Jun, S.B., Lee, C.J., Byun, K.M., Sung, J.H., Wilson, B.S., Rebsher, S.J., Oh, S.H., Kim, S.J.: Design for a simplified cochlear implant system. IEEE Trans. Biomed. Eng. **54**(6), 973–982 (2007)

10. Hamici, Z., Itti, R., Champier, J.: A high-efficiency power and data transmission system for biomedical implanted electronic devices. Meas. Sci. Technol. **7**, 192–201 (1996)

11. Sokal, N.O., Sokal, A.D.: Class E-A new class of high-efficiency tuned single-ended switching power amplifiers. IEEE J. Solid-State Circuits **SC-10**, 168–176 (1975)

12. Hamida, A.B., Samet, M., Lakhoua, N., Drira, M., Mouine, J.: Sound spectral processing based on fast Fourier transform applied to cochlear implant for the conception of a graphical spectrogram and for the generation of stimulation pulses. In: Proc. 24th Annu. Conf. IEEE Industrial Electronics Society, vol. 3, pp. 1388–1393 (1998)

13. Tang, Z., Smith, B., Schild, J.H., Peckham, P.H.: Data transmission from an implantable biotelemeter by load-shift keying using circuit configuration modulation. IEEE Trans. Biomed. Eng. **42**, 524–528 (1995)

14. Fernandez, C., Garcia, O., Prieto, R., Cobos, J.A., Gabriels, S., Van Der Borght, G.: Design issues of a core-less transformer for a contact-less applications. In: Proc. APEC02, pp. 339–345 (2002)

15. Merrill, D.R., Bikson, M., Jefferys, J.G.R.: Electrical stimulation of excitable tissue: design of efficacious and safe protocols. J. Neurosci. Methods **141**, 171–198 (2004)

16. Wyatt, J., Rizzo, J.: Ocular implants for the blind. IEEE Spectr. **33**, 47–53 (1996)

17. Seo, J.M., Kim, S.J., Chung, H., Kim, E.T., Yu, H.G., Yu, Y.S.: Biocompatibility of polyimide microelectrode array for retinal stimulation. Mater. Sci. Eng. C **24**, 185–189 (2004)

18. Hesse, L., Schanze, T., Wilms, M., Eger, M.: Implantation of retinal stimulation electrodes and recording of electrical stimulation response in the visual cortex of the cat. Graefes Arch. Clin. Exp. Ophthalmol. **238**, 840–845 (2000)

19. Chow, A.Y., Chow, V.Y.: Subretinal electrical stimulation of the rabbit retina. Neurosci. Lett. **225**, 13–16 (1997)

20. Shivdasani, M.N., Luu, C.D., Cicione, R., Fallon, J.B., Allen, P.J., Leuenberger, J., Suaning, G.J., Lovell, N.H., Shepherd, R.K., Williams, C.E.: Evaluation of stimulus parameter and electrode geometry for an effective suprachoroidal retinal prosthesis. J. Neural Eng. **7**, 036008 (2010)

21. Sivaprakasam, M., Liu, W., Wang, G., Weiland, J.D., Humayun, M.S.: Architecture tradeoffs in high-density microstimulators for retinal prosthesis. IEEE Trans. Circuits Syst. **52**(12), 2629–2642 (2005)

22. Zhou, J.A., Woo, S.J., Park, S., Kim, E.T., Seo, J.M., Chung, H., Kim, S.J.: A suprachoroidal electrical retinal stimulator design for long-term animal experiments and in vivo assessment of its feasibility and biocompatibility in rabbits. J. Biomed. Biotechnol. **2008**, 547428 (2008) 10 pp.

23. Uehara, A., Pan, Y., Kagawa, K., Tokuda, T., Ohta, J., Nunoshita, M.: Micro-sized photo-detecting stimulator array for retinal prosthesis by distributed sensor network approach. Sens. Actuators A **120**, 78–87 (2005)

24. Pan, Y.L., Tokuda, T., Uehara, A., Kagawa, K., Ohta, J., Nunoshita, M.: A flexible and extendible neural stimulation device with distributed multi-chip architecture for retinal prosthesis. Jpn. J. Appl. Phys. **44**, 2099–2103 (2005)

25. Ohta, J., Tokuda, T., Kagawa, K., Sugitani, S., Taniyama, M., Uehara, A., Terasawa, Y., Nakauchi, K., Fujikado, T., Tano, Y.: Laboratory investigation of microelectronics-based stimulators for large-scale suprachoroidal transretinal stimulation (STS). J. Neural Eng. **4**, S85–S91 (2007)

26. Wilson, B., Dorman, M.: Cochlear implants: a remarkable past and a brilliant future. Hear. Res. **242**, 3–21 (2008)

27. Tucci, D., Merson, M., Wilson, B.: A summary of the literature on global hearing impairment: current status and priorities for action. Otol. Neurotol. **31**(1), 31–41 (2010)
28. Lee, C.J.: An LCP-based cortical stimulator for pain control and a low cost but safe packaging using a new AC logic concept. Ph.D. Dissertation, School of Electrical Engineering and Computer Science, College of Engineering, Seoul National University, Aug. 2009. A patent pending

Printed by Books on Demand, Germany